Beyond Complementary Metal-Oxide-Semiconductor Nanoelectronics

Complementary metal-oxide-semiconductor (CMOS) nanoelectronics are approaching their physical limits, leading to issues such as excessive leakage and increased power consumption. New materials, computing methods, and state variables are being investigated to solve these issues. These emerging technologies embrace a wide range of cutting-edge developments aimed at surpassing the physical constraints of traditional CMOS devices. This research is vital for enabling future computing abilities, particularly in devices that could serve new functions in fields such as artificial intelligence and quantum information processing.

Beyond Complementary Metal-Oxide-Semiconductor Nanoelectronics offers an in-depth examination of the challenges associated with conventional CMOS technology, outlining the obstacles and issues encountered with traditional CMOS scaling, including power and energy consumption, as well as variability, and highlighting the necessity for alternative technologies. It presents possible successor technologies to CMOS, including emerging logic devices, memory chips, compute-in-memory architectures, new materials, and integration methods. These technologies leverage alternative physical effects, including spintronics, neuromorphic computing, and quantum phenomena, and introduce novel functionalities to enhance performance.

Providing an engaging and clear perspective on the future of nanoelectronics, this book will be a valuable reference for graduate students specializing in electrical engineering, computer engineering, materials science, and physics, as well as for university instructors delivering advanced coursework on nanoelectronics, very-large-scale integration (VLSI) design, and solid-state devices. Professional engineers and designers in the semiconductor field who wish to stay up to date on new technologies, as well as investigators in nanoelectronics and semiconductor technology, will also find it useful.

- Explains why conventional CMOS scaling is reaching its physical and technological limits and outlines new approaches necessary to improve performance.
- Highlights the importance of moving beyond electric-charge-based computing to explore alternative properties such as electron spin, for information storage and processing.
- Bridges the gap between devices and systems by integrating the core physics and materials science of devices with high-level architectural design and advanced packaging techniques.

Beyond Complementary Metal-Oxide-Semiconductor Nanoelectronics

Diversified Approaches to Transcending Physical Scaling Limits

Vinod Kumar Khanna

CRC Press
Taylor & Francis Group
Boca Raton London New York

CRC Press is an imprint of the
Taylor & Francis Group, an **informa** business

Front cover image: Jerome Mettling/Shutterstock

First edition published 2027
by CRC Press
2385 NW Executive Center Drive, Suite 320, Boca Raton FL 33431

and by CRC Press
4 Park Square, Milton Park, Abingdon, Oxon, OX14 4RN

CRC Press is an imprint of Taylor & Francis Group, LLC

Library of Congress Cataloging-in-Publication Data
Names: Khanna, Vinod Kumar, 1952- author
Title: Beyond complementary metal-oxide-semiconductor nanoelectronics : diversified approaches to transcending physical scaling limits / Vinod Kumar Khanna.
Description: Boca Raton : CRC Press, 2027. | Includes bibliographical references and index. |
Identifiers: LCCN 2026018398 (print) | LCCN 2026018399 (ebook) | ISBN 9781041271727 hbk | ISBN 9781041272519 pbk | ISBN 9781003753469 ebk
Subjects: LCSH: Semiconductors--Materials | Semiconductors--Technological innovations | Nanoelectronics | Logic devices | Spintronics | Integrated circuits--Very large scale integration
Classification: LCC TK7871.85 .K45 2026 (print) | LCC TK7871.85 (ebook)
LC record available at https://lccn.loc.gov/2026018398
LC ebook record available at https://lccn.loc.gov/2026018399

ISBN: 978-1-041-27172-7 (hbk)
ISBN: 978-1-041-27251-9 (pbk)
ISBN: 978-1-003-75346-9 (ebk)

DOI: 10.1201/9781003753469

Typeset in Times
by SPi Technologies India Pvt Ltd (Straive)

Dedication

To my parents, Late Shri Amarnath Khanna and Shrimati Pushpa Khanna, whose boundless love and unwavering encouragement have always been my source of inspiration.

To my grandsons Hansh and Avyaan, my daughter Aloka, and my wife Amita, who fill my days with joy and illuminate my world.

Contents

Preface

CMOS nanoelectronics is the bedrock of our digital world, powering its relentless march toward smaller, faster, and more energy-efficient devices. This technology forms the launchpad for breakthroughs in artificial intelligence and lightning-fast communications. Yet as the semiconductor industry pushes into the sub-5 nm frontier, it faces obstacles such as soaring manufacturing costs, increased gate leakage, and rising power demands. These hurdles are sparking a wave of innovation, from advanced 2.5D and 3D integration to the exploration of novel materials and architectures. While CMOS is poised to lead for years to come, the race to develop "Beyond CMOS" strategies is well underway.

In recent years, a quiet revolution has been reshaping the semiconductor world. Advanced packaging has emerged as the new frontier, shifting the spotlight from traditional chip scaling to inventive back-end integration. It is viewed as an alternative route to transcending scaling limits. As Moore's Law hits physical and economic limits, breakthrough technologies in integration and packaging offer new possibilities by stacking and linking chips in ways that boost performance, shrink size, and save energy. From 2.5D and 3D ICs to chiplets and hybrid bonding, these innovations are rewriting the rules, driven by the demands of AI and high-performance computing. Exciting trends like chip-on-wafer-on-substrate and co-packaged optics are knitting together diverse chiplets, supercharging memory bandwidth, and keeping systems cool. This book offers a comprehensive gateway into this dynamic landscape, emphasizing how new technologies can enhance and extend CMOS for future computing.

The book begins by examining the physical and technological limitations of CMOS nanoelectronics near the atomic scale, which act as impediments to the progress of CMOS nanoelectronics. With this background, nanoelectronics beyond CMOS technology is explored, defining its aims and objectives. The expanding portfolio of silicon and supplementary materials embraces both narrow and wide-bandgap semiconductors, together with dielectrics and topological insulators, carbon-based and two-dimensional nanomaterials, and ferroelectric and magnetic materials. Important CMOS and upcoming electronic logic devices include the fin field-effect transistor (FinFET), gate-all-around field-effect transistor, forksheet FET, complementary field-effect transistor; tunneling, negative-capacitance and ferroelectric field-effect transistors; single-electron transistor, 2D material FETs, and nanoelectromechanical systems (NEMs) switches. Members of the spintronic logic family are magneto-electric spin-orbit (MESO), magnetic tunnel junction, spin Hall effect devices, besides the tunneling spin-FET, the all-semiconductor/all-electric spin-FET, and skyrmionic logic. Among non-volatile memories, mention may be made of the resistive, magnetic, phase-change, and ferroelectric random-access memories. Non von-Neuman computing architectures that have attracted attention are in-memory computing, neuromorphic computing, and quantum computing. The role of beyond-CMOS nanoelectronics in quantum computing, as well as using beyond-CMOS nanoelectronics to build quantum computers, is discussed. Top-down nanofabrication techniques such as electron beam lithography (EBL), wide-aperture extreme ultraviolet lithography (EUVL), nanoimprint lithography (NIL), and block copolymer lithographies are prominent, as are reactive ion etching (RIE) and focused ion beam (FIB) milling. Bottom-up fabrication techniques such as atomic layer deposition (ALD), molecular beam epitaxy (MBE), metal-organic chemical vapor deposition (MOCVD), focused electron beam-induced deposition (FEBID), wet chemical synthesis, and self-assembly play significant roles. Homogeneous (monolithic) integration relies on methods such as advanced heteroepitaxy, M3D sequential integration, selective material deposition, and van der Waals integration. Heterogeneous integration utilizes techniques of BEOL integration, wafer bonding, chiplet/die-to-wafer assembly, micro-transfer printing, in-situ integration on engineered substrates, 3D integration with TSVs, and 2.5D interposers. Low-temperature fabrication processes are pertinent to both homogenous and heterogeneous integration. Looking toward the future, Boltzmann tyranny is a challenge. Overlay errors call for urgent attention as processing complications increase with the number of layers. The Poisson yield model proves inadequate for analysis.

This book is crafted for advanced undergraduate and graduate students, industry professionals tackling scaling dilemmas, and those pioneering "More than Moore" applications and unconventional device designs. As a vital resource for researchers delving into materials such as graphene and carbon nanotubes, as well as spintronics and quantum computing, this book aims to be a wellspring of insight and inspiration for all interested in this exciting advancement. It is earnestly hoped that the book will meet the needs of the target audience. Wishing all the readers a joyful expedition into the world of the post-CMOS age.

Vinod Kumar Khanna
Chandigarh, India

Acknowledgments

I am deeply thankful to Almighty God for the countless blessings in my life.

The present technological maturity of CMOS nanoelectronics is the culmination of the brilliance of many scientists whose breakthrough discoveries in physics have propelled the field forward. I owe a great debt to the trailblazers in nanoelectronics, whose vision and dedication have shaped the scenario of contemporary CMOS nanoelectronics, which has evolved from a specialized, low-power curiosity into the powerhouse that drives modern computing. Fueled by the tireless shrinking of transistors to boost speed, efficiency, and packing density, CMOS technology has shifted from micro-scale tinkering to the artful control of matter at the atomic scale. This evolution has leapt from flat, bulk-silicon devices to intricate 3D structures that push up against the limits of quantum physics, with the new horizon set on exploring technologies beyond traditional CMOS, the focal theme of this book.

My heartfelt thanks go to the commissioning and production editors and to the CRC Press team for their steadfast support and collaboration.

Above all, I am profoundly grateful to my family and my wife, whose resolute encouragement and motivation have been my guiding light.

Vinod Kumar Khanna
Chandigarh, India

About the Author

INTRODUCTION

Vinod Kumar Khanna is an independent researcher based in Chandigarh, India. A retired Chief Scientist at CSIR–CEERI, Pilani, and former Professor at AcSIR, Ghaziabad, he has held esteemed positions, including Emeritus Scientist at CSIR and Professor Emeritus at AcSIR.

ACADEMIC QUALIFICATIONS

Khanna's academic career is marked by a national scholarship from the Ministry of Education and Social Welfare, Government of India, awarded in 1970 for his outstanding Higher Secondary results. He received an M.Sc. degree in physics, specializing in electronics, from the University of Lucknow, Lucknow, Uttar Pradesh, in 1975, followed by a Ph.D. in physics from Kurukshetra University, Kurukshetra, Haryana, in 1988, where he explored the development and modeling of porous alumina humidity sensors.

RESEARCH AND TEACHING EXPERIENCE AND ACCOMPLISHMENTS

With a research career spanning four decades, from 1977 to 2017, he began as a Research Assistant in the Department of Physics at the University of Lucknow. In 1980, he joined CSIR-CEERI, Pilani, where he actively worked on a range of CSIR-funded research and development projects. His expertise covers power semiconductor devices, microelectronics, microelectromechanical systems (MEMS), and nanotechnology-based sensors and dosimeters. Among his notable achievements are the development of high-voltage and high-current rectifiers for railway traction, advanced TV deflection transistors, power Darlington transistors for AC motor drives, fast-switching thyristors, power (double-diffused metal-oxide-semiconductor field-effect transistors (DMOSFETs), and insulated-gate bipolar transistors (IGBTs). He developed techniques such as sealed-tube Ga/Al diffusion for deep junctions, surface electric-field control using edge beveling and contouring, and the design of floating-field limiting rings for large-area devices. His work included an in-depth characterization of minority-carrier lifetime in power semiconductor devices. He played a key role in creating P–I–N diode neutron dosimeters and P-channel metal-oxide-semiconductor field-effect transistor (PMOSFET)-based gamma-ray dosimeters. In sensor technology, he contributed to the development of nanoporous aluminum oxide humidity sensors, ion-sensitive field-effect transistor (ISFET)-based microsensors for biomedical, food, and environmental applications, microheater-embedded gas sensors for automotive electronics, MEMS acoustic sensors for satellite launch vehicles, and capacitive MEMS ultrasonic transducers for medical applications. As a faculty member at AcSIR, he coordinated the MEMS/IC Technology course in the advanced semiconductor electronics program and taught MEMS Technology to M.Tech students. As an adjunct faculty, BESU, Kolkata, he shared his expertise in MEMS Technology & Design with M.Tech (Mechatronics) students. He was also invited by IIT Jodhpur to deliver lectures on semiconductor fundamentals and technology to B.Tech students. He guided numerous B.Tech and M.Tech theses from institutions such as BITS Pilani, VIT Vellore, and Kurukshetra University, and supervised a Ph.D. thesis on MEMS acoustic sensor.

SEMICONDUCTOR FACILITY CREATION AND MAINTENANCE

He played a pivotal role in establishing and managing advanced facilities for diffusion and oxidation, edge beveling and contouring, reactive sputtering, and carrier lifetime measurement. As Head of the MEMS and Microsensors Group, he led the upkeep of a state-of-the-art six-inch MEMS fabrication facility for R&D and spearheaded the expansion of processing equipment at CSIR-CEERI.

SCIENTIFIC POSITIONS HELD

Over his distinguished career at CSIR-CEERI, from April 1980 until his retirement in November 2014, he rose through the ranks, earning a merit promotion along the way. He concluded his tenure as Chief Scientist, Professor (AcSIR), and Head of the MEMS and Microsensors Group. Following retirement, he continued to contribute as Emeritus Scientist at CSIR and Emeritus Professor at AcSIR until November 2017. Now settled in Chandigarh, he channels his passion into writing and reading.

MEMBERSHIP OF LEARNED SOCIETIES

He is recognized as a Fellow and Life Member of the Institution of Electronics and Telecommunication Engineers (IETE), India, and holds life memberships in the Indian Physics Association (IPA), the Semiconductor Society of India (SSI), and the Indo-French Technical Association (IFTA).

FOREIGN TRAVEL

He has presented research at international conferences, including the IEEE Industry Application Society Annual Meeting in Denver, Colorado, in 1986. His global collaborations have taken him to research assignments in Germany and Italy, and he represented India as part of a delegation to the Institute of Chemical Physics in Novosibirsk, Russia.

AWARDS AND HONORS

He has earned multiple CEERI Foundation Day Merit Team Awards for pioneering projects, including fast-switching thyristors, power Darlington transistors for transportation, P–I–N diode neutron dosimeters, and high-voltage TV deflection transistors. His research excellence was recognized with a prestigious CSIR-DAAD Fellowship in 2008 under the Indo-German Bilateral Exchange Programme for Senior Scientists. His global impact is reflected in his inclusion in the Stanford–Elsevier list of the world's top 2% scientists and his ranking as the #2 Scholar in Nanoelectronics by ScholarGPS.

RESEARCH PUBLICATIONS AND BOOKS

He has published 194 research papers in leading journals and conferences, authored 25 books, contributed six book chapters, and has five granted patents—including two from the United States.

About This Book

Beyond CMOS nanoelectronics aims to overcome the physical, power, and scaling barriers of conventional silicon technology by developing high-performance and energy-efficient devices. This field charts a bold course for the semiconductor industry, connecting fundamental physics to tomorrow's computers. It equips researchers and engineers with new tools to design energy-efficient systems, rather than simply downsize transistors. As Moore's law approaches its limits in physics, efficiency, and cost, research attention is shifting toward several vital new frontiers. In the past, packaging simply shielded delicate silicon components from threats like moisture and bumps. Now, it has transformed from a mere protective layer into a key force powering advances in semiconductor scaling and system performance.

Chapter 1 explores the core physical and thermodynamic limits of traditional charge-based MOSFETs, from short-channel effects to the notorious power-dissipation bottlenecks known as the "silicon ceiling." It traces the industry's trip from the relentless pursuit of "More Moore" to the horizons of "More than Moore," signaling the twilight of CMOS scaling and sparking the search for "beyond CMOS" solutions.

Chapter 2 elucidates the vast panorama of beyond-CMOS nanoelectronics, where the boundaries of research extend to novel materials such as high-mobility semiconductors, 2D layers, spintronics, and nanomaterials. It explores not only non-von Neumann architectures such as neuromorphic and quantum systems but also the rise of hybrid, co-optimized solutions that blend the best of multiple worlds. Beyond CMOS technology is a realm teeming with emerging materials, devices, and novel architectures, all poised to enhance or one day surpass the reign of traditional silicon-based CMOS. This domain ranges from the fundamental limits of scaling to emerging materials, inventive switching, and revolutionary computing designs.

Chapter 3 describes a new generation of materials that enable state-of-the-art logic and memory devices by exploiting quantum effects, electron spin, or molecular features, rather than relying solely on electron charge. These high-mobility and wide-bandgap materials are the backbone of beyond-CMOS nanoelectronics, powering devices with higher voltages, superfast switching, and remarkable thermal resilience compared to traditional silicon. Standout examples include gallium nitride, silicon carbide, diamond, and a range of two-dimensional materials. The 2D atomically thin wonders deliver exceptional carrier mobility and precise electrostatic control, making them ideal for ultra-small channels. Carbon nanotubes, with their unique one-dimensional structure, perform well in field-effect transistors, thanks to their superior thermal conductivity and the ability to fine-tune threshold voltages by tweaking their diameter or chirality. Meanwhile, spintronic and magnetic materials open new avenues by processing information through electron spin rather than charge.

Chapter 4 reconnoiters the remarkable progress from FinFETs to gate-all-around designs built with silicon nanowires and nanosheets, providing superior electrostatic control. Forksheet FETs are emerging as the leading-edge transistor design, poised to surpass nanosheets and redefine performance at 2 nm nodes and smaller. Complementary FETs are designed to significantly reduce the required footprint with vertical stacking of NFET and PFET devices. Tunneling field-effect transistors, known as steep-slope switches, utilize band-to-band tunneling to break through the 60 mV/dec subthreshold swing barrier of traditional CMOS, smoothing the path for ultra-low power electronics. Negative-capacitance and ferroelectric FETs bring ferroelectric materials into the gate stack, creating internal voltage amplification through negative capacitance and enabling much lower operating voltages. The chapter also spotlights 2D-material-based FETs, highlighting the promise of graphene, transition metal dichalcogenides, and other thin materials for building ultra-scaled channels on silicon platforms. NEM switches are known for their virtually zero leakage currents.

Chapter 5 introduces the fascinating world of spintronic logic devices, where information is processed via the spin of electrons rather than their charge. The chapter delves into spin-FETs, magnetic tunnel junctions, and the majority-gate logic that enables non-volatile computing. Unlike conventional logic gates, majority-gate devices decide their output by tallying the dominant spin direction among several magnetic inputs, laying the groundwork for more efficient and compact arithmetic circuits than those built with standard Boolean gates. Skyrmionic logic applies the unique physical properties and interactions of quasi-particles known as skyrmions to perform Boolean operations.

Chapter 6 delves into the non-volatile memory technologies, spotlighting resistive RAM, spin-transfer torque MRAM, phase-change memory, and ferroelectric memory, and showcasing their transformative impact on high-speed, high-density applications, neural networks that lead to neuromorphic and brain-inspired architectures.

Chapter 7 dives into the world of in-memory computing, revealing how devices like memristors encourage unconventional approaches. These breakthroughs enable massively parallel, hardware-driven neural networks that bring neuromorphic and brain-inspired architectures to life.

Chapter 8 examines nanofabrication, exploring how advanced top-down techniques like E-beam and EUV lithography, alongside bottom-up approaches such as self-assembly, craft features smaller than 10 nm.

Chapter 9 assesses the leap from standalone devices to fully integrated systems, looking into the promise of processing-in-memory, the rise of 3D integration, and the economic hurdles facing next-generation technologies.

Chapter 10 casts an eye on current realities and what lies ahead. According to experts, CMOS will remain unchallenged as the dominant technology until at least 2030 or 2040. In the meantime, the focus will shift toward augmenting CMOS for powerful accelerators tailored for AI and high-performance computing between 2024 and 2030.

Beyond-CMOS nanotechnologies are inaugurating revolutionary computing paradigms, shifting from traditional Boolean logic to brain-inspired and quantum approaches that excel in AI and big-data applications. Memory solutions like ReRAM and MRAM promise to erase the speed barrier between processors and memory, offering swift, non-volatile alternatives that could soon outpace flash and volatile RAM. By engaging new state variables such as electron spin or magnetic states, these devices dramatically cut power consumption, a game-changer for both mobile gadgets and sprawling data centers. The fusion of diverse materials, such as GaN for high-speed radio frequencies and MEMS for sensors, onto silicon chips is fueling a new generation of "mostly digital" devices tailored for 5G, IoT, and autonomous technologies. The global roadmap for devices and systems emphasizes the push for "Green materials," ensuring that innovation goes hand in hand with sustainability and environmental care. As readers turn the pages of this book, they embark on a thrilling excursion into a future with decreasing blurredness and rapidly coming into focus.

Abbreviations, Acronyms, Chemical Symbols, and Mathematical Notation

A

A	Ampere, area, a constant
A°	Angstrom unit = 10^{-10} m
AAO	Anodic aluminum oxide
ADAS	Advanced driver assistance system
ADCs	Analog-to-digital converters
aF	Attofarad
AFM	Antiferromagnetic
Ag	Silver
AI	Artificial intelligence
AIMC	Analog in-memory computing
aJ	Attojoule
ALD	Atomic-layer deposition
AlGaAs	Aluminum gallium arsenide
AlGaN	Aluminum gallium arsenide
AlN	Aluminum nitride
Al_2O_3	Aluminum oxide (alumina) or AlO_x
AlSb	Aluminum antimonide
AlScN	Aluminum scandium nitride
ANNs	Artificial neural networks
APAM	Atomic-precision advanced manufacturing
ART	Aspect ratio trapping
AS-ALD	Area-selective atomic layer deposition
ASIC	Application-specific integrated circuit
Au	Gold

B

$BaTiO_3$	Barium titanate (BTO)
BC	Beyond CMOS
BE	bottom electrode
BEOL	Back-end-of-line
$BiFeO_3$	Bismuth ferrite (BFO)
$(Bi, Sb)_2Se_3$	Bismuth antimony selenide (Bismuth antimony selenium alloy)
$Bi_{1-x}Sb_x$	Bismuth-antimony alloys

Bi_2Se_3	Bismuth selenide
Bi_2SeO_5	Bismuth selenite (bismuth selenium pentoxide)
Bi_2Te_3	Bismuth telluride
Bi_2TeO_5	Bismuth tellurite (bismuth tellurium oxide)
BJT	Bipolar junction transistor
BL	Bit line
BOM	Bill of materials
BOX	Buried oxide
BP	Black phosphorus
BPNR	Black phosphorus nanoribbon
BTBT	Band-to-band tunneling

C

C	Coulomb
°C	Degree Centigrade
C60	Buckminsterfullerene
CAM	Content-addressable memory
CBRAM	Conductive bridge RAM
CCTO	Calcium copper titanate ($CaCu_3Ti_4O_{12}$)
CCVD	Catalytic chemical vapor deposition
CD	Critical dimension
CdSe	Cadmium selenide
CdTe	Cadmium telluride
CeO_2	Cerium dioxide, Ceria, Cerium (IV) oxide, Ceric oxide
CFET	Complementary field-effect transistor
CiM	Computing-in-memory
CIPS	Copper indium phosphide sulfide
CLM	Channel length modulation
cm	Centimeter
CMOS	Complementary metal-oxide-semiconductor
CMP	Chemical mechanical polishing
CMR	Colossal magnetoresistance
CNN	Cellular neural network
CNT	Carbon nanotube
CNTFET	Carbon nanotube field-effect transistor
Co	Cobalt
CoFeB	Cobalt-Iron-Boron (ferromagnetic alloy)
Co_2MnSi	Cobalt manganese silicon
CoO	Cobalt (II) oxide or cobalt monoxide
CPU	Central processing unit
CPW	Coplanar waveguide
$Cr_2Ge_2Te_6$	Chromium germanium telluride (CGT)
CrI_3	Chromium (III) iodide (*or* chromium triiodide)
CrO_2	Chromium (IV) oxide (chromium dioxide or magtrieve)
Cryo-CMOS	Cryogenic complementary metal-oxide-semiconductor

$Cs_2AgBiBr_6$	Cesium silver bismuth bromide
CSD	Chemical solution deposition
3C-SiC	Cubic silicon carbide
Cu	Copper
$CuInP_2S_6$	Copper indium thiophosphate (CIPS)
$CuInS_2$	Copper indium sulfide
CuMnAs	Copper manganese arsenide
CuO	Copper (II) oxide or Cupric oxide
CVD	Chemical vapor deposition

D

1D, 2D	1-Dimensional, 2-Dimensional, 2.5-Dimensional, 3-Dimensional
2.5D, 3D, 3.5D	3.5-Dimensional
DACs	Digital-to-analog converters
DBs	Dangling bonds
DC	Direct current
DDT	Dodecanethiol
dec	Decade
DG	Double-gate (MOSFET)
DIBL	Drain-induced barrier lowering
DMS	Dilute magnetic semiconductor
DNA	Deoxyribonucleic acid
DoM	Dielectric-on-metal
DRAM	Dynamic random-access memory
DUV	Deep ultraviolet
D2W	Die-to-wafer (assembly)

E

EBL	Electron beam lithography
ECRAM	Electrochemical random-access memory
EDP	Energy-delay product
ELO	Epitaxial lateral overgrowth
EMIB	Embedded multi-die interconnect bridge
eNVMs	Emerging non-volatile memories
ESHS	Environmental, safety, health, and sustainability
EUVL	Extreme ultraviolet lithography
eV	Electron volt

F

F	Farad
Fabs	Fabrication plants *or* Fabrication facilities
FD-SOI	Fully depleted silicon-on-insulator
FE	Ferroelectric
Fe	Iron

FeCAP	Ferroelectric memcapacitor
$FeCl_3$	Iron(III) chloride, ferric chloride
FeFET	Ferroelectric field-effect transistor
Fe_nGeTe_2	Iron germanium telluride (FGT)
Fe_2O_3	Iron (III) oxide (ferric oxide, iron trioxide, hematite)
Fe_3O_4	Iron (II, III) oxide (*ferrous* ferric oxide or magnetite)
FEOL	Front-end-of-line
FeRAM	Ferroelectric random-access memory or FRAM
FeSFET	Ferroelectric semiconductor field-effect transistor
fF	Femtofarad
fJ	FemtoJoule
FM	Ferromagnetic (materials)
FSFET	Forksheet FET
1F-1T	A ferroelectric transistor *(F)* with a current-limiting *transistor (T)*
FTJ	ferroelectric tunnel junction

G

GAAFET	Gate-all-around FET
GAA-NSFET	Gate-all-around nanosheet FET
GAA-NWFET	Gate-all-around nanowire FET
GaAs	Gallium arsenide
GaN	Gallium nitride
GaN-on-Si	Gallium nitride-on-silicon
Ga_2O_3	Gallium(III) oxide
GaSb	Gallium antimonide
GdOCl	Gadolinium oxychloride
Ge	Germanium
GeH_4	Germane
GeOI	Germanium-on-insulator
Ge-on-Si	Germanium-on-silicon
GeSn	Germanium-tin
$Ge_{0.96}Sn_{0.04}$	Germanium-tin alloy with 96% Ge and 4% Si
GHz	Gigahertz
GND	Ground
GNEM	Graphene nanoelectromechanical (switch)
GNR	Graphene nanoribbon
GO	Graphene oxide
GP	Ground-plane

H

h-BN	Hexagonal boron nitride
HBT	Heterojunction bipolar transistor
HCI	Hot carrier injection
HEMT	High-electron-mobility transistor

Hf	Hafnium
HfO_2	Hafnium oxide
HfSiON	Hafnium silicon oxynitride
HgTe	Mercury telluride
HK/MG	High *k*/Metal gate
HM	Heavy-metal
HRS	High resistance state
4H-SiC	4-Hexagonal silicon carbide
HT	High temperature
Hz	Hertz
HZO	Hafnium zirconium oxide, $Hf_{1-x}Zr_xO_2$

I

IBE	Ion beam etching
IBM	International Business Machines Corporation
IC	Integrated circuit
IG	Independent-gate
IMC	In-memory computing
IMF	Interfacial misfit
IMP	Implication (logic)
InAlAs	Indium aluminum arsenide
InAs	Indium arsenide
INC	International nanotechnology conference on communication and cooperation
InGaAs	Indium gallium arsenide
InP	Indium phosphide
InSb	Indium antimonide
α-In_2Se_3	Alpha-phase indium selenide
I/O	Input/Output
IoT	Internet of Things
IP	Intellectual property
IPWGN	International planning working group on nanoelectronics
Ir	Iridium
IRDS	International roadmap for devices and systems
IS	In-situ (a compute model)
ISHE	Inverse spin Hall effect
ISOC	Inverse spin-orbit coupling (effect)
ITO	Indium tin oxide
ITRS	International technology roadmap for semiconductors

J

J	Joule
JJ	Josephson junction
JLFET	Junctionless FET

K

KGD	Known good die (yield)
kW	kilowatt

L

L1/L2 caches	Level-1/Level-2 caches
$LaFeO_3$	Lanthanum ferrite (or Lanthanum iron oxide)
$LaMnO_3$	Lanthanum manganite (or Lanthanum manganate)
$LaNiO_3$	Lanthanum nickel oxide
LaOCl	Lanthanum oxychloride (lanthanum oxide chloride)
$La_{0.7}Sr_{0.3}MnO_3$	Lanthanum strontium manganite (Lanthanum strontium (LSMO) manganese oxide)
LB	Langmuir-Blodgett
LDD	Lightly doped drain
LED	Light-emitting diode
LGAA	Lateral gate-all-around
LIF	Leaky integrate-and-fire (model)
LiM	Logic-in-memory
$LiNbO_3$	Lithium niobate
L-K equation	Landau-Khalatnikov equation
LN	Logic network
LRS	Low resistance state
LSI	Large-scale integration
LT	Low temperature
LTP	Long-term potentiation

M

mA	Milliampere
MAC	Multiply-and-accumulate (operations)
$MAPbI_3$	Methylammonium lead iodide
MBE	Molecular beam epitaxy
MCUs	Microcontrollers
M3D	Monolithic 3D (integration)
ME	Magnetoelectric (effect)
MEK	Methyl ethyl ketone
MEMS	Microelectromechanical systems
MESO	Magneto-electric spin-orbit (logic)
MFIS	Metal/ferroelectric/insulator/semiconductor
MgO	Magnesium oxide
MHz	Megahertz
MIDA	Metal-induced dopant activation
mid-IR	mid-infrared
MIFL	Multi-interface free layer
MIV	Monolithic interlayer via

mm	Millimeter
Mn	Manganese
Mn_2Au	Manganese gold
MNIST	Modified national institute of standards and technology
Mn_3O_4	Manganese(II, III) oxide (hausmannite)
MOCVD	Metal-organic chemical vapor deposition
MoS_2	Molybdenum disulfide
MOSFET	Metal-oxide-semiconductor field-effect transistor
$MoTe_2$	Molybdenum ditelluride
MOVPE	Metal-organic vapor phase epitaxy
MPCVD	Microwave plasma chemical vapor deposition
MRAM	Magnetic random-access memory
ms	Millisecond
MTJ	Magnetic tunnel junction
MtM	More-than-Moore
MTP	Micro-transfer printing
MV	Megavolt
mV	Millivolt
MVM	Matrix-vector multiplication
MWIR	Mid-wavelength infrared

N

NA	Numerical aperture
nA	Nanoampere
Nb	Niobium
NBG	Narrow bandgap (Semiconductors)
NC	Negative capacitance
NCFET	Negative capacitance field-effect transistor
NEMS	Nanoelectromechanical systems
NFET	N-channel field-effect transistor
NH_3	Ammonia
NHE	Nanoheteroepitaxy
Ni	Nickel
NIL	Nanoimprint lithography
NiO	Nickel (II) oxide
NISQ	Noisy intermediate-scale quantum
nm	Nanometer
NML	Nanomagnetic logic
NMOSFET	N-channel metal-oxide-semiconductor field-effect transistor
NO	Nitric oxide
N_2O	Nitrous oxide
NRI	Nanoelectronics research initiative
NS	Nanosheet
NV	Nitrogen-vacancy
NVM	Non-volatile memory

NW	Nanowire
nW	Nanowatt

O

OI	Optical interconnect
OTMS	Octadecyltrimethoxysilane

P

PCL	Phase-change layer
PCM	Phase-change memory, phase-change material
Pd	Palladium
PDK	Process design kit
PDMS	Polydimethylsiloxane
PDN	Pull-down network
PDP	Power-delay product
PECVD	Plasma-enhanced chemical vapor deposition
PEDOT: PSS	Poly(3,4-ethylenedioxythiophene) polystyrene sulfonate
PFET	P-channel field-effect transistor
PGMs	Platinum-group metals
PIC	Photonic integrated circuit
pJ	PicoJoule
PLD	Pulsed laser deposition
PMMA	Poly(methyl methacrylate)
PMOSFET	P-channel metal-oxide-semiconductor field-effect transistor
PPA	Power, Performance, and Area
PPC	Polypropylene carbonate
ps	Picosecond
Pt	Platinum
PUFs	Physically unclonable functions
PUN	Pull-up network
PVD	Physical vapor deposition
P(VDF-TrFE)	Poly(vinylidene fluoride-co-trifluoroethylene) copolymer
PZT	Lead zirconate titanate, chemical formula: $Pb(Zr_xTi_{1-x})O_3$, $0 \leq x \leq 1$

Q

QALE	Quasi-atomic layer etching
QCA or QDCA	Quantum-dot cellular automata
QDs	Quantum dots
QEC	Quantum error correction
Q factor	Quality factor
QPCs	Quantum point contacts
QPU	Quantum processing unit
QSHI	Quantum spin Hall insulator

R

RA	Resistance-area (product)
RAM	Random-access memory
ReRAM	Resistive random-access memory or RRAM
ReS_2	Rhenium disulfide
RF	Radio frequency
RFET	Reconfigurable transistor
RHEED	Reflection high-energy electron diffraction
RMG	Rapid melt growth
Ru	Ruthenium

S

S	Siemen
s	Second
SA	Sense amplifier
SAG	Selective area growth
SAM	Self-assembled monolayer
SBT	Strontium bismuth tantalate
Sb_2Te_3	Antimony telluride
SCEs	Short-channel effects
SCM	Storage class memory
SEG	selective epitaxial growth
SET	Single-electron transistor
sFET	Spin field-effect transistor
SFL	Synthetic free layer
SG-FET	Split-gate FET
SHE	Spin Hall effect
Si	Silicon
SiC	Silicon carbide
$SiCl_2H_2$	Dichlorosilane (DCS)
SiCOI	Silicon-carbide-on-insulator
$Si_{1-x}Ge_x$	Silicon-germanium alloy
$Si_{0.16}Ge_{0.84}$	Silicon-germanium alloy with 16% Si and 84% Ge
SiH_4	Silane
Si_2H_6	Disilane
Si-NPA	Silicon nanoporous pillar array
SiNW	Silicon nanowire
SiO_2	Silicon dioxide
SiON	Silicon oxynitride
SiP	System-in-package
SL	Source line
SNNs	Spiking neural networks
SnTe	Tin telluride
SO	Spin-orbit
SOC	Spin-orbit coupling

SoC	System-on-a-chip
SOD	Silicon-on-diamond
SOI	Silicon-on-insulator
SOLES	Silicon on lattice-engineered substrates
SOT	Spin-orbit-torque
SQUIDs	Superconducting quantum interference devices
SRAM	Static random-access memory
SRDP	Spike-rate-dependent plasticity
$SrTiO_3$	Strontium titanate (or strontium titanium oxide)
STB	Spin-tunnel barrier
STDP	Spike-timing-dependent plasticity
STFET	Spin tunneling field-effect transistor
STM	Scanning tunneling microscopy
STO	Strontium titanate
STP	Short-term potentiation
STT	Spin-transfer torque
STT-MRAM	Spin-transfer torque magnetic random-access memory
SWCNT	Single-walled carbon nanotube
SyAF	Synthetic antiferromagnets

T

Ta	Tantalum
TaN	Tantalum nitride
TaO_x	Tantalum oxide
TASE	Template-assisted selective epitaxy
1T1C	One-transistor and one-capacitor
TCA	Thermal cyclic annealing
TDD	Threading dislocation density
TDs	Threading dislocations
TE	Top electrode
TFET	Tunnel field-effect transistor
TFT	Thin-film transistor
THz	Terahertz
TI	Topological insulator
TiN	Titanium nitride
TMAH	Tetramethylammonium hydroxide
TMDCs/TMDs	Transition metal dichalcogenides
TMR	Tunnel magnetoresistance
TPa	Tera Pascal
1T1R	One-transistor and one-resistor
4T2R	4 transistors, 2 resistors
TRI	Time-reversal invariant (band structure)
1T1S	One-transistor and one-selector
TSS	Topological surface states
TSV	Through-silicon via
TTL	Transistor-transistor logic

U

UHV	Ultrahigh vacuum

V

V	Volt
VCMA	Voltage-controlled magnetic anisotropy
vdW	van der Waals
VGAA	Vertical gate-all-around
VLSI	Very-large-scale integration
VMM	Vector-matrix multiplication
VSe_2	Vanadium diselenide

W

W	Watt, Tungsten
WBG	Wide-bandgap semiconductors
WKB	Wentzel-Kramers-Brillouin (approximation)
WL	Word line
WoW	Wafer-on-wafer (bonding)
WS_2	Tungsten disulfide
WSe_2	Tungsten diselenide
WTe_2	Tungsten ditelluride

Y

YSZ	Yttrium-stabilized zirconia

Z

zJ	zeptojoule
ZnO	Zinc oxide
ZnS	Zin sulfide
ZrO_2	Zirconium dioxide (zirconia)
ZTO	Zinc tin oxide

Greek Letters

κ

κ^2	Linear area scaling factor

μ

μA	Microampere
μC	MicroCoulomb
μF	Microfarad

μm	Micrometer (micron)
μS	microSiemens
μTP	Micro-transfer printing
μW	Microwatt

Γ

Γ-valley	Gamma-valley

Ω

Ω	Ohm
kΩ	Kiloohm

Mathematical Symbols

A

A_v	Voltage amplification factor

C

C	Capacitance
C_{dep}	Depletion capacitance
C_{eff}	Effective gate capacitance
C_{FE}	Ferroelectric capacitor
C_G	Gate capacitance
C_{GD}	Gate-to-drain capacitance
C_{GS}	Gate-to-source capacitance
C_i	Gate insulator capacitance
C_{Load}	Load capacitance
C_{MOS}	MOSFET capacitor
C_{Ox}	Oxide capacitance per unit area

D

D_0	Average defect density

E

E	Electric field, Switching energy
e	Elementary charge
E_b	Minimum energy barrier for magnetization reversal
$E_{Breakdown}$	Breakdown electric field
E_C	Charging energy
E_c	Critical electric field, coercive field
E_{CMOS}	Energy consumed by CMOS
E_G	Bandgap

E_{Min}	Minimum energy dissipation
E_{Neuro}	Energy consumed by neuromorphic chip
E_{Sw}	Switching energy

F

F	Feature size
f	Frequency
f_1, f_2	Old and new operating frequency

G

G	Electrical conductance
G_0	Conductance quantum
G_{AP}	Anti-parallel conductance
G_{ij}	Conductance of the memory cell located at the intersection of the i-th row and j-th column
G_{P}	Parallel conductance

H

H	Magnetic field, height
h	Planck's constant
$\hbar$	Reduced Planck's constant ($h/2\pi$)

I

I	Current
I_{DS}	Drain-source current
I_j	Output current in the j-th column
I_{L}	Current through the left branch
I_{Off}	Off current
I_{On}	On current
I_{R}	Current through the right branch
I_{Sum}	Total accumulated current

J

J	Current density
J_{c}	Charge current
J_{s}	spin current

K

K	Kelvin
k	Dielectric constant, Thermal conductivity
k_{B}	Boltzmann constant

L

L	Channel length
$L-\Delta L$	Effective channel length
L_1, L_2	Initial and new processor technology nodes (characteristic feature sizes)
L_{2D}/L_{3D}	Average wire length in the 2D/3D integrated circuit

M

M	Magnetization
m	Ideality factor
m^*	Effective electron mass
m_e	Electron rest mass
$\mathrm{MW}_{\mathrm{Max}}$	Maximum memory window

N

n	Subthreshold slope factor, Time exponent
N_1, N_2	Initial and expected number of transistors
N_A	Acceptor doping concentration

P

P	Power, polarization, pitch
P(A)	Probability of obtaining a good Chiplet A
P_{Avg}	Average power consumed per transistor
P_{Dynamic}	Dynamic power dissipation
P_{Static}	Static power dissipation
P_r	Remanent polarization

Q

q	Electronic charge

R

R	Resistance, the wafer radius
R_{AP}	Resistance in anti-parallel state
r_j	Junction depth
R_{Off}	Off-resistance
R_{On}	On-resistance
R_P	Resistance in parallel state

S

S or SS	Subthreshold swing

T

T	Temperature
t	Switching time
T_2	Coherence time (dephasing or transverse relaxation time)
t_{Delay}	Total propagation delay
T_{op}	Operation time
t_{Ox}	Thickness of the oxide
t_{Switch}	Average switching time for each magnet

V

V_D	Drain voltage
v_d	Drift velocity
V_{DD}	Drain supply
V_{DS}	Drain-source voltage
V_{DS}	Drain-source saturation voltage
V_{GS}	Gate-source voltage
V_i	Input voltage applied to the *i*-th row
V_{In} or V_{IN}	Input voltage
V_{Out} or V_{OUT}	Output voltage
V_S	Source voltage
v_{Sat}	Saturation velocity
$V_{T,Long}$	Threshold voltage of long-channel MOSFET
V_{Th} or V_T	Threshold Voltage

W

W	Width

X

x_d	Depletion region width

Y

Y	Yield
Y_0	A non-random yield component

Greek Letters

Δ

Δ	Thermal stability factor
ΔL	Linear displacement, Change in length
$\Delta\theta_{Radians}$	Rotational error in radians.
ΔQ	Total switchable charge
ΔV_T	Shift in the threshold voltage

ϵ

ϵ	Dielectric constant, lattice mismatch percentage

η

η_D	Spin detection efficiency
η_S	Spin injection efficiency

θ

θ_{SH}	Spin Hall angle

λ

λ	Channel-length modulation parameter

μ

μ	Carrier mobility
μ_h	Hole mobility
μ_n	Electron mobility

ρ

ρ_{3D}	Volumetric transistor density
$\rho_{effcetive}$	Effective resistivity
$\rho_{Nominal}$	Standard resistivity

σ

σ	Standard deviation
σ_{2D}	Two-dimensional electrical conductivity
σ_{SH}	Spin Hall conductivity
σ_{xx}	Longitudinal conductivity

τ

τ_d	Switching delay

Φ

Φ_B	Potential barrier height

1 Physical and Technological Limitations of CMOS Nanoelectronics

1.1 INTRODUCTION

Complementary metal-oxide-semiconductor (CMOS) nanoelectronics pertain to semiconductor devices with dimensions ranging from 1 to 100 nm, including contemporary FinFETs. As these devices are miniaturized down to atomic scales, they confront essential physical and technological challenges that jeopardize the continuation of Moore's Law. Physical limitations stem from the intrinsic laws of nature and the discrete characteristics of matter, exemplified by quantum tunneling, short-channel effects (SCEs), atomic limits, thermal effects, and statistical variability. Technological constraints encompass obstacles related to manufacturing and the cost-effectiveness of scaling, including lithographic constraints, interconnect bottlenecks, manufacturing costs, and reliability. In response to these challenges, the industry is moving toward innovative materials and designs.

1.2 ORIGINS OF CMOS TECHNOLOGY

Frank Wanlass and Chih-Tang Sah invented CMOS at Fairchild Semiconductor in 1963 (Wanlass and Sah 1963). Wanlass created the initial configuration of a logical circuit that integrated P-channel and N-channel MOS transistors to significantly reduce power consumption. In contrast, Sah made substantial contributions to its evolution, and together they shared their findings at a conference in 1963. Frank Wanlass created and secured a patent for the fundamental CMOS circuit design, which was granted in 1967 (Wanlass 1967).

1.3 CORE PRINCIPLES OF CMOS AND ITS ADVANTAGES

CMOS technology uses a complementary pair of P-type (PMOS) and N-type (NMOS) transistors to implement logic gates (Kang and Leblebici 2002). Its complementary design is shown in Figure 1.1(a). This design ensures that when a logic gate is in a stable state, one transistor is always turned off, resulting in extremely low power usage. To understand how one transistor is always off, we examine the equivalent circuit of the CMOS arrangement sketched in Figure 1.1(b). As shown, in a CMOS circuit, a pair of MOSFETs (an N-channel and a P-channel) is arranged

DOI: 10.1201/9781003753469-1

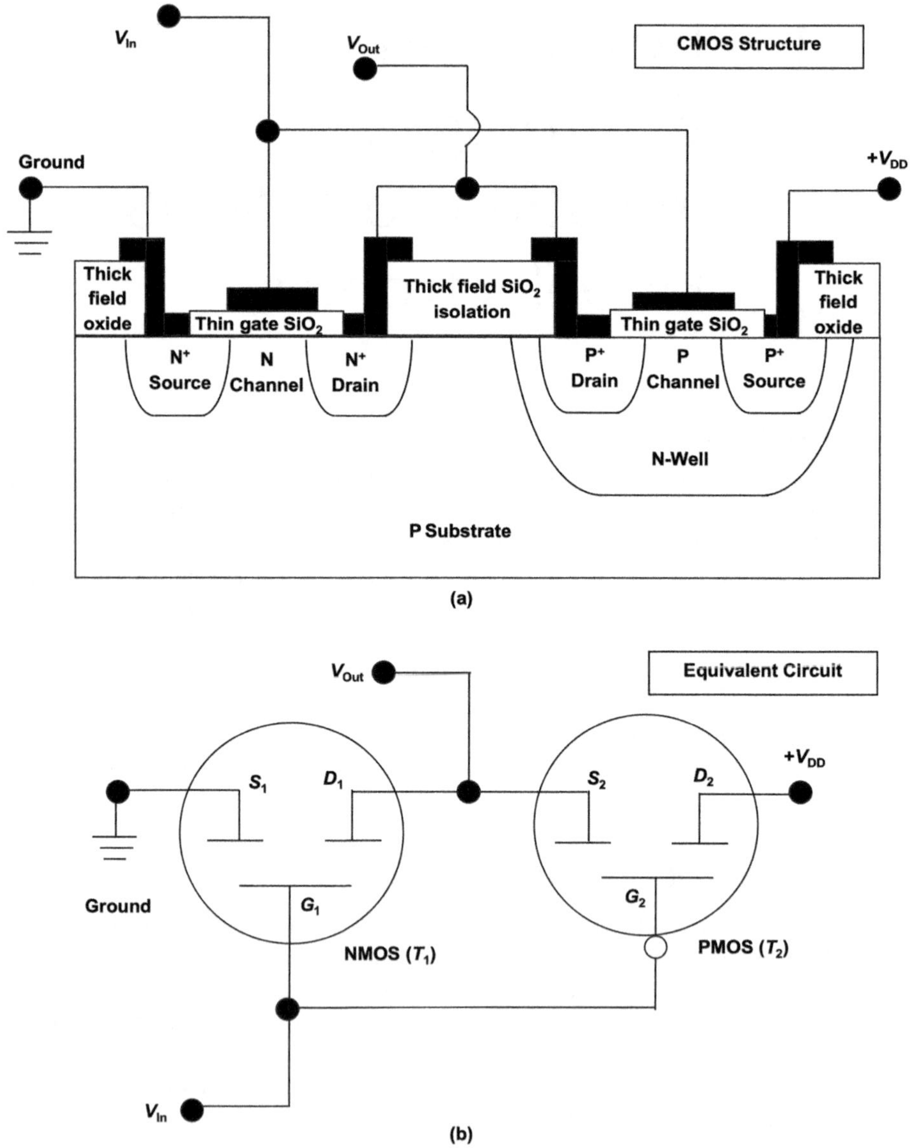

FIGURE 1.1 CMOS structure, equivalent circuit, and advantages: (a) Cross-sectional diagram shows a P-substrate in which an N-channel MOSFET is fabricated by N^+ source and drain diffusions. Adjoining this NMOSFET, an N-well is created, and a P-channel MOSFET is fabricated by P^+ source and drain diffusions in the N-well. Thin gate SiO_2, thick field oxide, gate metal, source, and drain contact metals are indicated. The source of NMOSFET is grounded, the drain of the NMOSFET is shorted with source of PMOSFET, and the drain of PMOSFET is connected to $+V_{DD}$ supply. Input signal V_{in} is applied to the shorted gates of NMOSFET and PMOSFET, and the output signal V_{Out} is drawn from the connected drains of the MOSFETs. (b) The equivalent circuit of CMOS structure consists of an NMOS transistor (T_1) with source S_1, gate G_1, and drain D_1 and a NPMOS transistor (T_2) with source S_2, gate G_2, and drain D_2. Terminal D_1 is connected with S_2, S_1 is grounded, and D_2 is joined to $+V_{DD}$. V_{in} is fed to the shorted gate terminals G_1, G_2 of T_1, T_2. V_{Out} is taken from interconnected D_1, S_2 terminal.

in series, with the P-channel device' source connected to a positive supply, $+V_{DD}$, and the N-channel device's source connected to ground. The gates of both devices are connected to a standard input, and their drain terminals are combined to form a standard output.

When the input voltage is at a high level, $+V_{DD}$ volts, the gate G_2 of MOSFET T_2 is at zero potential relative to its source S_2, causing T_2 to be OFF with a resistance of $R_{OFF} = 10^{10}\ \Omega$. Meanwhile, gate G_1 of MOSFET T_1 is at a positive potential relative to its source S_1, indicating that T_1 is ON with a resistance of $R_{ON} = 1\ \text{k}\Omega$. In this scenario, the output will be approximately 0 V.

Conversely, when the input is at 0 V, the gate G_2 of MOSFET T_2 experiences a negative potential with respect to its source S_2, making MOSFET T_2 ON with $R_{ON} = 1\ \text{k}\Omega$. At the same time, the gate G_1 of MOSFET T_1 is at 0 V relative to its source S_1, which keeps T_1 OFF with a resistance of $R_{OFF} = 10^{10}\ \Omega$. These resistances form a potential divider, leading to an output of approximately $+V_{DD}$ volts.

In CMOS, complementary pull-up and pull-down operations alternate, activating in opposition, ensuring no static power is wasted (Figure 1.2).

The CMOS architectural design offers several notable benefits compared to other logic families like transistor-transistor logic (TTL) or NMOS logic (Figure 1.3(a)). As already noted, supremacy of CMOS is assured by its minimal static power usage. In a stable state (when the output is either high or low logic), one transistor remains off, preventing direct current from flowing from the power supply to ground. Power consumption predominantly occurs during brief state transitions, resulting in dynamic power dissipation. This power usage, too, is negligible. Its low power consumption makes CMOS particularly suitable for contemporary battery-operated

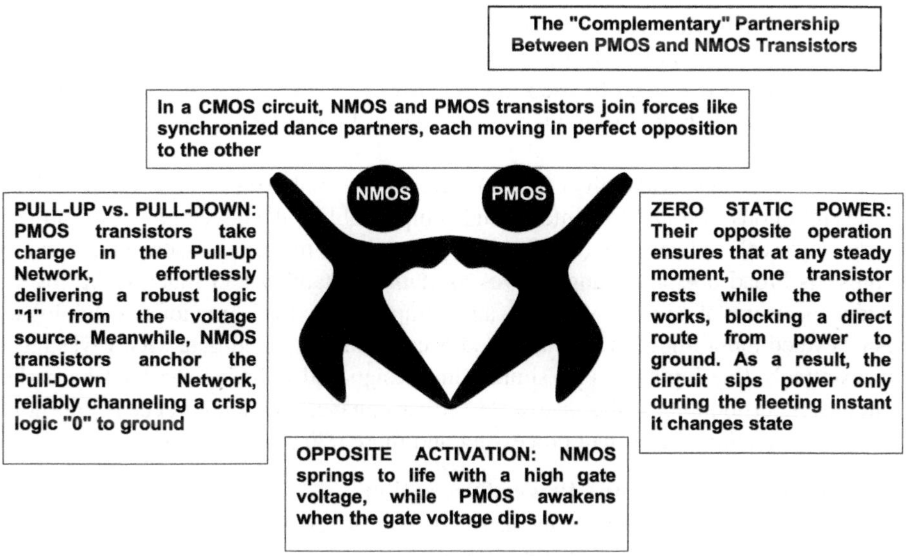

FIGURE 1.2 Complementary concept in CMOS is exhibited by pull-up vs. pull-down operation, through opposite activation, resulting in zero static power.

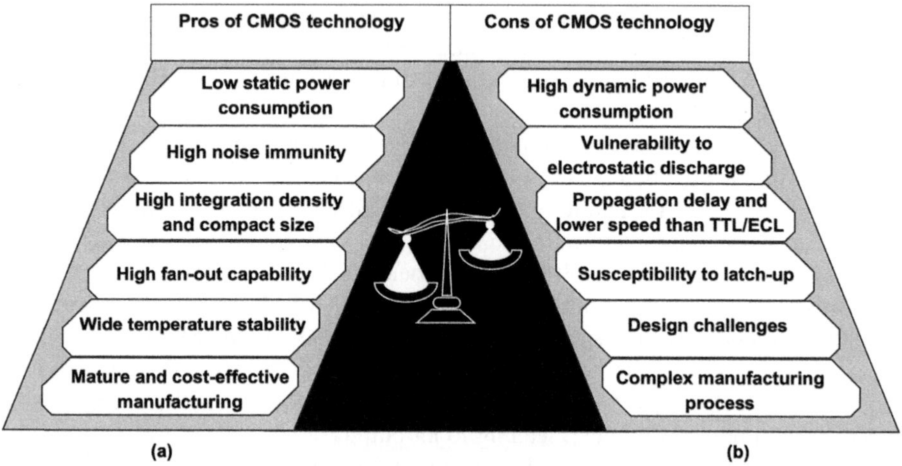

FIGURE 1.3 Advantages and disadvantages of CMOS technology: (a) Advantages of CMOS technology are low power consumption, high immunity to noise, high integration density, high fan-out capability, broad operating temperature range, and standardized production line. (b) Disadvantages of CMOS technology are high dynamic power consumption, vulnerability to electrostatic discharge, propagation delay and lower speed than transistor-transistor logic (TTL) or ECL, susceptibility to latch-up where a parasitic, low-impedance path forms between power supply rails, design challenges, and a complex manufacturing process requiring expensive, cleanroom facilities.

and high-density devices. Besides power reduction, CMOS provides a strong noise resilience: CMOS circuits exhibit high resistance to voltage variations and electrical interference, ensuring consistent performance. The output voltage approaches the supply rails (V_{DD} and GND), guaranteeing a clear differentiation between logic "0" and "1." Still, it is crucial to keep the drawbacks of CMOS technology in mind, as highlighted in the diagram (Figure 1.3(b)), to avoid potential failures when designing applications.

High density and scalability of CMOS technology enable the integration of millions or even billions of transistors into a single chip, enabling the development of intricate very-large-scale integration (VLSI) systems such as microprocessors and memory chips. Its broad operating range serves to maintain stability and functionality over a wide spectrum of supply voltages and temperatures than previous technologies. An elevated input impedance is achieved because the insulated-gate MOSFETs result in extremely low input current, simplifying design and reducing the load on earlier circuits. Transitioning to the nanoscale further enhances switching speeds and noise tolerance, enabling faster data processing and more reliable performance.

Example 1.1 Dynamic Power Dissipation

A CMOS inverter operating at the nanoscale runs at 250 MHz with a supply voltage (V_{DD}) of 0.8 V. The total load capacitance (C_{Load}), which includes gate, interconnect, and parasitic capacitances, is estimated at 30 fF (femtofarads). Determine the dynamic power dissipation ($P_{Dynamic}$) for this individual CMOS inverter.

Solution: In CMOS circuits, dynamic power dissipation primarily arises from the charging and discharging of the load capacitance, represented by the formula:

$$P_{\text{Dynamic}} = C_{\text{Load}} V V_{\text{DD}}^2 fP \tag{1.1}$$

where C_{Load} indicates the load capacitance, V_{DD} signifies the supply voltage, and f denotes the operating frequency.
Identifying the provided values: $C_{\text{Load}} = 30\ \text{fF} = 30 \times 10^{-15}$ F, $V_{\text{DD}} = 0.8$ V, $f = 250\ \text{MHz} = 250 \times 10^6$ Hz. Inserting the values into the formula and carrying out the computation:

$$\begin{aligned} P_{\text{Dynamic}} &= 30 \times 10^{-15}\ \text{F} \times (0.8)^2\ \text{V}^2 \times 250 \times 10^6 \\ \text{Hz} &= 30 \times 0.64 \times 250 \times 10^{-15+6}\ \text{FV}^2 \\ \text{Hz} &= 4800 \times 10^{-9} \\ \text{W} &= 4.8 \times 10^{-6} \\ \text{W} &= 4.8 \times 10^{-6} \times 10^6\ \mu\text{W} = 4.8\ \mu\text{W} \end{aligned} \tag{1.2}$$

since

$$\text{FV}^2\ \text{Hz} = \frac{\text{C}}{\text{V}}\text{V}^2\frac{1}{\text{s}} = \frac{\text{C}}{\text{s}}\text{V} = \text{A} \times \text{V} = \text{W} \tag{1.3}$$

The dynamic power dissipation for the single nanoscale CMOS inverter is 4.8 μW.
Connection to "Beyond CMOS": Although this power figure for a single gate may appear minimal, a conventional processor comprises billions of transistors. The significant cumulative power consumption and challenges in lowering the supply voltage—due to the fundamental limit of 60 mV/dec subthreshold slope—are essential factors driving research into "beyond CMOS" technologies aimed at enhancing computational efficiency and reducing switching energy.

1.4 DAWN OF THE CMOS NANOELECTRONICS ERA

The shift to CMOS nanoelectronics occurred several decades later, mainly during the late 1990s and early 2000s, as feature sizes reached the nanoscale (Wind et al. 1995, Wong et al. 1999, Xiang et al. 2000, Veendrick 2001). Around the mid-2000s, CMOS technology shrunk to sub-100 nm, marking the beginning of the nanoelectronics era. This evolution to nanoelectronics was a gradual progression influenced by Moore's Law, as engineers consistently reduced transistor size to fit hundreds of thousands, and ultimately billions, onto a single chip.

Example 1.2 Moore's Law and Dimensional Scaling

A semiconductor firm has a processor containing 2 billion transistors, produced using a 22 nm technology node. Following Moore's law (which, for this scenario, states that the number of transistors doubles every 2 years), the firm intends to launch a new chip in 2 years, using a more advanced technology node that reduces linear dimensions by a factor of 0.7.

Calculate: The anticipated number of transistors in the new processor. The estimated feature size (technology node) of the latest processor. The estimated reduction in the chip's area, assuming the overall design size continues to occupy the same percentage of the die area.

Solution: Given parameters: Initial transistors $N_1 = 2$ billion. Initial technology node $L_1 = 22$ nm, Time interval = 2 years. Transistor doubling period (according to Moore's Law) = 2 years. Linear scaling factor $\kappa = 0.7$
Expected number of transistors (N_2): According to the simplified version of Moore's Law, the number of transistors doubles every two years.

$$N_2 = N_1 \times 2 = 2 \text{ billion} \times 2 = 4 \text{ billion transistors} \tag{1.4}$$

Estimated feature size of the new processor (L_2): The linear dimensions will be reduced by a factor of 0.7 for the upcoming technology node.

$$L_2 = L_1 \times \kappa = 22 \text{ nm} \times 0.7 = 15.4 \text{ nm} \tag{1.5}$$

In practice, this value would typically be rounded to the nearest standard node, likely 16 nm or 14 nm, but the calculation gives the theoretical scaled value.
Estimated reduction in the chip's area: The area changes by the square of the linear scaling factor (κ^2).

$$\text{Area scaling factor} = \kappa^2 = (0.7)^2 = 0.49 \tag{1.6}$$

The new area will be around 49% of the original area.

$$\text{Area reduction} = (1 - 0.49) \times 100\% = 0.51 \times 100\% = 51\% \tag{1.7}$$

The new technology node provides an approximate 51% reduction in area per circuit.
Beyond CMOS Connection: This example highlights the key principle of CMOS scaling that has propelled the industry for many years. However, as dimensions approach the single-digit nanometer range (as observed in L_2), physical limitations, such as increased leakage currents and short-channel effects, make traditional CMOS technologies increasingly challenging to maintain. "Beyond CMOS" nanoelectronics investigates alternative techniques, e.g., spintronics, quantum dots, and 2D materials, to either enhance or substitute conventional silicon devices once these scaling constraints are reached.

Example 1.3 Moore's Law and Operating Frequency

In a historical interpretation of Moore's law (often presented in a simplified manner), the dimensions of devices shrink by roughly a factor of 0.84 each year, enhancing device density and performance. This size reduction generally results in a corresponding rise in operating frequency. Given that the current CMOS technology node operates at an average frequency (f_1) of 3 GHz, with a characteristic feature size (L_1) of 45 nm, what would the anticipated new operating frequency (f_2) be if the feature size reduces to the next central node's target of 32 nm (L_2), assuming frequency changes inversely with linear device dimensions?

Solution: The operating frequency is inversely related to the linear dimension of the device ($f \propto 1/L$). Setting up the ratio, we have:

$$\frac{f_2}{f_1} = \frac{L_1}{L_2} \tag{1.8}$$

Rearranging this, we solve for the new frequency (f_2):

$$f_2 = f_1\left(\frac{L_1}{L_2}\right) \tag{1.9}$$

By substituting the given values: $f_1 = 3$ GHz, $L_1 = 45$ nm, and $L_2 = 32$nm, we find

$$f_2 = 3\text{ GHz}\left(\frac{45\text{ nm}}{32\text{ nm}}\right) = 3\text{ GHz} \times 1.40625 \tag{1.10}$$

Performing the calculation yields:

$$f_2 = 3\text{ GHz} \times 1.40625 = 4.21875\text{ GHz} \tag{1.11}$$

Thus, the expected new operating frequency for the 32 nm node would be around 4.22 GHz. This kind of scaling analysis illustrates the historical improvements achieved through miniaturization of CMOS devices and the physical limitations that drive exploration into "beyond CMOS" technologies such as single-electron transistors and quantum computing approaches.

1.5 SERVICES OF CMOS NANOELECTRONICS TO THE SEMICONDUCTOR INDUSTRY

CMOS nanoelectronics has been foundational to the semiconductor sector for many years, mainly because of its remarkable features, such as ultra-low power consumption, outstanding scalability, high integration density, strong noise immunity, and reliability. Additionally, it benefits from a well-established, cost-effective manufacturing ecosystem that has enabled continuous Moore's law advancement, driving everything from simple computing tasks, such as the real-time clock on a motherboard, to advanced microprocessors and artificial intelligence (AI) accelerators.

CMOS nanoelectronics offer a wide array of essential services and functionalities vital to the semiconductor sector, particularly through the design and manufacturing of integrated circuits (ICs) that boast high density, efficiency, and performance. Here are a few contributions of CMOS nanoelectronics to the semiconductor field:

1.5.1 Design and Manufacturing Services

IC manufacturing is the core process of producing various ICs, from microprocessors to memory chips, utilizing a stable and scalable method. Transistor miniaturization (scaling) is the ongoing reduction of transistor dimensions to the nanometer scale, in accordance with Moore's Law, to achieve higher packing density and enhanced performance. VLSI design is facilitating the creation of intricate circuits containing millions or even billions of transistors on a single chip. System-on-a-chip (SoC) integration consists of merging all essential electronic system components (CPU, memory, etc.) into one CMOS die, lowering the bill of materials (BOM) cost and enhancing efficiency. Application-specific integrated circuit (ASIC) prototyping and production involves manufacturing tailored chips for particular applications, benefiting from the energy efficiency and high-performance characteristics of CMOS technology. Advanced packaging and heterogeneous integration are implementing 2.5D and 3D stacking methods along with chiplets to transcend conventional 2D scaling and boost overall system performance. Wafer manufacturing and processing are delivering the full infrastructure and methodologies required for producing silicon wafers utilized in chip production. Photolithography metrology provides accurate measurement and control solutions for deep ultraviolet (DUV) and extreme ultraviolet (EUV) lithography techniques used for nanoscale pattern creation.

1.5.2 Functional and Performance Services

The low power consumption of CMOS circuits is achieved by engineering circuits that use minimal power when not in active use (near-zero static power), prolonging battery durability in portable devices. Their high switching speeds facilitate rapid transistor operations, which is key for high-performance computing systems and real-time processing. The high noise resistance ensures robust circuits that are less affected by electrical noise, guaranteeing dependable performance in diverse settings. A wide operating voltage range allows adaptable usage in various applications, from energy-efficient wearables to higher-power industrial systems.

1.5.3 Specialized and Emerging Services

CMOS image sensor production involves fabricating image sensors that are widely utilized in digital cameras, smartphones, and medical imaging technologies. Analog and mixed-signal circuit design merges both digital and analog capabilities on a singular CMOS chip, which is crucial for communication systems (RF circuits, transceivers) and sensor data processing. Integration with emerging technologies supports research and development for hybrid technologies, such as coordinating CMOS with quantum computing control circuits or nanomagnetic memories. Development of AI Accelerators is delivering optimized CMOS architectures specifically designed for machine learning and AI tasks. Biomedical and biosensor applications include manufacturing ultra-low-power devices for medical implants, microfluidic setups, and extensive diagnostic biosensor arrays. Automotive electronics production is engaged in developing reliable and power-efficient microcontrollers and sensors for advanced driver assistance systems (ADAS), navigation, and entertainment systems. Industrial automation control is creating microcontrollers and sensors to oversee industrial robots, IoT gadgets, and smart factory systems for improved efficiency and precision. Intellectual property (IP) and process design kit (PDK) creation is crafting essential IP blocks and standardized PDKs to streamline effective and reliable chip design across the sector.

1.6 IMPEDIMENTS TO THE PROGRESS OF CMOS NANOELECTRONICS NEAR THE ATOMIC SCALE

The ongoing scaling of CMOS devices into the nanometer range has faced substantial physical, material, power, and economic obstacles. Picture a long-channel MOSFET as a broad, tranquil river, where electrons glide smoothly from source to drain. Here, the depletion regions at each end are mere ripples compared to the channel's length and never touch (Figure 1.4). But as we shrink the channel length, sometimes dipping below 1 μm, the river narrows until the banks nearly collide, and the once orderly flow turns unpredictable. The real difference between long- and short-channel MOSFETs is not just their size but the way the electric fields inside begin to play by a whole new set of rules.

In short channels, the drain electric field has a significant impact on the channel, aiding in charge depletion and channel inversion even at reduced gate voltages. As a result, the threshold voltage decreases, turning on the transistor at lower voltages, thereby increasing subthreshold leakage current. As the length of the transistor channel approaches that of the depletion region, issues such as Drain-Induced Barrier Lowering (DIBL) and "punch-through" emerge, complicating effective transistor off-state operation. The punch-through effect is explained in Figure 1.5.

With extremely thin gate oxides and lower threshold voltages, quantum-mechanical tunneling occurs, resulting in increased current leakage (both off-state and gate leakage) and elevated static power consumption. The inability to proportionally reduce the supply voltage (V_{DD}) with device size increases power density, making heat dissipation a significant concern. At the nanoscale, minor variations in manufacturing processes (such as doping levels or gate lengths) significantly affect device performance, leading to inconsistencies and reduced yield. Traditional materials like silicon dioxide have reached their fundamental limits in dielectric strength and conductivity, prompting the adoption of new materials, such as high-κ dielectrics and metal gates, to control leakage. Nanoscale devices are more vulnerable to issues such as soft errors (caused by energetic particles), time-dependent dielectric breakdown, and electromigration, which can lead to device degradation and eventual failure. The cost of establishing foundries and manufacturing chips at advanced nodes (such as 2 nm) has grown dramatically due to the need for advanced lithography techniques (such as EUV lithography) and intricate testing, which affects the profitability of continued scaling. Instead of outright failure, the industry has responded with creative solutions to prolong CMOS longevity, including the introduction of new materials (such as high-k dielectrics and metal gates), innovative device architectures (such as FinFETs, gate-all-around transistors, and nanosheets), and novel integration methods (such as 3D integration). Currently, the leading technology nodes for commercial CMOS nanoelectronics are the 3 and 2 nm nodes, with ongoing research and development aimed at advancing beyond the 1 nm node. The phrase "technology node" denotes a particular generation of manufacturing technology characterized by specific design guidelines and performance standards, rather than indicating a precise physical measurement such as gate length. CMOS continues to be the leading technology, but researchers are actively investigating alternative "beyond CMOS" devices (such as carbon nanotubes (CNTs), graphene, and spintronics) to complement or potentially replace conventional transistors in the future as silicon's physical limits are ultimately approached.

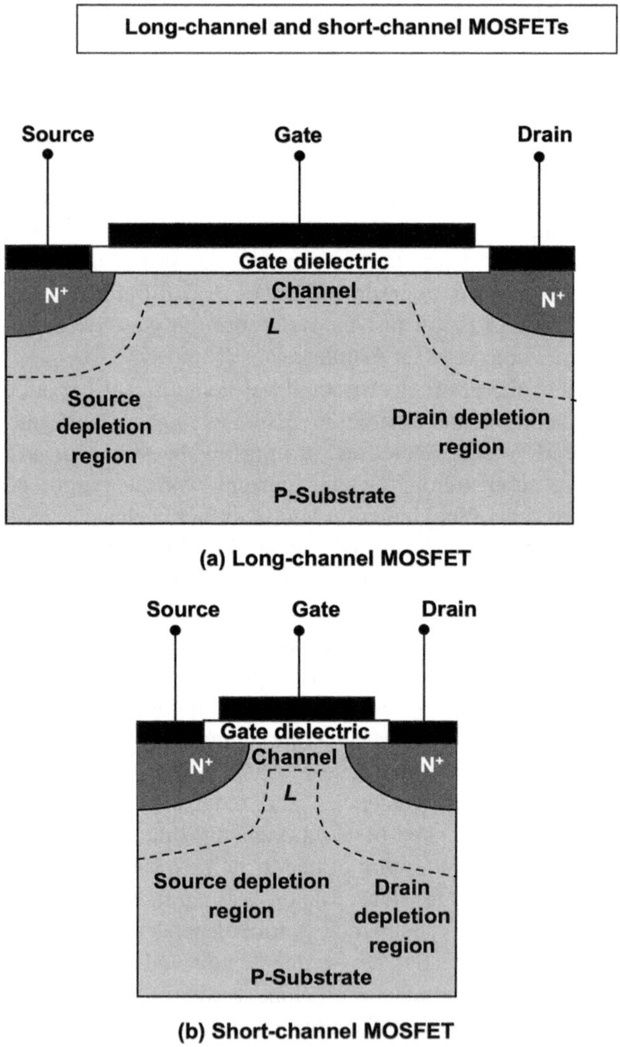

FIGURE 1.4 Long-channel and short-channel MOSFETs. MOSFETs are classified as long-channel or short-channel devices based on the ratio of channel length (L) to the depth of the source and drain depletion regions. Long-channel MOSFETs have a channel length much greater than the depletion width, while short-channel MOSFETs have a channel length comparable to the depletion width.

1.6.1 Threshold Voltage Reduction (V_{Th} Roll-Off)

In long-channel devices, the threshold voltage (V_{Th}) tends to remain invariable and stable. However, as the channel length is shortened, the threshold voltage decreases. This occurrence is known as V_{Th} roll-off. This happens because the depletion region from the source begins to overlap with that of the drain, reducing the gate's effectiveness in controlling the channel (Ogura et al. 1980).

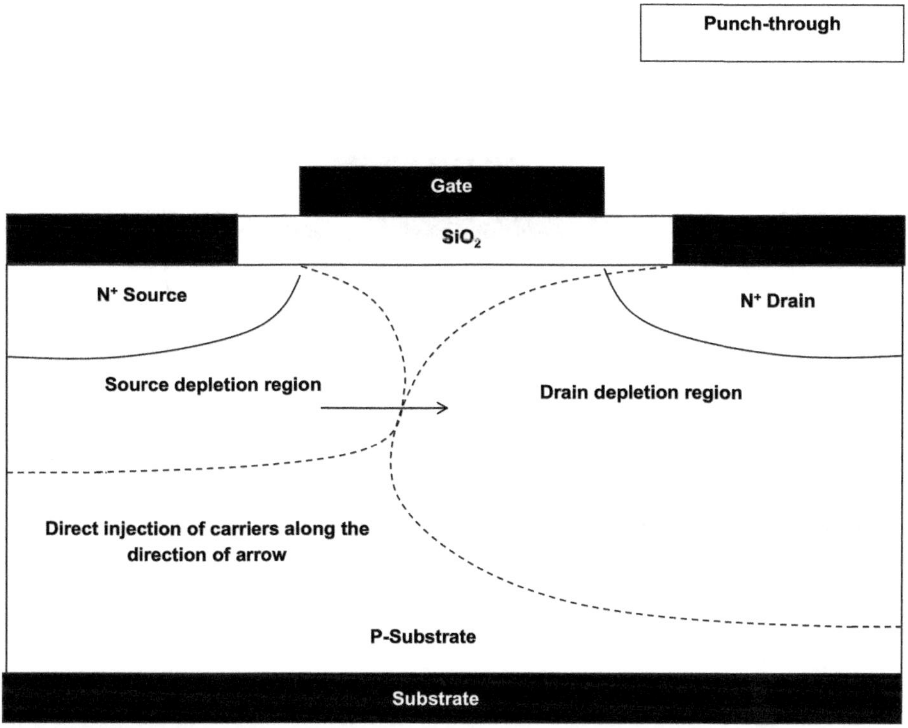

FIGURE 1.5 Source-drain punch-through. This occurs when the drain depletion region extends across the channel and merges with the source depletion region, especially in short-channel MOSFETs under high drain bias. As a result, uncontrolled high-density current flows in the substrate, increasing subthreshold leakage, power consumption, and the risk of device breakdown.

A viable approach to lessen the impact of V_{Th} roll-off involves using a lightly doped drain (LDD): introducing a lightly doped region between the source/drain and the channel reduces the electric field at the junction, helping sustain the threshold voltage and minimize roll-off. Another solution strategy is to use high-permittivity (high-k) gate dielectrics, such as HfO_2, improves gate control over the channel, offsetting the diminished gate-channel coupling seen in short-channel devices.

Example 1.4 Threshold Voltage Shift

An N-channel MOSFET has the following specifications: Channel length (L): 0.5 μm, Junction depth (r_j): 0.1 μm, Substrate doping concentration (N_A): 1×10^{16} cm^{-3}, Oxide capacitance per unit area (C_{Ox}): 6.9×10^{-8} F/cm^2, and source/drain depletion region widths (x_d): 0.2 μm (for simplicity, we will assume equal widths with $V_{DS} = 0$). The permittivity of silicon ϵ_{Si} is 1.04×10^{-12} F/cm, and the elementary charge e is 1.6×10^{-19} C. Calculate the shift in the threshold voltage ΔV_T resulting from the short-channel effect.

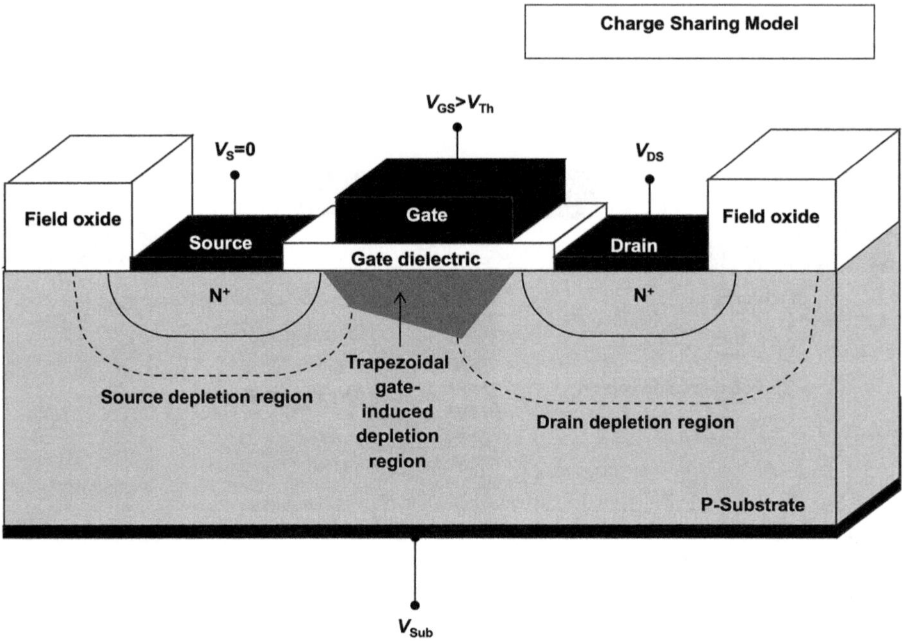

FIGURE 1.6 The charge-sharing model of a MOSFET. Diagram shows the N⁺ diffusions for source and drain, and source and drain contact metals, the gate dielectric, the gate metal, the field oxide, the source voltage V_S, the gate-source voltage V_{GS} > threshold voltage V_{Th}, the drain-source voltage V_{DS}, the source depletion region, drain depletion region, and the trapezium-shaped gate-induced depletion region produced by the encroachment of the source and drain depletion regions under the gate.

Solution: We can use the trapezoidal approximation model, often called the charge-sharing model and illustrated in Figure 1.6, which provides a formula:

$$\Delta V_T = -\frac{eN_A x_d}{C_{Ox}}\left\{1-\sqrt{1+\frac{2r_j}{x_d}}+\frac{r_j}{L}\left(\sqrt{1+\frac{2r_j}{x_d}}-1\right)\right\} \tag{1.12}$$

It is worth noting that a more straightforward and commonly used version, assuming semi-circular source/drain depletion regions, is given by

$$\Delta V_T = V_{T,\ \mathrm{Long}}\frac{r_j}{L}\left(\sqrt{1+\frac{2r_j}{x_d}}-1\right) \tag{1.13}$$

or similar alternatives depending on specific model assumptions. For this numerical problem, we will apply a well-known simplified expression:

$$\Delta V_T = -\frac{eN_A x_d}{C_{Ox}}\left\{1-\frac{r_j}{L}\left(\sqrt{1+\frac{2x_d}{r_j}}\right)\right\} \tag{1.14}$$

This equation estimates the proportion of the bulk charge managed by the gate, assuming that the unsupported sections near the source and drain form trapezoids and semicircles. Now,

$$
\begin{aligned}
-\frac{eN_A x_d}{C_{Ox}} &= -\frac{1.6\times10^{-19}\,\text{C}\times1\times10^{16}\,\text{cm}^{-3}\times0.2\times10^{-4}\,\text{cm}}{6.9\times10^{-8}\,\text{F/cm}^2}\\
&= -\frac{1.6\times1\times0.2}{6.9}\times10^{-19+16-4+8}\frac{\text{C}\times\text{cm}^{-3}\times\text{cm}\times\text{cm}^2}{\text{F}} = -\frac{0.32}{6.9}\times10^1\,\frac{\text{C}}{\frac{\text{C}}{\text{V}}}\\
&= -\frac{3.2}{6.9}\,\text{V} = -0.464\,\text{V}
\end{aligned}
\tag{1.15}
$$

$$
\begin{aligned}
1-\frac{r_j}{L}\left(\sqrt{1+\frac{2x_d}{r_j}}\right) &= 1-\frac{0.1\times10^{-4}\,\text{cm}}{0.5\times10^{-4}\,\text{cm}}\left(\sqrt{1+\frac{2\times0.2\times10^{-4}\,\text{cm}}{0.1\times10^{-4}\,\text{cm}}}\right)\\
&= 1-0.2\sqrt{1+\frac{0.4}{0.1}} = 1-0.2\sqrt{1+4} = 1-0.2\sqrt{5}\\
&= 1-0.2\times2.236 = 1-0.4472 = 0.5528
\end{aligned}
\tag{1.16}
$$

$$
\begin{aligned}
\therefore \Delta V_T &= -0.464\,\text{V}\times0.5528 = -0.256499\,\text{V} = -0.256499\times1000\,\text{mV}\\
&= -256.499\,\text{mV} \approx -256.5\,\text{mV}
\end{aligned}
\tag{1.17}
$$

Thus, the estimated change in the threshold voltage ΔV_T is about −256.5 mV. This decrease in threshold voltage is due to the short-channel effect, known as V_T roll-off.

1.6.2 Drain-Induced Barrier Lowering (DIBL)

In short-channel MOSFETs, an increase in the drain voltage can diminish the potential barrier at the source-channel junction, permitting carriers to enter the channel even when the gate voltage is below the threshold (Troutman 1979, Chamberlain and Ramanan 1986). This DIBL effect can be understood from Figure 1.7. It results in elevated off-state currents and worsens subthreshold characteristics. This effect can be overcome by reducing the drain-source overlap. Minimizing the overlap between the drain and the channel can help alleviate DIBL. FinFET designs offer enhanced electrostatic control due to the gate wrapping around the channel, which helps to lessen DIBL. Scaling of gate oxide is also helpful. Employing a thinner gate oxide enhances gate control and mitigates the effects of drain-induced electric fields.

Example 1.5 DIBL

A nanoscale FinFET device is evaluated at two distinct drain voltages (V_{DS}). The associated threshold voltages (V_{Th}) are determined from the curves of subthreshold current versus gate voltage ($I_D - V_{GS}$). The given values are as follows: for $V_{DS,Low} = 0.05$ V, the threshold voltage is $V_{Th,Low} = 0.45$ V, and for $V_{DS,High} = 1.0$ V, the threshold voltage is $V_{Th,High} = 0.25$ V. Calculate the DIBL coefficient for this device in millivolts per volt (mV/V).

Solution: The DIBL effect leads to a reduction in the threshold voltage as the drain voltage increases. The DIBL coefficient is determined by taking the absolute value of the

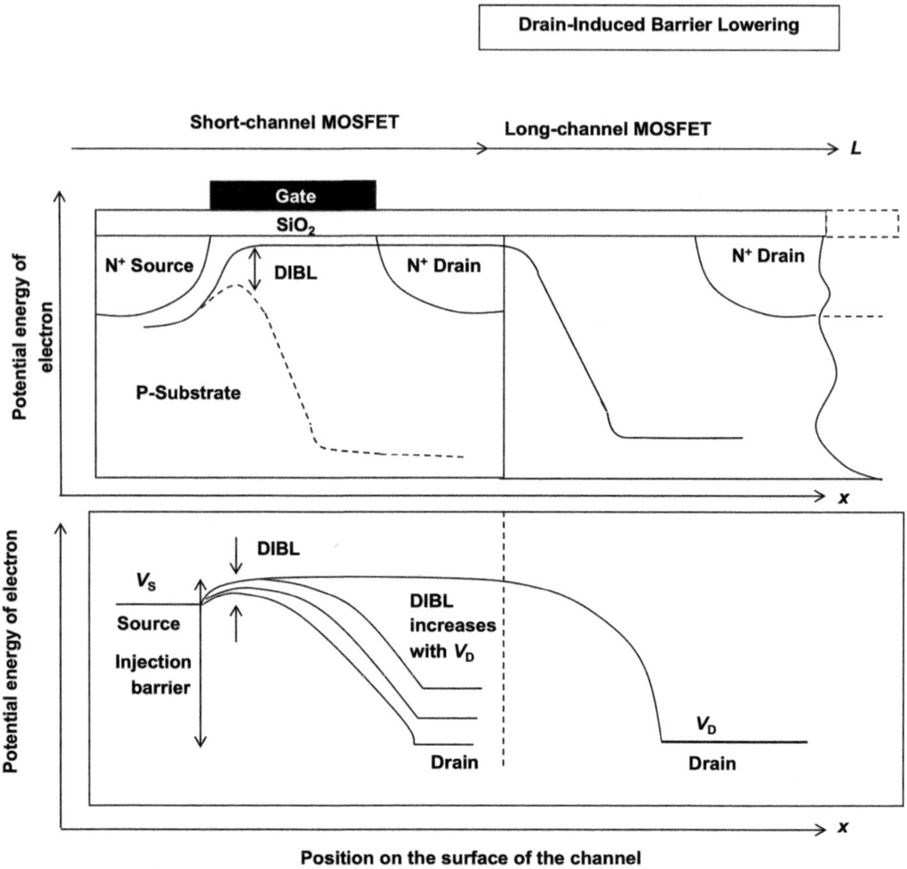

FIGURE 1.7 Drain-induced barrier lowering in a short-channel MOSFET. In both top and bottom diagrams, potential energy of the electron is plotted on the Y-axis against position x on the surface of the channel on the X-axis. Top diagram shows N^+ source and N^+ drain of the short-channel MOSFET along with SiO_2 insulator and gate metal. The N^+ drain of the long-channel MOSFET is shown farther away from the N^+ source in the same diagram. The P-substrate is indicated. The potential vs. position curves of the two MOSFETs are drawn below their structures. DIBL effect is represented by the distance between the top flat portions of the two curves. The bottom diagram illustrates how the injection barrier at the source end decreases with increasing drain voltage V_D with respect to source voltage V_S. Accordingly, the magnitudes of DIBL effect vary too. The greater the difference between V_D and V_S, more pronounced is the DIBL.

change in threshold voltage ΔV_{Th} divided by the change in drain voltage ΔV_{DS}. The equation for calculating the DIBL coefficient is:

$$\text{DIBL} = \left|\frac{\Delta V_{Th}}{\Delta V_{DS}}\right| = \left|\frac{V_{Th,High} - V_{Th,Low}}{V_{DS,High} - V_{DS,Low}}\right| \tag{1.18}$$

Calculating the change in threshold voltage ΔV_{Th},

$$\Delta V_{Th} = V_{Th,High} - V_{Th,Low} = 0.25\text{ V} - 0.45\text{ V} = -0.20\text{ V} \tag{1.19}$$

Calculating the change in drain-source voltage ΔV_{DS},

$$\Delta V_{DS} = V_{DS,High} - V_{DS,Low} = 1.0 \text{ V} - 0.05 \text{ V} = 0.95 \text{ V} \tag{1.20}$$

Hence,

$$\text{DIBL} = \left|\frac{\Delta V_{Th}}{\Delta V_{DS}}\right| = \left|\frac{-0.20 \text{ V}}{0.95 \text{ V}}\right| = 0.2105\text{V/V} = 0.2105 \times 1000\text{mV/V} = 210.5\text{mV/V} \tag{1.21}$$

The DIBL coefficient for the FinFET device is roughly 210.5 mV/V. (A DIBL value that is lower, for instance, 70 mV/V or below, typically signifies improved management of short-channel effects in nanoscale devices).

1.6.3 Subthreshold Slope Degradation

In a perfect MOSFET, the subthreshold slope (which indicates how sharply the device switches from off to on) remains constant. Nonetheless, in short-channel devices, this slope becomes flatter due to increased leakage currents, particularly at lower gate voltages (Shappir et al. 2003). This decline can lead to higher-power consumption when the device is in the off-state. Enhanced doping of the channel is frequently pursued as a potential remedy. Increasing channel doping improves gate control over carriers, thereby reducing the subthreshold slope. Multi-gate transistors featuring several gates, such as Tri-gate or Gate-All-Around devices, offer superior electrostatic control over the channel, thereby improving the subthreshold slope.

Example 1.6 Ideal Subthreshold Swing

Calculate the fundamental, theoretical minimum subthreshold swing in millivolts per decade (mV/dec) for a typical MOSFET functioning at room temperature (300 K). Given constants: Boltzmann constant (k_B): 1.38×10^{-23} J/K, Elementary charge (q): 1.60×10^{-19} C, Temperature (T): 300 K.

Solution: The theoretical subthreshold swing is defined by the thermal voltage (V_T) and can be determined using the formula:

$$S = 2.3\left(\frac{k_B T}{q}\right) \tag{1.22}$$

The factor 2.3 converts from the natural logarithm to a base-10 logarithm, since subthreshold swing is expressed in voltage per decade (a tenfold change in current).

To calculate the thermal voltage (V_T):

$$V_T = \frac{k_B T}{q} = \frac{1.38 \times 10^{-23} \text{ J/K } 300 \text{ K}}{1.6 \times 10^{-19} \text{ C}} = \frac{1.38 \times 300 \times 10^{-23+19}}{1.6} \text{ J/C} = 258.75 \times 10^{-4} \text{ V}$$
$$= 0.025875 \text{ V} \tag{1.23}$$

To find the ideal subthreshold swing (S):

$$S = 2.3\left(\frac{k_B T}{q}\right) = 2.3 \times 0.025875\ \mathrm{V} = 2.3 \times 0.025875 \times 1000\ \mathrm{mV} = 59.5125\ \mathrm{mV} \tag{1.24}$$

The fundamental, theoretical minimum subthreshold swing for a conventional CMOS device at room temperature is roughly 59.5 mV/dec.

Significance (Beyond CMOS Link): This value of 59.5 mV/dec represents a fundamental physical limit for traditional MOSFETs that use thermionic emission (diffusion) for current flow in the subthreshold region. Actual devices tend to exhibit a slightly higher SS (approximately 70–100 mV/dec) due to non-ideal behaviors and short-channel effects. Research is being conducted on "Beyond CMOS" devices, such as tunnel field-effect transistors (TFETs), which operate on an alternative principle (quantum-mechanical tunneling) and could achieve a subthreshold swing lower than 59.5 mV/dec, enabling sharper turn-off characteristics and significantly reduced power consumption.

Example 1.7 Subthreshold Swing in a Short-Channel Device

An ideal long-channel N-channel MOSFET at room temperature ($T = 300$ K) exhibits a subthreshold slope (SS) of 60mV/dec. A short-channel variant of the same device is produced. Due to short-channel effects, particularly charge-sharing and drain-induced barrier lowering (DIBL), its subthreshold performance is compromised. The device characteristics yield a subthreshold slope factor (n) of 1.25. Calculate the actual subthreshold slope (in mV/dec) of this short-channel device at room temperature.

Solution: The relationship for the subthreshold slope, also referred to as the subthreshold swing, is expressed by the equation:

$$S_{\mathrm{Actual}} = n \times \frac{kT}{q} \times \ln(10) = n \times \text{Ideal SS} \tag{1.25}$$

where n represents the subthreshold slope factor (which ideally equals 1, but often exceeds 1 in real or short-channel devices). k is the Boltzmann constant (1.38×10^{-23} J/K). T denotes the absolute temperature (in Kelvin). q signifies the elementary charge (1.602×10^{-19} C). The term $kT/q \times \ln(10)$ approximates the ideal subthreshold slope at a specific temperature; it is approximately 60 mV/dec at $T = 300$ K.

Recognizing the provided values: Subthreshold slope factor, $n = 1.25$. Ideal subthreshold slope at room temperature (300 K), Ideal SS ~ 60mV/dec.

Utilizing the formula:

$$S_{\mathrm{Actual}} = n \times \text{Ideal SS} \tag{1.26}$$

$$S_{\mathrm{Actual}} = 1.25 \times 60\mathrm{mV/decade} = 75\mathrm{mV/decade} \tag{1.27}$$

The actual subthreshold slope of the short-channel device amounts to 75 mV/dec. This measurement exceeds the ideal (60 mV/dec), indicating a less steep turn-off characteristic and increased off-state leakage current.

Example 1.8 Subthreshold Leakage Current

A nanoscale CMOS transistor features a gate length (L_G) of 10 nm. At this dimension, a significant concern regarding static power is the subthreshold leakage current (I_{Off}). Let's consider the following parameters for an individual NMOS transistor within a logic circuit: Subthreshold swing (S): 70 mV/dec, Off-current requirement $\left(I_{Off,Spec}\right)$: 1 nA, Operating temperature (T): 300 K, Supply voltage (V_{DD}): 0.8 V, and Threshold voltage (V_T): 0.3 V (nominal). Compute the maximum allowable change in the threshold voltage ΔV_T, while ensuring that the off current for this specific transistor does not exceed twice the specified threshold $\left(2I_{Off,Spec}\right)$ due to manufacturing process variations.

Solution: The subthreshold current can be simplified to $I_{off} \propto 10^{-V_T/S}$, or, more precisely, the relationship between current change and voltage change can be expressed using the subthreshold swing. The subthreshold swing (S) is the voltage required to change the drain current by one order of magnitude (a factor of 10). We can establish a link between the target current change and the voltage variation:

$$\frac{I_{off,Actual}}{I_{off,Spec}} = 10^{\Delta V_T/S} \tag{1.28}$$

Our goal is to have the actual current as $2 \times I_{Off,spec}$, which implies

$$\frac{I_{off,Actual}}{I_{off,Spec}} = 2 \tag{1.29}$$

Thus, we derive the equation

$$2 = 10^{\Delta V_T/S} \tag{1.30}$$

to solve for ΔV_T:

$$\log_{10} 2 = \Delta V_T/S \tag{1.31}$$

This leads to the expression

$$\Delta V_T = S \log_{10} 2 \tag{1.32}$$

By substituting the known values, we have $S = 70$ mV/dec and $\log_{10}(2) \sim 0.301$ to find

$$\Delta V_T = 70\,\text{mV} \times 0.301 = 21.07\text{mV} \tag{1.33}$$

The final result indicates that the threshold voltage variation must be controlled during manufacturing to remain within approximately 21 mV, satisfying the specified off-current requirements for all transistors. This emphasizes the importance of managing process variations as a significant challenge in nanoscale numerical and manufacturing applications.

1.6.4 Gate Leakage Current Increase due to Gate Oxide Thickness Reduction

In addition to subthreshold leakage, gate leakage current is another contributor to static power loss in transistors (Chen and Guo 2006). But the two currents differ in their flow paths, the physical processes that drive them, and their sensitivity to temperature changes. While subthreshold leakage current travels between the drain and the source, passing through the semiconductor channel, gate leakage current flows from the gate terminal to the channel, source, or drain via the gate oxide (insulator). Subthreshold leakage current is highly temperature-dependent, rising exponentially with increasing temperature; however, gate leakage current shows weak sensitivity to temperature changes (almost unaffected).

A thinner gate oxide typically improves the subthreshold slope (lowering it) by providing better gate control. However, a reduced gate oxide thickness results in a substantial, exponential increase in gate leakage current due to carrier tunneling via Fowler-Nordheim and direct tunneling through the insulating layer.

As technology advances to the nanometer scale, gate leakage becomes predominant. When the thickness of the gate oxide decreases to just a few atomic layers, electrons can tunnel directly through the fragile oxide barrier, even when the device is in the "off" state. This leads to considerable power loss, presenting a significant obstacle in nanoelectronics. The direct tunneling current density (J) through a very thin gate oxide, the capacitance can be estimated using a simplified WKB (Wentzel-Kramers-Brillouin) approximation formula:

$$J \propto \exp\left(-\frac{4\pi\sqrt{2m^{*}\Phi_{B}}}{h}t_{Ox}\right) \tag{1.34}$$

where m^{*} indicates the effective electron mass in the oxide (approximately $0.5 \times m_{e}$ for SiO_2, with m_e being the free electron mass, Φ_B represents the potential barrier height (approximately 3.1 eV for SiO_2), h is Planck's constant (6.626×10^{-34} J·s), and t_{Ox} is the thickness of the oxide. For simplification purposes, the constant term can be condensed into a factor K so that the relationship becomes

$$J \propto \exp(-Kt_{Ox}) \tag{1.35}$$

where $K \sim 1.2 \times 10^{10}$ m^{-1} (utilizing standard physical constants).

Example 1.9 Gate Leakage Current

When reducing the gate oxide thickness from $t_{Ox1} = 2.0$ nm to $t_{Ox2} = 1.0$ nm, how many times does the gate leakage current density J increase?

Answer: We can determine the ratio of the new current density (J_2) to the initial current density (J_1):

$$\frac{J_2}{J_1} = \frac{\exp(-Kt_{Ox2})}{\exp(-Kt_{Ox1})} = \exp\{-K(t_{Ox2} - t_{Ox1})\} \tag{1.36}$$

Substituting the provided values into the equation:

$K = 1.2 \times 10^{10}$ m^{-1}, $t_{Ox1} = 2.0$ nm $= 2.0 \times 10^{-9}$ m, $t_{Ox2} = 1.0$ nm $= 1.0 \times 10^{-9}$ m

Determining the difference in thickness:

$$t_{Ox2} - t_{Ox1} = 1.0 \times 10^{-9}\ \text{m} - 2.0 \times 10^{-9}\ \text{m} = -1.0 \times 10^{-9}\ \text{m} \tag{1.37}$$

Computing the exponent term:

$$\begin{aligned} \exp\{-K(t_{Ox2} - t_{Ox1})\} &= \exp\left\{\left(-1.2 \times 10^{10}\ \text{m}^{-1}\right) \times \left(-1.0 \times 10^{-9}\ \text{m}\right)\right\} \\ &= \exp\left(1.2 \times 10^{-9}\right) = \exp(12) \end{aligned} \tag{1.38}$$

Calculating the ratio:

$$\frac{J_2}{J_1} = \exp(12) = 162{,}754.79 \tag{1.39}$$

By reducing the gate oxide thickness from 2.0 to 1.0 nm, the gate leakage current density increases by roughly a factor of 162,754. This substantial rise underscores why gate leakage and power dissipation pose fundamental challenges for ongoing CMOS scaling at nanoscale dimensions, prompting the exploration of solutions beyond CMOS technology.

1.6.5 Velocity Saturation

As the channel length is reduced, charge carriers (mainly electrons) attain higher velocities due to the stronger electric field between the source and drain. However, beyond a certain threshold, their velocity saturates due to carrier scattering (Hwang et al. 2009). This imposes a limit on the drive current and diminishes the device's overall performance. Solutions entail using strained silicon and high-mobility materials. Introducing strain into the silicon structure enhances carrier mobility, helping sustain elevated carrier velocities even in the presence of strong electric fields. Employing high-mobility channel materials such as Gallium Arsenide (GaAs) or Indium Gallium Arsenide (InGaAs), which offer superior electron mobility compared to silicon, can alleviate the effects of velocity saturation.

Example 1.10 Velocity Saturation

A silicon (Si) N-channel MOSFET features a channel length L of 0.1 μm (100 nm). The mobility of electrons at low fields μ in the channel is recorded as 500 $\text{cm}^2/(\text{V s})$, while the saturation velocity of electrons (v_{Sat}) in silicon is around 1×10^7 cm/s. Determine the critical electric field (E_c) at which the velocity of electrons is expected to start saturating in this component.

Solution: The critical electric field marks the transition of carrier velocity from a linear relationship with the electric field ($v_d = \mu E$) to a constant, saturated speed (v_{Sat}). The connection between the critical electric field E_c, low-field mobility, and saturation velocity v_{Sat} is represented by the equation:

$$E_c = \frac{v_{Sat}}{\mu} \tag{1.40}$$

Provided values: Saturation velocity, $v_{Sat} = 1 \times 10^7\ \text{cm/s}$; Low-field mobility, $\mu = 500\ \text{cm}^2/(\text{V s})$.

Inserting the values into the equation:

$$E_c = \frac{1 \times 10^7 \mathrm{cm/s}}{500\ \mathrm{cm^2/V\ s}} = 20{,}000 \mathrm{V/cm} = 20 \mathrm{kV/cm} \tag{1.41}$$

The critical electric field (E_c) for this component is 20 kV/cm (or 2×10^4 V/cm). In a short-channel device measuring 0.1 μm (100 nm), the average electric field throughout the channel, even at low drain voltages (for instance, a V_{DS} of 0.5 V) would result in an average field of = 0.5 V/(0.1×10^{-4}) cm = 5×10^4 V/cm = 50 kV/cm, which can easily surpass this critical field, leading to significant velocity saturation effects on the device's performance. This phenomenon imposes a limit on the maximum possible drain current and alters the $I_D - V_{GS}$ relationship from a square-law to a more linear form.

1.6.6 Channel Length Modulation

In MOSFETs with shorter channels, the actual channel length may vary with an increase in drain voltage due to the growth of the depletion region (Duffy et al. 2000). This occurrence is referred to as channel length modulation (CLM). The phenomenon is illustrated in Figure 1.8. It results in a rise in drain current even when

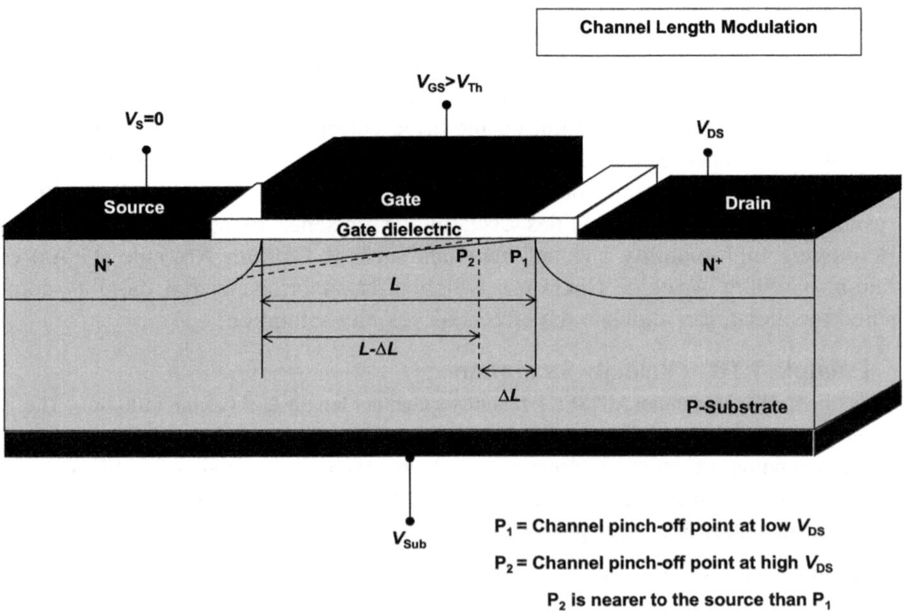

FIGURE 1.8 Channel length modulation in a short-channel MOSFET. It is a phenomenon in MOSFETs where the effective channel length decreases as the drain-source voltage (V_{DS}) increases, primarily in the saturation region when the channel pinches off near the drain. As V_{DS} increases beyond the pinch-off voltage ($V_{DS,\,sat}$), the depletion region at the drain expands, moving the pinch-off point closer to the source. The effective length of the inversion layer becomes $L - \Delta L$, where L is the drawn channel length. Because channel resistance is proportional to length, reducing the length decreases resistance and causes a slight increase in drain current I_{DS}, even in saturation.

the MOSFET has entered saturation, thereby influencing the output resistance of the device. LDD structure employing lightly doped drain regions aids in lowering the electric field at the drain, which can help alleviate CLM. In halo implantation, extra doping is introduced around the source and drain. This helps manage the growth of the depletion region, thereby diminishing the effects of CLM.

Example 1.11 Channel Length Variation

A short-channel N-channel MOSFET is operating in the saturation region with a drain current of $I_D = 0.5$ mA at $V_{DS} = 1.5$ V. For this device, the channel-length modulation parameter is $\lambda = 0.5\ V^{-1}$. Determine the new drain current $I_{D,New}$, when the drain-source voltage V_{DS} is increased to 2.0 V while keeping the gate-source voltage V_{GS} and other parameters unchanged.

Solution: To find the drain current in the saturation region by factoring in channel length modulation (CLM), we use the modified square-law equation:

$$I_D = \frac{1}{2}\mu_n C_{Ox}\frac{W}{L}(V_{GS} - V_{Th})^2(1 + \lambda V_{DS}) \tag{1.42}$$

This expression can be simplified to

$$I_D = K(1 + \lambda V_{DS}) \tag{1.43}$$

where K includes all constant parameters.

The critical insight for this problem is to use the current ratio to calculate the new current value without needing all specific parameters. The ratio can be established as:

$$\frac{I_{D,\ New}}{I_{D,\ Original}} = \frac{K(1 + \lambda V_{DS,New})}{K(1 + \lambda V_{DS,Original})} \tag{1.44}$$

This can be rewritten as:

$$\frac{I_{D,\ New}}{I_{D,\ Original}} = \frac{1 + \lambda V_{DS,New}}{1 + \lambda V_{DS,Original}} \tag{1.45}$$

Now, substituting the given values yields:

$$I_{D,\ Original} = 0.5\ \text{mA} \tag{1.46}$$

$$V_{DS,Original} = 1.5\ \text{V} \tag{1.47}$$

$$V_{DS,New} = 2.0\ \text{V} \tag{1.48}$$

$$\lambda = 0.5\ \text{V}^{-1} \tag{1.49}$$

After plugging these values into the ratio, we have:

$$\frac{I_{D,\ New}}{0.5\ \text{mA}} = \frac{1 + 0.5\ \text{V}^{-1}\ 2.0\ \text{V}}{1 + 0.5\ \text{V}^{-1}\ 1.5\ \text{V}} = \frac{1 + 1}{1 + 0.75} \tag{1.50}$$

Calculating the new current gives us:

$$I_{\mathrm{D,\ New}} = \frac{2}{1.75} \times 0.5\ \mathrm{mA} \tag{1.51}$$

This simplifies to:

$$I_{\mathrm{D,\ New}} \approx 0.5714\ \mathrm{mA} \tag{1.52}$$

The updated drain current is approximately 0.5714 mA. This increase from 0.5 mA illustrates the influence of channel length modulation in a short-channel device, where the current's dependence on V_{DS} persists even in the saturation region.

1.6.7 Hot Carrier Injection (HCI)

In short-channel MOSFETs, carriers can acquire considerable energy due to the strong electric fields present near the drain. These energetic carriers, known as hot carriers, become trapped in the gate oxide, resulting in permanent damage, device deterioration, and changes in threshold voltage over time (Yu-Zhang and Ting-Wei 1988). Figure 1.9 offers a glimpse of how hot-electron effects unfold.

Approaches to mitigate HCI effects are based on drain engineering, using strained silicon and growing a non-uniform gate oxide, thicker at the drain end. The LDD configuration helps to lower the electric field close to the drain, which in turn mitigates the hot carrier phenomenon. Utilizing strained silicon not only enhances carrier mobility but also lessens scattering, leading to a reduction in the production of hot carriers. Implementing a non-uniform gate oxide that is thicker in the vicinity of the drain helps diminish the chances of hot carrier injection into the oxide.

Example 1.12 Hot Carrier Degradation

A short-channel NMOSFET undergoes an accelerated hot carrier stress test. The shift in threshold voltage (V_{Th}) after 100 seconds of stress is recorded as 15 mV. After 1000 seconds, the shift rises to 30 mV. The degradation follows a power-law relationship with respect to stress time (t): $V_{\mathrm{Th}} = At^n$, where A is a constant and n is the time exponent. Determine the time exponent n of the degradation.

Answer: Setting up two equations based on the given measurements.

$$\text{At time } t_1 = 100 \text{ seconds, Threshold voltage shift} \Delta V_{\mathrm{Th},1} = 15\ \mathrm{mV} \tag{1.53}$$

$$\text{At time } t_2 = 1000 \text{ seconds, Threshold voltage shift} \Delta V_{\mathrm{Th},2} = 30\ \mathrm{mV} \tag{1.54}$$

The equations are

$$15\ \mathrm{mV} = A(100\ \text{seconds})^n \tag{1.55}$$

$$30\ \mathrm{mV} = A(1000\ \text{seconds})^n \tag{1.56}$$

Then, dividing the second equation by the first to eliminate the constant A.

$$\frac{30\ \mathrm{mV}}{15\ \mathrm{mV}} = \frac{A(1000\ \text{seconds})^n}{A(100\ \text{seconds})^n} \tag{1.57}$$

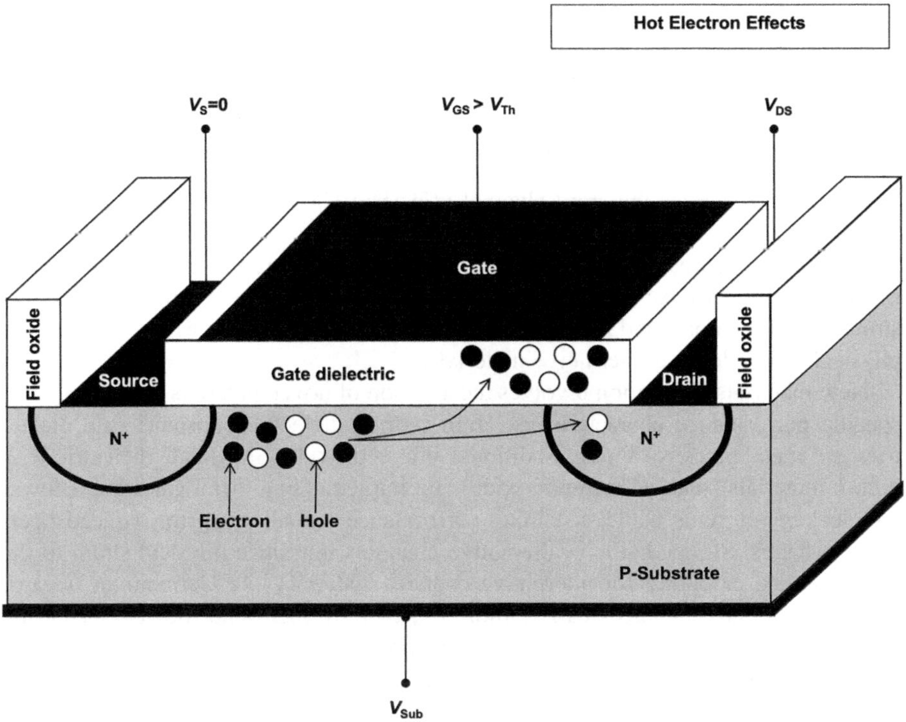

FIGURE 1.9 Hot-electron effects in a short-channel MOSFET. A short-channel MOSFET is shown with N^+ source, N^+ drain, and P-substrate. The gate oxide, gate metal, and field oxide are clearly marked. As the drain-to-source voltage increases, the electric field raises electron drift velocity and kinetic energy. Through impact ionization, these energetic carriers can release bound electrons, generating new electron-hole pairs. Hot electrons may overcome the Si–SiO_2 energy barrier (approximately 3.2 eV) and become trapped in the oxide layer. In an N-channel MOSFET, trapped negative charges increase the threshold voltage, requiring a higher voltage to activate the transistor. In a P-channel MOSFET, the threshold voltage decreases.

or,

$$\frac{30}{15} = (10)^n \tag{1.58}$$

or,

$$2 = (10)^n \tag{1.59}$$

Lastly, solving for n by applying the base-10 logarithm to both sides.

$$\log_{10}(2) = \log_{10}(10)^n = n \log_{10}(10) = n \times 1 = n \tag{1.60}$$

or,

$$n = \log_{10}(2) = 0.30103 \tag{1.61}$$

The resulting time exponent (n) for the hot carrier degradation in this device is approximately 0.30103. This value can be utilized to estimate the device's lifespan under typical operating conditions.

1.7 DISCUSSION AND CONCLUSIONS

Table 1.1 summarizes the short-channel effects in CMOS, and Table 1.2 presents possible solutions to these short-channel effects.

Methods to address the challenges of CMOS nanoelectronics can be categorized mainly into two groups: enhancing silicon CMOS performance through material and structural innovations, and investigating "post-CMOS" technologies that rely on new physical principles and alternative materials.

New materials integration involves the addition of novel materials aiming to boost specific performance characteristics. High-κ (high-dielectric constant) gate dielectrics and metal gates: swapping traditional silicon dioxide (SiO_2) gate dielectric with high-k materials (such as hafnium oxide) and implementing metal gates helps lower gate leakage currents while sustaining performance and allowing for reduced operating voltages. Strained silicon/alternative channels introduce physical strain to the silicon lattice or opting for alternative channel materials like Germanium (Ge) or III–V compounds enhances carrier mobility, thereby increasing device speed and performance. Ongoing research is investigating the potential of CNTs and two-dimensional materials such as molybdenum disulfide (MoS_2) or graphene as channels in highly scaled FETs to surpass the fundamental limitations of silicon.

Approaches dealing with device structures concentrate on improving the performance of current silicon-based CMOS technology by tackling problems such as leakage currents, power loss, and short-channel effects (SCEs). A significant advancement has been the shift from conventional planar transistors to three-dimensional designs through advanced transistor architectures. FinFETs (Fin field-effect transistors) utilize a fin-shaped silicon channel with a gate that encircles it, offering improved channel control, which leads to reduced leakage currents and SCEs. Nanosheet/Gate-all-around (GAA) FETs are regarded as the next advancement after FinFETs. These designs incorporate a gate that entirely surrounds the channel, providing superior electrostatic control and scalability down to sub-2 nm technology nodes. Fully depleted silicon-on-insulator (FD-SOI) technique employs an extremely thin silicon layer positioned on an insulating substrate to enhance performance and minimize leakage.

As the fundamental physical constraints of silicon-based FETs are approached (near the 2 nm node), studies are actively investigating fundamentally new devices that leverage different physics. These are intended to both augment and possibly supplant CMOS in the future of information processing. Several non-FET logic alternatives are proposed. Memristors are devices that alter their resistance based on the history of the current that passes through them, making them suitable for innovative memory and logic uses. Spintronics technology harnesses the electron's intrinsic spin, in addition to its electric charge, for data storage and processing. Nanomagnetic logic approach utilizes magnetic interactions to perform logic operations. Superconducting logic concept involves materials that demonstrate zero electrical resistance at extremely low temperatures.

TABLE 1.1
Causes Leading to Short-Channel Effects in CMOS

Sl. No.	Limitation	Explanation
1	Decrease in threshold voltage	Resulting from the drain field infiltrating the channel, which diminishes gate control and facilitates the activation of the transistor.
2	Drain-induced barrier lowering	The electric field from the drain "penetrating" the channel in short devices expands its depletion region, which consequently decreases the potential barrier at the source, lowering the required gate voltage (threshold voltage) for current flow and resulting in higher leakage.
3	Deterioration of subthreshold slope	Occurs when smaller devices experience diminished electrostatic control, resulting in greater off-current, elevated power consumption, and lower performance. This is mainly due to issues such as drain-induced barrier lowering (DIBL), channel doping effects, source/drain contact challenges (energy filtering), and hot carrier injection, which make the transition from "off" to "on" less sharp compared to the ideal 60 mV/dec.
4	Increase in gate leakage current due to reduction in gate oxide thickness	Decreasing the thickness of the gate oxide (t_{ox}) in miniaturized transistors significantly raises the gate leakage current mainly due to direct tunneling, a quantum-mechanical phenomenon that allows electrons to traverse the thin barrier. This effect is exacerbated by the significantly stronger electric field across the extremely thin oxide layer, which affects device performance and power consumption.
5	Saturation of velocity	Happens due to the presence of intense electric fields in the short channel that enable electrons to acquire sufficient energy to interact with the crystal lattice (resulting in the emission of optical phonons). This heightened scattering effectively restricts the speed of the carriers to a maximum level (saturation velocity, v_{Sat}), which halts any further increase in velocity even when electric fields are raised.
6	Modulation of channel length	The expansion of the depletion region at the drain side occurs as the drain voltage V_{DS} rises, which shortens the conductive channel and boosts the drain current I_{DS} when in saturation. This effect makes the channel seem narrower and less resistive, a phenomenon that is more noticeable in smaller devices.
7	Hot carrier injection (HCI)	This phenomenon is induced by intense electric fields close to the drain, particularly in miniaturized devices, which propel electrons and holes to elevated kinetic energies (become "hot"). These high-energy carriers can subsequently surpass the silicon-oxide barrier and become entrapped within the gate dielectric, leading to performance degradation by generating interface and oxide traps.

TABLE 1.2
Pointwise Solutions to Short-Channel Effects in CMOS

Sl. No.	Short-Channel Effect	Remedies and Mitigation Techniques
1	Decrease in threshold voltage	Reinforce channel doping levels, implement a thicker gate oxide layer, incorporate halo/pocket implant techniques, minimize junction depth: Adopt sophisticated device architectures: FinFETs (Fin field-effect transistors) and double-gate (DG) MOSFETs, enhance channel length optimization, introduce body bias (back-gate bias)
2	Drain-induced barrier lowering	Shallow source/drain junctions, enhanced channel doping, halo/pocket implants, backwards (retrograde) doping, thinner gate oxide, high-*k* dielectric materials, configurations: silicon-on-insulator (SOI) devices, multi-gate devices (FinFETs, GAA FETs)
3	Deterioration of subthreshold slope	FinFETs and GAAFETs (Gate-all-around), tunnel field-effect transistors (TFETs), negative capacitance field-effect transistors (NC-FETs), high-*k* dielectric materials, enhanced doping techniques, engineering of work function, shorter and shallower source/drain junctions, minimized channel length
4	Increase in gate leakage current due to reduction in gate oxide thickness	High-*k* dielectrics, optimization of interfacial layers (such as SiON) to reduce defects, the hydrogen/deuterium isotope effect for passivating dangling bonds, molecular dipole layers used to adjust the work function and lessen tunneling current
5	Saturation of velocity	Improved materials (expanded bandgaps), developing advanced device architectures (FinFETs, GAAFETs), enhancing doping/alloying techniques, minimizing scattering (temperature regulation, surface treatments), and employing advanced modeling (Monte Carlo) to forecast/lessen effects for high-frequency applications
6	Modulation of channel length	Multi-gate transistors (FinFETs, GAAFETs), SOI technology, junctionless FETs (JLFETs), high-*k* dielectrics, halo doping, shallow source/drain junctions, engineered spacer layers
7	Hot carrier injection (HCI)	Lightly doped drain (LDD), SOI, FinFETs/GAA, extended channel length, graded drain Junctions, nitrided oxides through annealing in NH_3, N_2O, or NO, strained silicon, Gate dielectrics like HfSiON with barrier layers (TiN).

In addition to enhancements at the single-device level, architectural design and system-level approaches are vital. Three-dimensional (3D) integration/chip stacking consists of stacking multiple layers of devices or chips atop one another, while utilizing through-silicon vias (TSVs) boosts integration density and functionality without solely depending on horizontal scaling. On-chip optical communication (nanophotonics) substitutes conventional metallic interconnects with optical communication pathways, which can mitigate the "interconnect bottleneck" (signal delays, cross-talk, and power usage) as devices become more closely positioned. Circuit design techniques based on employing dynamic voltage scaling or crafting circuits for subthreshold operation can assist in managing power loss and leakage currents at a broader system level. The next chapter deals with the scenario of technologies that seek to assuage the bewildering situation in CMOS.

REFERENCES

Chamberlain S. G. and S. Ramanan 1986 Drain-induced barrier-lowering analysis in VSLI MOSFET devices using two-dimensional numerical simulations, *IEEE Transactions on Electron Devices*, 33(11): 1745–1753, https://doi.org/10.1109/T-ED.1986.22737

Chen Z. and J. Guo 2006 Dramatic reduction of gate leakage current of ultrathin oxides through oxide structure modification, *Solid-State Electronics*, 50: 1004–1011, https://doi.org/10.1016/j.sse.2006.04.045

Duffy C., M. Kucic, A. C. Low and P. Hasler 2000 An exponential model of channel-length modulation applied towards floating-gate circuits, *Proceedings of the 43rd IEEE Midwest Symposium on Circuits and Systems (Cat.No.CH37144)*, Lansing, MI, 8–11 August, pp. 1044–1047, vol. 3, https://doi.org/10.1109/MWSCAS.2000.951395

Hwang S., H. Im, M. Song, K. Ishida, T. Hiramoto and T. Sakurai 2009 Velocity saturation effects in a short channel Si-MOSFET and its small signal characteristics, *Journal of The Korean Physical Society*, 55(2): 581–584, https://doi.org/10.3938/jkps.55.581

Kang S.-M. and Y. Leblebici 2002 *CMOS Digital Integrated Circuit: Analysis* and *Design*, 3rd ed. McGraw-Hill, New York, 1008.

Ogura S., P. J. Tsang, W. W. Walker, D. L. Critchlow and J. F. Shepard, 1980 Design and characteristics of the lightly doped drain-source (LDD) insulated gate field-effect transistor, *IEEE Transactions on Electron Devices*, 27(8):1359–1367, https://doi.org/10.1109/T-ED.1980.20040

Shappir A., Y. Shacham-Diamand, E. Lusky, I. Bloom and B. Eitan 2003 Subthreshold slope degradation model for localized-charge-trapping based non-volatile memory devices, *Solid-State Electronics*, 47(5): 937–941, https://doi.org/10.1016/S0038-1101(02)00454-9

Troutman R. R. 1979 VLSI limitations from drain-induced barrier lowering, *IEEE Journal of Solid-State Circuits*, 14(2):383–391, https://doi.org/10.1109/JSSC.1979.1051189

Veendrick H. 2001 100 nm CMOS technology. A design perspective, *2001 IEEE International Symposium on Circuits and Systems (ISCAS)*, Sydney, NSW, Australia, 9 May, pp. 10.2.1–10.2.12, https://doi.org/10.1109/TUTCAS.2001.946981

Wanlass F. and C. Sah 1963 Nanowatt logic using field-effect metal-oxide semiconductor triodes, *1963 IEEE International Solid-State Circuits Conference. Digest of Technical Papers*, Philadelphia, PA, 20–22 February, pp. 32–33, https://doi.org/10.1109/ISSCC.1963.1157450

Wanlass F. M. 1967 Low stand-by power complementary field effect circuitry, US3356858A, US Patent Filing Date: June 18, 1963, Grant Date: December 5, 1967.

Wind J., Y. Taur, Y. Lee, Y. Mii, R. G. Viswanathan, J. J. Bucchignano, A. T. Pomerene, R. M. Sicina, K. R. Milkove, J. W. Stiebritz, R. A. Roy, C. K. Hu, M. P. Manny, S. Cohen and W. Chen 1995 Lithography and fabrication processes for sub-100 nm scale complementary metal-oxide semiconductor. *Journal of Vacuum Science & Technology B*, 13(6): 2688–2695, https://doi.org/10.1116/1.588931

Wong H.-S. P., D. J. Frank, P. M. Solomon, C. H. J. Wann and J. J. Welser 1999 Nanoscale CMOS, *Proceedings of the IEEE*, 87(4): 537–570, https://doi.org/10.1109/5.752515

Xiang Q., C. Woo, E. Paton, J. Foster, B. Yu, and M.-R. Lin 2000 Deep sub-100 nm CMOS with ultralow gate sheet resistance by NiSi, *2000 Symposium on VLSI Technology. Digest of Technical Papers (Cat. No.00CH37104)*, Honolulu, HI, 13–15 June, pp. 76–77, https://doi.org/10.1109/VLSIT.2000.852776

Yu-Zhang C. and T. Ting-Wei 1988 Numerical simulation of avalanche hot-carrier injection in short-channel MOSFET's, *IEEE Transactions on Electron Devices*, 35(12): 2180–2188, https://doi.org/10.1109/16.8792

2 Exploring Nanoelectronics Beyond CMOS Technology

2.1 INTRODUCTION

In the previous chapter, we saw how CMOS, which has been the keystone technology supporting digital computing, has been facing scaling difficulties to cope with the ever-growing demands of the modern age. This prompts us to think about: what is beyond CMOS? "Beyond CMOS" refers to the exploration of alternative digital logic technologies aimed at overcoming the physical constraints, such as heat and quantum tunneling, that conventional CMOS transistors face as they approach atomic dimensions (under 7 nm) (Hutchby et al. 2002, Bernstein et al. 2010). Featured as a chapter in the International Roadmap for Devices and Systems (IRDS), the "Beyond CMOS (BC)" roadmap explores the frontier of emerging devices and materials, uncovering their promise and the hurdles they must overcome before gaining industry acceptance. It offers an IRDS perspective on device technologies, serving as a vital link from traditional CMOS to the exciting frontier of nanoelectronics beyond CMOS scaling (Das et al. 2021, Chen 2022). Research in alternative nanoelectronic technologies beyond CMOS is an extensive effort aimed at discovering new devices and circuits that can either supplement or eventually replace conventional silicon CMOS as it nears its fundamental physical scaling limits. The emphasis has shifted from merely finding substitutes to a hybrid methodology that seeks to enhance system performance and energy efficiency while introducing new functionalities. This chapter seeks to enquire about the commencing date of beyond-CMOS nanoelectronics, outline its aims and objectives, and define its scope.

2.2 WHEN DID BEYOND-CMOS NANOELECTRONICS FORMALLY START?

The idea of beyond-CMOS nanoelectronics began to gain traction in serious conversations and research plans in the early 2000s, particularly around 2001–2002, as the industry recognized the impending physical limits of traditional CMOS scaling. In 2001, the International Technology Roadmap for Semiconductors (ITRS) launched an "Emerging Technologies" section. During the mid-2000s, research initiatives focused on "beyond-CMOS" technologies received significant funding in the United States. The journey truly began with the pivotal launch of the Nanoelectronics Research Initiative (NRI) in 2005. By 2010–2011, the phrase "Beyond CMOS" was officially included as a key area in major industry roadmaps, including the ITRS and

DOI: 10.1201/9781003753469-2

its successor, the IRDS. In essence, while the fundamental concepts have been around for a while, the concentrated global effort in scientific and industrial exploration of "beyond CMOS" solutions began earnestly around 2010. With backing from the ITRS Emerging Research Devices chapter and the International Nanotechnology Conference on Communication and Cooperation (INC) initiative, the International Planning Working Group on Nanoelectronics (IPWGN) came together to spotlight gaps in technology research funding and programs to streamline efforts and harness technology for smarter, more effective research program planning (Galatsis et al. 2015).

2.3 AIMS AND OBJECTIVES OF BEYOND-CMOS NANOELECTRONICS

The main impetus for advancing beyond-CMOS nanoelectronics is the need to surpass the inherent power and performance constraints of conventional silicon CMOS technology, as Moore's Law scaling nears its physical and economic limits (Nikonov and Young 2013). A key factor related to power is reducing high-power consumption. While single CMOS transistors have become increasingly energy-efficient, the immense number of transistors present in contemporary chips, in the billions, leads to considerable overall power dissipation and heat generation, which has posed a challenge for electronics manufacturers since approximately 2005. Another vital factor is the existence of energy efficiency gap, a significant disparity in energy efficiency between digital computers and biological brains, particularly for applications like neuromorphic or brain-inspired computing. For instance, the human brain functions on roughly 20 W, whereas a supercomputer can demand about 200 kW. Research into beyond-CMOS aims to close this gap by facilitating more energy-efficient alternative computing models. Leakage power and off-state current must also receive due consideration. As CMOS devices are miniaturized to the nanoscale, controlling leakage current and off-state power use becomes progressively more challenging, which adversely affects the battery life of mobile and smart devices. Further, voltage-scaling complications arise. To reduce power consumption, supply voltages have been lowered; however, this introduces challenges in maintaining performance and ensuring compatibility with new devices that may require varying voltage levels.

Research into beyond-CMOS technologies is underway to develop systems that effectively balance performance and energy requirements. As CMOS dimensional scaling reaches fundamental limits, inventive devices and architectures are essential to continue improving chip efficiency and performance. Functional specialization is emphasized. Various applications necessitate specific device functionalities beyond a "one-size-fits-all" solution. Devices beyond CMOS can provide characteristics such as non-volatility and low-voltage operation that are better suited for tasks such as memory, sensing, and neuromorphic computing. Moreover, the von Neumann bottleneck is a serious problem in conventional computing architectures that separate memory and processing units, leading to constant data movement that is time- and energy-inefficient, particularly for AI applications. It must be mitigated, encouraging the development of novel information processing models that are inherently energy-efficient to facilitate tasks related to decision-making and pattern recognition.

A leading objective of beyond-CMOS initiatives is to reduce energy consumption below the fundamental thermal threshold (Landauer limit) and the limitations of scaled CMOS. Established methodologies are employed to compare new devices against the limits of CMOS, emphasizing particular metrics, a few of which are elaborated below. Switching energy (E) is the energy required to transition a logic device from one state to another. Switching time (t) is the time required for a device to switch between states. Power-delay product (PDP) is the product of power consumption and delay, serving as a standard measure of energy efficiency. Device density/area indicates the number of devices that can be accommodated in a specific area, typically measured in nanometers (for instance, advanced CMOS technology currently has gate lengths smaller than 3 nm).

Example 2.1 A primary motivation for research into "beyond CMOS" technologies is the issue of heat generation, often referred to as the "heat death" of CMOS scaling. The fundamental constraint on the energy required to switch a single bit of information at room temperature is the Landauer limit. Compute the minimum theoretical energy needed to erase one bit of information at room temperature (300 K), and assess how it compares to the average energy usage during a switching event in a modern CMOS transistor—key Constants: Boltzmann constant $(k_{\mathrm{B}}) : 1.38 \times 10^{-23}\ \mathrm{J/K}$, Room Temperature (T): 300 K.

Solution: The equation for the Landauer Limit, the minimum energy dissipation (E_{Min}) for an irreversible operation involving a 1-bit change, is expressed as:

$$E_{\mathrm{Min}} = k_{\mathrm{B}} T \ln(2) \tag{2.1}$$

Inputting the values and computing:

$$\begin{gathered} E_{\mathrm{Min}} = 1.38 \times 10^{-23}\ \mathrm{J/K} \times 300\ \mathrm{K} \ln(2) = 1.38 \times 300 \times 10^{-23} \times \ln(2) \\ \mathrm{J} = 1.38 \times 300 \times 10^{-23} \times 0.69315\ \mathrm{J} = 2.8696 \times 10^{-21}\ \mathrm{J} \approx 2.87 \times 10^{-21}\ \mathrm{J} \end{gathered} \tag{2.2}$$

Comparing with CMOS: A typical switching event for a high-performance, cutting-edge CMOS transistor (at around the 5 nm node) may dissipate energy in the range of 10^{-15} to 10^{-14} J (1 to 10 fJ or femtoJoules). Therefore, the theoretically minimal energy is several orders of magnitude (approximately 1 million times) lower than that consumed by current advanced CMOS devices. Technologies "beyond CMOS," such as spintronics and neuromorphic circuits, seek to narrow this disparity by leveraging alternative physical mechanisms, e.g., electron spin rather than charge, to approach this fundamental limit, thereby achieving much higher energy efficiency.

Example 2.2 Consider a contemporary microprocessor with a set chip area, where the transistor count increases every two years, as stated by Moore's Law. Yet, power usage per chip must stay constant to control heat generation. A chip in year X has N transistors and operates with power P. In year $X + 2$, the chip contains $2N$ transistors. If the total power P remains unchanged, what must happen to the average power per transistor (P_{Avg})?

Solution: In the year X, power is $P = N \times P_{\text{Avg}, X}$
In the year $X + 2$, power is: $P = 2N \times P_{\text{Avg},X+2}$
Since the total power P is unchanged, we can equate the two expressions:

$$N \times P_{\text{Avg},\ X} = 2N \times P_{\text{Avg},\ X+2} \tag{2.3}$$

Solving for the new average power per transistor $P_{\text{Avg},X+2}$

$$P_{\text{Avg},\ X+2} = \frac{N \times P_{\text{Avg},\ X}}{2N} = \frac{P_{\text{Avg},\ X}}{2} \tag{2.4}$$

To keep the total chip power constant (and thus manageable heat), the average power consumed by each transistor must be reduced by half with each successive generation, where the transistor count doubles. This steep reduction in per-transistor power (referred to as Dennard scaling) has become impractical, leading to the emergence of the "power wall" that necessitates the investigation of beyond-CMOS technologies such as spintronics and carbon nanotubes to achieve enhanced energy efficiency. The challenge often tackled in "beyond CMOS" research is the "power wall," where increased transistor density leads to excessive heat generation, even as supply voltages are lowered.

Example 2.3 A traditional silicon CMOS logic gate operating near its physical capacity generally has a switching energy of about 10^{-14} J (10 fJ). It operates at a supply voltage (V_{DD}) of around 0.7 V. A newly proposed carbon nanotube field-effect transistor (CNTFET) gate is engineered to operate at a reduced supply voltage of 0.4 V, with an intended switching energy of 10^{-16} J (0.1 fJ). Determine the possible power reduction of the CNTFET device in comparison to the CMOS device, assuming both run at a switching frequency of 1 GHz (10^9 operations per second), and discuss the importance of the energy reduction goal.

Solution: A conceptual approach for CMOS power calculation is based on the formula: Power (P) = Energy (E) × Frequency (f). In this context, Energy (E) denotes the energy consumed during each switching action, while frequency (f) denotes the number of these actions. This concept is particularly relevant to dynamic power consumption, which refers to the power required to charge and discharge the load capacitance (C) at the nodes within the circuit.

$$P_{\text{CMOS}} = 10^{-14}\ \text{J} \times 10^9\ \text{s}^{-1} = 10^{-5}\ \text{J/s} = 10^{-5}\ \text{W} = 10^{-5} \times 10^6\ \mu\text{W} = 10\ \mu\text{W} \tag{2.5}$$

$$P_{\text{CNTFET}} = 10^{-16}\ \text{J} \times 10^9\ \text{s}^{-1} = 10^{-7}\ \text{J/s} = 10^{-7}\ \text{W} = 10^{-7} \times 10^6\ \mu\text{W} = 0.1\ \mu\text{W} \tag{2.6}$$

Power reduction ratio:

$$\frac{P_{\text{CMOS}}}{P_{\text{CNTFET}}} = \frac{10\ \mu\text{W}}{0.1\ \mu\text{W}} = 100 \tag{2.7}$$

Significance: The target value represents a 100-fold decrease in power consumption per gate. This substantial energy reduction is a key driver for research beyond CMOS technology, as it addresses heat dissipation, which limits performance scaling in conventional CMOS devices.

Example 2.4 Consider a microchip with a surface area of 1 cm^2 that has a maximum power dissipation capacity of 100 W (which is typical for high-performance cooling systems). If one transistor operates at a specific voltage and frequency and dissipates an average of 1 nW (1×10^{-9} W), determine how many transistors can be placed on the chip before reaching the power dissipation limit.

Solution: The total power limit is 100 W, and the power used by each transistor is 1 nW = 1×10^{-9} W. To find the number of transistors (N), we divide the total power limit by the power consumed by each transistor:

$$N = \frac{\text{Total power limit}}{\text{Power per transistor}} \tag{2.8}$$

Accordingly,

$$N = \frac{100\ \text{W}}{1 \times 10^{-9}\ \text{W}} \tag{2.9}$$

results in

$$N = 100 \times 10^{9} \text{transistors} = 100 \text{ billion transistors} \tag{2.10}$$

This calculation indicates that there is a finite limit to the number of transistors that can be placed on a chip based solely on power dissipation, even with advanced cooling techniques. In practice, contemporary high-performance chips are made possible by innovative design strategies and by having numerous transistors switched off at any given time. The research into "beyond CMOS" aims to create new devices (such as tunnel field-effect transistors (TFETs), spintronic devices, or neuromorphic devices) that are fundamentally more energy-efficient, potentially reducing power per transistor by several orders of magnitude, thereby allowing for the incorporation of trillions of components without surpassing thermal limitations.

In context, the primary challenge motivating the pursuit of "beyond CMOS" solutions is the growing power density in integrated circuits as transistors continue to shrink. Conventional CMOS devices face a fundamental constraint in the steepness of their current turn-off (the subthreshold slope), which hinders significant reductions in operating voltage and, consequently, power consumption. This results in chips producing more heat than can be effectively dissipated.

Example 2.5 You are assessing the scenario of two distinct transistor technologies intended for low-power applications: Device A, a standard CMOS transistor operating at its scaling limit, and Device B, a TFET, considered a promising alternative to CMOS. The essential metrics for each device, evaluated at the circuit level, are listed in Table 2.1.

Calculate and compare the energy-delay product (EDP) of both devices to determine which is more energy-efficient for the specified application. The device with the lower EDP is typically considered more efficient.

TABLE 2.1
Device Metrics

Metric	Device A (CMOS)	Device B (TFET)
Switching energy (E_{Sw})	1 fJ (femptojoule)	0.3 fJ
Switching delay (τ_{d})	1 ps (picosecond)	2 ps

Solution: The formula for EDP calculation is:

$$\text{EDP} = E_{\text{Sw}} \times \tau_{\text{d}} \tag{2.11}$$

Computing the EDP for device A (CMOS):

$$\text{EDP}|_{\text{A}} = E_{\text{SwA}} \times \tau_{\text{dA}} = 1\text{ fJ} \times 1\text{ ps} = 1\text{ fj ps} \tag{2.12}$$

Computing the EDP for device B (TFET):

$$\text{EDP}|_{\text{B}} = E_{\text{SwB}} \times \tau_{\text{dB}} = 0.3\text{ fJ} \times 2\text{ ps} = 0.6\text{ fj ps} \tag{2.13}$$

Comparing the outcomes:

$$\text{EDP}|_{\text{A}} = 1\text{ fj ps against EDP}|_{\text{B}} = 0.6\text{ fj ps} \tag{2.14}$$

and assessing, we find

$$0.6\text{ fj ps} < 1\text{ fj ps} \tag{2.15}$$

Obviously, device B (TFET) exhibits a lesser EDP than device A (CMOS). This numerical result indicates that the TFET is more energy-efficient in this particular scenario, despite being marginally slower, a significant factor in exploring these devices as potential substitutes or complements to conventional CMOS nanoelectronics.

2.4 DEFINITION AND SCOPE OF BEYOND-CMOS NANOELECTRONICS

Emerging technologies that go beyond CMOS nanoelectronics encompass a wide range of innovations aimed at surpassing the physical constraints of traditional CMOS devices. These technologies leverage alternative physical effects, including spintronics, neuromorphic computing, and quantum phenomena, and introduce novel functionalities to enhance performance and achieve greater density, transcending CMOS limitations. However, these technologies are still in early development phases and exhibit significant variability.

"Nanoelectronics that go beyond CMOS" refers to the exploration and development of innovative digital logic and memory technologies that overcome the inherent physical and economic scaling limitations of standard CMOS technology, which has historically underpinned Moore's law. This domain investigates alternative materials (such as carbon nanotubes, graphene, or magnetic substances) and fundamentally different operational principles—like utilizing electron spin (spintronics), quantum phenomena, or light (photonics) instead of only electron charge—to facilitate ongoing enhancements in computing capabilities, energy efficiency, and device density. The aim is not necessarily to completely replace CMOS but to augment it with specialized, high-performance hybrid systems for future applications such as AI hardware, quantum computing, and advanced sensing—areas where traditional silicon faces challenges.

The "Beyond CMOS" chapter of the IEEE 2023 International Roadmap for Devices and Systems (IRDS) spotlights a wave of breakthrough technologies poised to push past the boundaries of traditional CMOS, which is nearing its physical and economic limits. This year's roadmap shines a light on several pivotal areas and innovations (IEEE 2023):

(i) Strategic goals and paradigms

Augmenting CMOS: Although CMOS will continue to lead, beyond CMOS technologies are envisioned as powerful hardware accelerators, supercharging its capabilities—especially in the realm of neuro-inspired computing.

Energy Efficiency: A central motivation is to slash the enormous energy appetite of today's computing, with a sharp focus on AI and edge applications.

Sustainability: The 2023 edition spotlights "Green materials"—innovative substances that promise both lower economic impact and higher performance—as a catalyst for building a circular semiconductor economy.

(ii) Upcoming device technologies: The roadmap surveys and highlights a lineup of promising contenders for the future of logic and memory beyond CMOS:

Spintronics: Uses electron spin for ultra-low-power information processing.

2D Materials: Explores transition-metal dichalcogenides, such as molybdenum disulfide (MoS_2), unlocking ultra-thin channel materials ideal for stacked nanosheet architectures.

Switching Mechanisms: Investigates tunnel FETs, ferroelectric FETs, and negative capacitance FETs as innovative approaches to overcoming voltage-scaling barriers.

Novel Memory: Advanced MRAM and phase-change memory are being developed to emulate synaptic and neuronal functions, paving the way for neuromorphic computing.

(iii) Key challenges:

Heterogeneous Integration: Managing the complex packaging of diverse materials (e.g., 2D materials onto silicon).

Metrology: Integrating novel nanoscale materials and 3D structures brings new measurement challenges to the forefront.

Programming Models: Shifting to non-Von Neumann architectures, such as neuro-inspired chips, demands bold rethinking of software and programming approaches.

Figure 2.1 presents a panoramic view of the beyond-CMOS journey, with key features and aspects of materials, logic devices, memories, computing architectures, fabrication, and packaging techniques.

Table 2.2 compares CMOS and beyond-CMOS technologies.

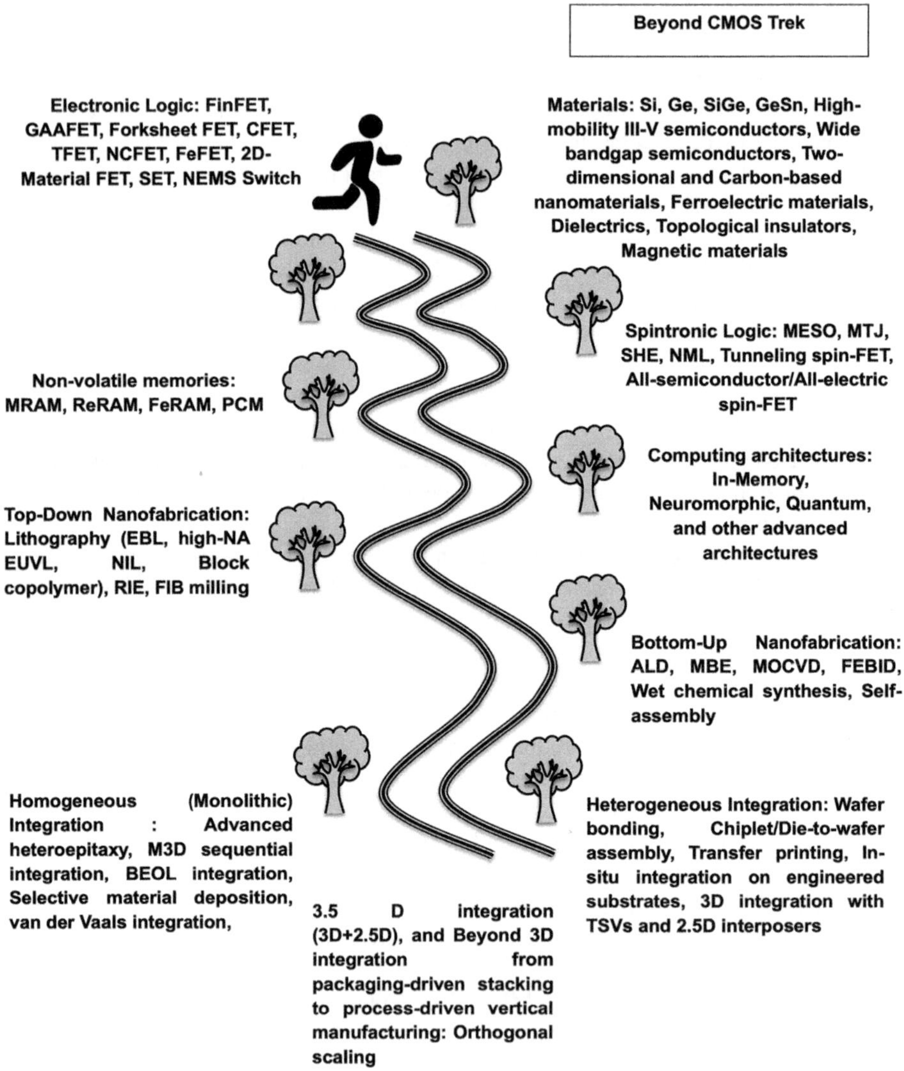

FIGURE 2.1 The beyond-CMOS journey. It marks the industry's shift from traditional silicon-based MOSFET scaling to advanced materials, devices, and architectures that address physical and thermal limitations. This evolution is driven by the need for greater performance and energy efficiency. Key developments include the adoption of III-V compounds and 2D materials such as graphene and WSe_2 to enhance carrier mobility and enable further miniaturization. Innovations in logic devices, including tunneling, negative capacitance, and ferroelectric FETs, as well as nanomagnetic logic (NML) and spintronics, are central to this progress. The integration of advanced non-volatile memories, such as magnetic RAM and resistive RAM, directly onto CMOS platforms helps reduce energy consumption for data transfers. Additionally, the industry is moving beyond the classic von Neumann architecture and non-Boolean logic, embracing neuromorphic and quantum computing to better support AI and deep learning workloads. Advances in nanofabrication and new integration and packaging methods are also critical. Overall, the journey is defined by a transition into a hybrid era, in which traditional CMOS is enhanced by emerging technologies to meet the demands of edge computing and AI.

TABLE 2.2
CMOS Nanoelectronics and Beyond-CMOS Nanoelectronics

Sl. No.	Feature	CMOS Nanoelectronics	Beyond CMOS Nanoelectronics
1	Core technology	Utilizes the traditional silicon-based MOSFET (metal-oxide-semiconductor field-effect transistor).	Investigates a variety of materials and devices beyond silicon, encompassing spintronics, nanomagnetics, and two-dimensional materials such as graphene and carbon nanotubes.
2	Information carrier	Employs the electrical charge of electrons to depict and process binary information (0s and 1s).	Incorporates alternative computational state variables, such as electron spin (in spintronics), magnetic flux (in superconducting devices), or physical states (in memristors).
3	Performance scaling	Adheres to Moore's law, which historically resulted in the doubling of transistors on a chip approximately every two years. This trend is approaching its physical constraints due to higher leakage currents and power density.	Strives to facilitate ongoing progress in computing speed and efficiency once the scaling defined by Moore's law concludes.
4	Speed	Although advancements in scaling have enhanced the switching speed over the years, it is ultimately constrained by electron movement and thermal effects.	This section examines devices and architectures that aim for increased computational speed and reduced latency, including quantum computing and certain types of optical computing.
5	Memory	On-chip cache is primarily dominated by volatile memory technologies such as SRAM, while non-volatile flash memory serves as the main storage solution. There is an ongoing effort to discover a cohesive, high-speed, high-density, non-volatile memory alternative.	A significant focus is on creating fast, high-capacity, and non-volatile memory technologies to supplement or replace traditional options. Notable examples of these technologies include magnetic RAM (MRAM), resistive RAM (ReRAM), and phase-change memory (PCM).
6	Manufacturing	This industry relies on a well-established and standardized fabrication process, enabling mass production with high yields.	It necessitates new and often intricate fabrication techniques that currently do not align with existing CMOS process flows. Key challenges involve material integration and variations in manufacturing processes.
7	Computing architecture	Usually structured on the von Neumann model, which includes distinct processing and memory components.	It explores innovative architectural models, such as neuromorphic or biologically inspired computing, to harness the unique characteristics of new devices for extensive parallel processing.

(*Continued*)

TABLE 2.2 (Continued)

Sl. No.	Feature	CMOS Nanoelectronics	Beyond CMOS Nanoelectronics
8	Technological maturity	Represents a well-established and mature technology that serves as the backbone of contemporary digital electronics.	It is currently in the research and development phase, facing numerous critical obstacles before it can achieve widespread commercial acceptance.
9	Hybrid integration	Increasingly adopted in a "More-than-Moore" strategy, where additional functionalities like sensors are incorporated alongside the primary CMOS logic.	It is projected to be integrated with current CMOS platforms soon to develop hybrid systems that merge new capabilities with the proven advantages of CMOS.
10	Power consumption	High leakage currents and elevated power dissipation present significant challenges as devices continue to miniaturize, leading to the phenomenon known as "dark silicon."	The goal is to enhance power efficiency by utilizing less power-hungry mechanisms, although managing overall system power remains a complex issue.
11	Examples	Examples include advanced silicon-based MOSFETs like FinFETs and Gate-All-Around (GAA) transistors.	Spintronic devices (which leverage spin), neuromorphic computing (inspired by the brain), quantum computing, and innovative tunneling FETs (Tunnel FETs) are noteworthy advancements.
12	Future role	It will continue to be a key technology but will require enhancements and eventual replacement as it approaches its intrinsic limits.	This area of research is vital for enabling future computing abilities, focusing on devices that could serve new functions in fields such as AI and quantum information processing.

Collaboration between design and technology researchers in nanoelectronics, particularly in the beyond-CMOS field, is a lively mix of different terminologies, working styles, and a lack of shared priorities. Sotomayor Torres et al. (2012) unveiled a groundbreaking approach to evaluating next-generation non-CMOS devices, such as spintronic, ferroelectric, and TFETs. Instead of focusing solely on traditional logic, the study explores how these devices perform on complex, non-Boolean tasks, such as neuromorphic computing with cellular neural networks (CNNs). The results were striking: spintronic devices deliver up to three times better energy-delay performance in CNNs, even though they lag in conventional Boolean circuits. This discovery sparked a call for inventive circuit designs that can harness these strengths, setting the stage for a new era of efficient, specialized computing.

Environmental, safety, health, and sustainability (ESHS) now stand at the forefront of the semiconductor industry's priorities. The "beyond CMOS" chapter in the

International Roadmap for Devices and Systems spotlights "Green materials," underscoring their vital role in driving sustainable innovation throughout semiconductor manufacturing (Ueda et al. 2025).

The semiconductor industry is advancing through four primary strategies (Figure 2.2): more Moore, which focuses on miniaturization; more-than-Moore, which emphasizes functional diversity; beyond CMOS, which explores novel technologies; and extended CMOS, which integrates these approaches. These strategies collectively sustain performance improvements as conventional single-chip scaling becomes less effective, thereby supporting progress in AI, 5G, and the Internet of Things.

2.5 ORGANIZATIONAL PLAN OF THE BOOK

In alignment with major focus areas in nanoelectronics, the book is structured into the following chapters.

2.5.1 Silicon and Supplemental Materials

Chapter 3 discusses the limitations of traditional silicon-based materials in nanoscale devices, highlighting the need for innovative substitutes with improved characteristics. It focuses on various high-mobility materials, including III–V semiconductors (such as InGaAs and InSb), which offer superior electron mobility, essential for high-speed, energy-efficient applications. The integration of germanium and SiGe with CMOS technology enhances performance through strained channels. Additionally, carbon-based materials like carbon nanotubes and graphene nanoribbons show promise for low-power, high-performance devices. Two-dimensional (2D) materials, including graphene and transition-metal dichalcogenides (TMDs), are also emphasized for their potential in transistors and memory, owing to their unique properties and ability to maintain high mobility at the nanoscale, addressing challenges in 3D semiconductors. Overall, these materials are presented as key alternatives to silicon for advancing nanoscale electronic devices.

The chapter also discusses ferroelectric materials, including hafnates and perovskites like lead zirconate titanate (PZT), which are being examined for potential use in ferroelectric-based memory devices and negative capacitance field-effect transistors (NCFETs), which in turn could enhance switching performance and reduce power consumption. Various novel dielectrics and insulators are surveyed with applications in advanced electronics. High-*k* gate dielectrics such as HfO_2, Al_2O_3, and ZrO_2 are highlighted for their ability to minimize leakage currents and enable lower operating voltages, which are crucial for power management in nanoscale devices. Additionally, topological insulators are explored for their unique properties, offering low-dissipation conduction pathways that could benefit low-energy electronics and quantum computing applications. Lastly, magnetic materials are noted for their role in emerging memory technologies such as magnetoresistive random access memory (MRAM) and spintronic devices, which leverage electron spin for data processing.

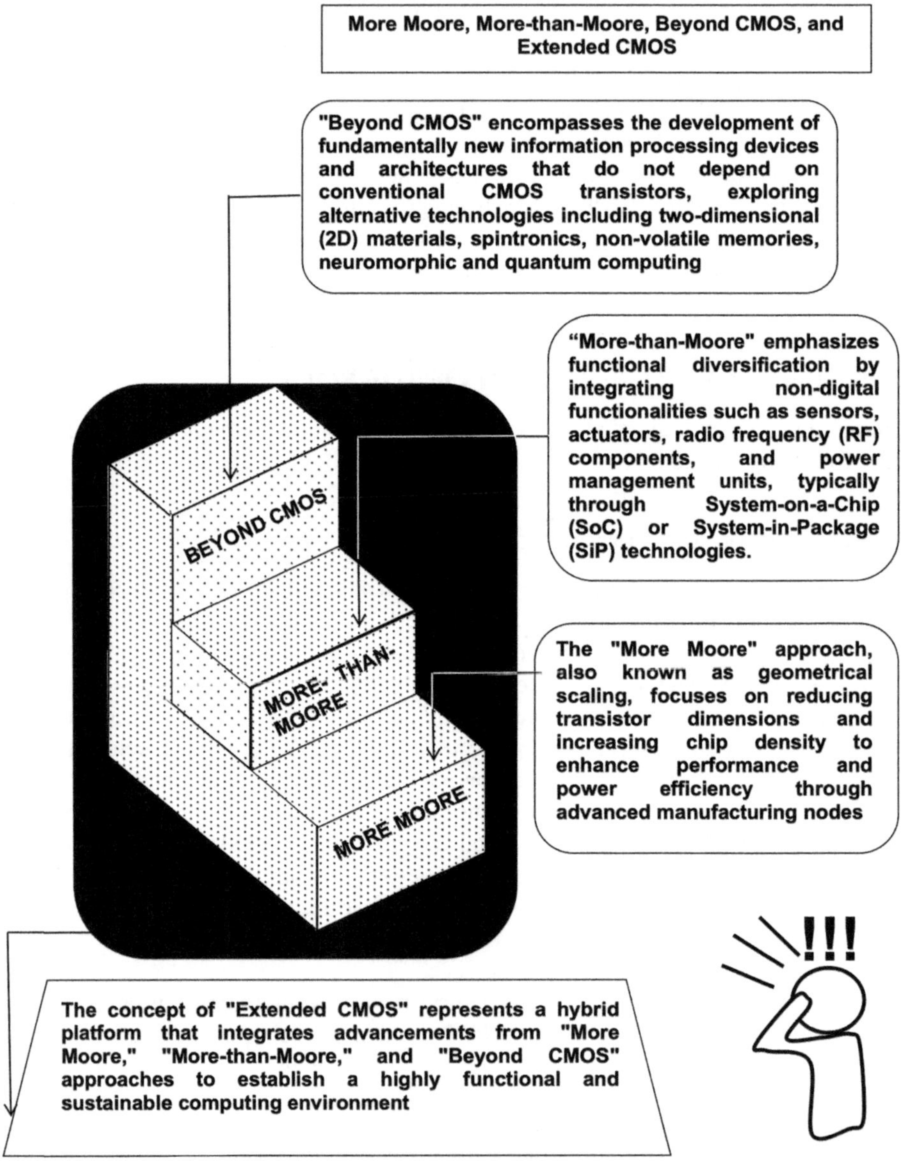

FIGURE 2.2 More Moore, more-than-Moore, beyond CMOS, and extended CMOS. The "More Moore" approach, also known as geometrical scaling, focuses on reducing transistor dimensions and increasing chip density to enhance performance and power efficiency as manufacturing nodes advance. In contrast, "More-than-Moore" emphasizes functional diversification by integrating non-digital components such as sensors, actuators, radio-frequency (RF) components, and power management units, typically through system-on-a-chip (SoC) or system-in-package (SiP) technologies. "Beyond CMOS" encompasses the development of fundamentally new information processing devices and architectures that do not rely on conventional CMOS transistors, exploring alternative technologies, including quantum computing, spintronics, neuromorphic computing, and two-dimensional (2D) materials. The concept of "extended CMOS" represents a hybrid platform that integrates advancements from "more Moore," "more-than-Moore," and "beyond CMOS" to establish a highly functional and sustainable computing environment.

2.5.2 CMOS and Ancillary Electronic Logic Devices

Chapter 4 discusses the evolution of transistor technology as conventional planar MOSFETs are being replaced by innovative architectures to enhance gate control and tackle scaling challenges. Key developments include Multi-gate Devices such as FinFETs and Gate-All-Around FETs, which improve gate control, reduce leakage, and minimize short-channel effects. Vertical GAA FETs have a perpendicular channel orientation, enabling higher integration density and smaller footprints. TFETs utilize band-to-band tunneling to improve power efficiency but face challenges such as low on-currents and hysteresis. NCFETs aim to surpass the traditional limits on subthreshold swing by leveraging negative capacitance to enable more efficient switching. Ferroelectric FETs (FeFETs) incorporate ferroelectric materials to enable non-volatile memory (NVM) applications and reduce power consumption. Nanowire-based devices, such as silicon and carbon nanotube FETs, provide better electrostatic control and carrier transport. 2D Material FETs use materials such as MoS_2 and ReS_2, showcasing potential for high-performance, low-power devices with excellent electrostatic control. These advancements highlight transformative approaches in semiconductor technology aimed at enhancing performance and efficiency.

2.5.3 Upcoming Spintronic Logic Devices

Chapter 5 discusses emerging logic device technologies that leverage magnetic properties to enable efficient data processing. Magneto-electric spin-orbit (MESO) logic devices change magnetization in response to an electric field and convert this change into electrical signals via spin-orbit coupling, functioning as logic gates. Magnetic tunnel junction (MTJ) logic devices utilize voltage-controlled magnetic switching, offering low power consumption and non-volatility for memory and logic tasks. NML uses nanomagnets for data storage and processing, enabling high-density integration. Additionally, devices based on the Spin Hall effect (SHE) operate by accumulating spin-polarized electrons for logic operations, emphasizing high energy efficiency in spin logic applications.

2.5.4 Non-Volatile Memories in the Beyond-CMOS Era

Chapter 6 discusses four advanced NVM technologies—ReRAM, MRAM, PCM, and FeRAM—and highlights their advantages over traditional flash memory, including improved density, faster access speeds, and lower power consumption. ReRAM operates by altering the electrical resistance of a material by forming or dissolving conductive filaments. MRAM relies on magnetic polarization to represent binary data. PCM uses Joule heating to transition a material between crystalline and amorphous states, ensuring data retention. FeRAM stores information using the electrical polarization of ferroelectric materials, which can be modified by applying voltage. These technologies represent significant advancements in memory storage solutions.

2.5.5 Emerging Computing Paradigms and Architectures

Chapter 7 discusses three innovative computational approaches beyond CMOS nanoelectronics that diverge from traditional computing models: neuromorphic computing, IMC), and quantum computing. Neuromorphic computing is inspired by the human brain; this approach uses devices such as memristors to mimic neural networks, thereby enhancing AI and machine learning capabilities. IMC integrates memory and processing units to reduce data movement and improve power efficiency by enabling data to be processed directly in RAM, bypassing the need for separate processors. Quantum computing utilizes quantum properties such as superposition and entanglement. Quantum computing processes information using qubits, enabling parallel computation for certain types of problems. These models leverage advanced materials and phenomena to enhance computing efficiency beyond conventional paradigms.

2.5.6 Nanofabrication Techniques for Post-CMOS Processing

Chapter 8 outlines two primary areas of focus: advanced lithography techniques and material deposition methods. Advanced lithography techniques are essential for creating intricate patterns below 10 nm on wafers. Key methods include extreme ultraviolet lithography (EUVL), which uses short-wavelength light (13.5 nm) to achieve finer patterns for cutting-edge nodes (7 nm and below). Nanoimprint lithography (NIL) is a mechanical approach that shapes polymers using a mold, offering high resolution and cost-effectiveness for non-CMOS applications. Other techniques, such as electron-beam lithography and block-copolymer lithography, are also being explored to achieve higher resolution. Novel material deposition techniques for beyond-CMOS devices require materials such as III–V semiconductors and graphene. Key deposition methods include atomic layer deposition (ALD), which allows precise, layer-by-layer growth of materials, crucial for creating high-κ gate insulators and uniform layers on complex structures. Molecular beam epitaxy (MBE) is used to grow high-quality crystalline layers of novel materials on silicon substrates, which are important for hybrid chips. Metal-organic chemical vapor deposition (MOCVD) focuses on producing high-quality films of III–V semiconductors, essential for optoelectronic devices. These advancements are pivotal for the continued evolution of materials and designs in the semiconductor industry.

2.5.7 Post-CMOS Integration and Advanced Packaging

Chapter 9 discusses the challenges and opportunities arising from integrating various materials and architectures in semiconductor manufacturing, particularly through 3D integration techniques. Rather than focusing on simply miniaturizing existing designs, the "Beyond CMOS" approach emphasizes creating a unified, three-dimensional system with integrated materials and devices. Heterogeneous integration merges diverse devices and functions onto a single platform, typically using methods such as through-silicon vias (TSVs) to vertically interconnect stacked layers. Two bonding techniques are highlighted: wafer-on-wafer (WoW) bonding, which joins full wafers for high throughput, and chip-on-wafer (CoW) bonding, which offers greater flexibility. Wafer bonding/layer transfer enables the merging of different component layers and the creation of vertical interconnections. Monolithic 3D ICs, which form devices in layers on a single wafer, aim to simplify interconnections but face fabrication challenges. Advanced packaging methods, such as 3D stacked ICs, seek to improve performance and manage thermal issues through effective heterogeneous integration. The chapter also mentions back-end-of-line (BEOL) integration, which introduces new materials in later manufacturing stages, enhancing functionality while minimizing disruption. Low-temperature fabrication processes are vital for integrating thermally sensitive materials with existing CMOS technologies without damage. Additional techniques, such as self-assembly, are explored as a potential low-cost, high-density fabrication approach that uses chemical processes to arrange nanoscale components into functional structures, although it remains an area of ongoing research.

2.5.8 Future Outlook and Research Trends

Beyond-CMOS nanoelectronics present opportunities, such as greater device density, reduced power consumption, and improved functionality (such as neuromorphic computing), yet face obstacles, including high production costs, device variability, reliability issues, complex integration, and the nascent stage of many promising technologies, such as spintronic devices. To address these challenges, considerable research is needed in areas such as device physics, materials science, system architecture, and manufacturing processes to transcend the limitations of current CMOS scaling.

2.6 DISCUSSION AND CONCLUSIONS

Following the discussion of CMOS challenges in the preceding chapter, this chapter establishes the framework for nanoelectronics beyond CMOS. Experts agree that CMOS will not be entirely replaced by any one technology. Instead, the future of computing points toward a vibrant mix of technologies working together in a hybrid ecosystem.

The sequence of ensuing chapters is materials, then logic devices (electronic and spintronics), followed by memories, computing architectures, fabrication processes, and packaging. We will delve into the planned chapters in succession, starting with the materials in the following chapter.

REFERENCES

Bernstein K., R. K. Cavin, W. Porod, A. Seabaugh and J. Welser 2010 Device and architecture outlook for beyond CMOS switches, *Proceedings of the IEEE*, 98(12): 2169–2184, https://doi.org/10.1109/JPROC.2010.2066530

Chen A. 2022 Beyond-CMOS roadmap—From Boolean logic to neuro-inspired computing, *Japanese Journal of Applied Physics*, 61(SM1003), https://doi.org/10.35848/1347-4065/ac5d86

Das S., A. Chen and M. Marinella 2021 Beyond CMOS, *2021 IEEE International Roadmap for Devices and Systems Outbriefs*, Santa Clara, CA, USA, 30 November, pp. 1–129, https://doi.org/10.1109/IRDS54852.2021.00011

Galatsis K., P. Gargini, T. Hiramoto, D. Beernaert, R. DeKeersmaecker and J. Pelka 2015 Nanoelectronics research gaps and recommendations: A report from the International

Planning Working Group on Nanoelectronics (IPWGN) [Commentary], *IEEE Technology and Society Magazine*, 34(2): 21–30, https://doi.org/10.1109/MTS.2015.2425811

Hutchby J. A., G. I. Bourianoff, V. V. Zhirnov and J. E. Brewer 2002 Extending the road beyond CMOS, *IEEE Circuits and Devices Magazine*, 18(2): 28–41, https://doi.org/10.1109/101.994856

IEEE 2023 Beyond CMOS, International Roadmap for Devices and Systems: 2023 Update, IEEE, 2023. [Online]. Available: https://irds.ieee.org/editions/2023

Nikonov D. E. and I. A. Young 2013 Overview of beyond-CMOS Devices and a uniform methodology for their benchmarking, *Proceedings of the IEEE*, 101(12): 2498–2533, https://doi.org/10.1109/JPROC.2013.2252317

Sotomayor Torres C. M., J. Ahopelto, M. W. M. Graef, R. M. Popp and W. Rosenstiel 2012 Beyond CMOS – benchmarking for future technologies, *2012 Design, Automation & Test in Europe Conference & Exhibition (DATE)*, Dresden, Germany, 2012, pp. 129–134, https://doi.org/10.1109/DATE.2012.6176445

Ueda A., H. Akinaga, S. Agarwal, J. A. Hagmann, S. Das, M. J. Marinella and A. Chen 2025 Green materials in semiconductors: perspective from the IRDS beyond-CMOS roadmap, *Nanotechnology*, 36(14), https://doi.org/10.1088/1361-6528/adb041

3 The Expanding Portfolio of Silicon and Supplementary Materials

3.1 INTRODUCTION

As traditional silicon (Si) scaling reaches fundamental limits, the field of nanoelectronics is investigating a variety of emerging materials to facilitate new computing paradigms. These novel materials possess distinct characteristics, including atomic-scale control, enhanced electrostatic tunability, and quantum-mechanical effects that conventional materials cannot provide. The main objective of these emerging materials is not to entirely replace CMOS but rather to work in conjunction with or enhance current platforms, aiming to create systems that are both more energy-efficient and powerful, particularly for advanced computing architectures such as neuromorphic or quantum computing. These materials serve as the basis for developing devices that utilize new state variables, moving beyond the simple electron charge used in CMOS. We shall explore materials for beyond-CMOS nanoelectronics under the broad categories of narrow- and wide-bandgap semiconductors, carbon-based and two-dimensional nanomaterials, dielectrics and topological insulators, ferroelectric materials, spintronic and magnetic materials. Figures 3.1–3.5 showcase the essential materials that define each class. Formidable challenges arise from the intricate processes of material synthesis, sophisticated fabrication methods, and the need for seamless integration with existing Si technology to advance from research prototypes to commercially viable solutions.

3.2 NARROW BANDGAP (NBG) SEMICONDUCTORS

3.2.1 Silicon

Si serves as the fundamental and overwhelmingly prevalent material in contemporary nanoelectronics. It remains the leading material primarily due to its exceptional semiconducting characteristics, excellent thermal stability, plentiful availability, affordability, low cost, ability to form a high-quality native oxide layer (SiO_2), and a well-developed manufacturing infrastructure that has evolved over many years. The entire modern microchip sector, which includes cutting-edge processors with dimensions measured in mere nanometers (such as gate-all-around nanowires and nanosheet FETs), relies on Si and its compatibility with CMOS technology.

DOI: 10.1201/9781003753469-3

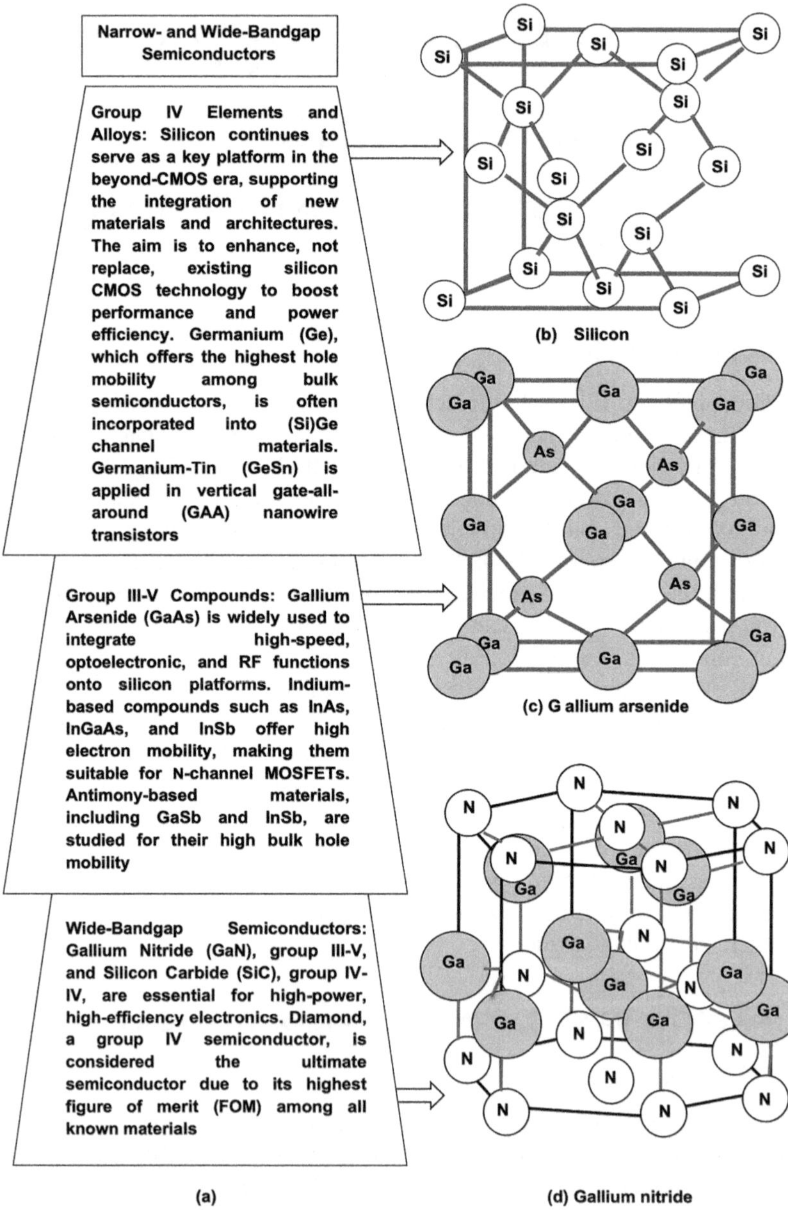

FIGURE 3.1 (a) A quick view of narrow- and wide-bandgap semiconductors. These consist of group IV elements and alloys: Silicon, Germanium (Ge), Germanium-Tin (GeSn); group III–V compounds: gallium arsenide (GaAs), indium-based compounds such as InAs, InGaAs, and InSb; antimony-based materials, including GaSb and InSb; wide-bandgap semiconductors: gallium nitride (GaN), group III–V, and silicon carbide (SiC), group IV–IV, and diamond, a group IV semiconductor. Diagrams on the right-hand side show atomic structures of silicon (b), gallium arsenide (c) and gallium nitride (d) lattices.

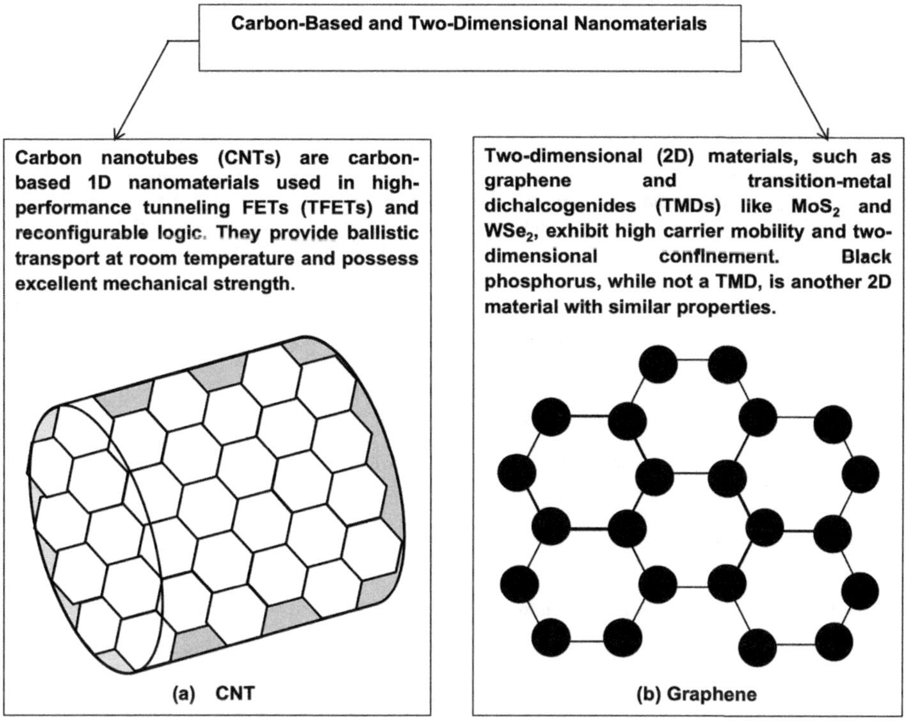

FIGURE 3.2 Carbon-based and two-dimensional nanomaterials for CMOS technology. The main materials are carbon nanotubes (CNTs); two-dimensional (2D) materials, such as graphene and transition-metal dichalcogenides (TMDs) like MoS_2 and WSe_2; and black phosphorus (not a TMD). A carbon nanotube (a) is a one-dimensional cylindrical structure created by rolling a graphene (b) sheet. Graphene is a single-atom-thick, two-dimensional sheet of carbon atoms.

Key reasons for Si's dominance are:

Its Abundance in Earth's Crust and Low Cost: As the second-most abundant element in Earth's crust (found in sand), Si is extremely reasonably priced and readily accessible for the mass production of highly pure wafers.

Semiconducting Properties: Si, being a semiconductor, allows precise control of its conductivity through a method known as doping (the introduction of impurities such as phosphorus or boron). Its main properties are presented in Figure 3.6(a) (ioffe-Si 2004, An et al. 2023).

Ease of Processing: Si readily forms a high-quality, stable insulator, silicon dioxide (SiO_2), which is crucial for the development of gate dielectrics in transistors. The established fabrication infrastructure (fabs) enables machining of Si with nanometer-level accuracy using techniques such as photolithography and etching.

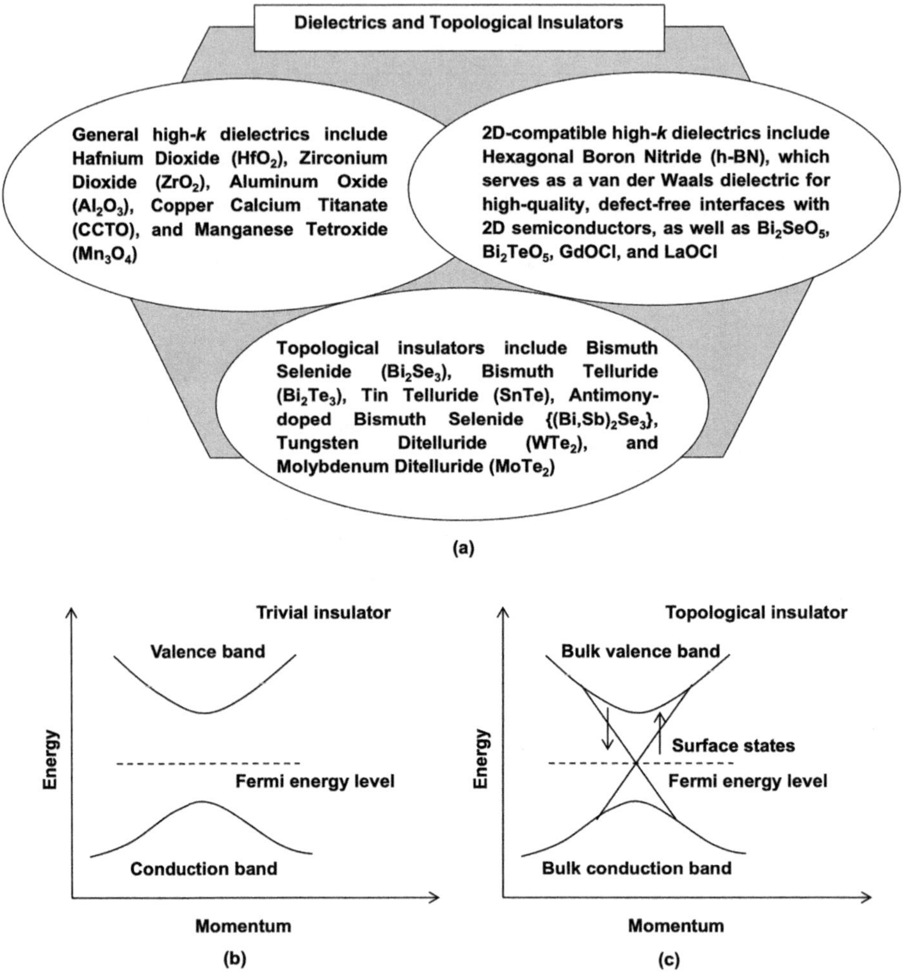

FIGURE 3.3 Types of insulators: (a) important dielectrics and topological insulators. General high-*k* dielectrics include HfO_2, ZrO_2, Al_2O_3, CCTO, and Mn_3O_4. 2D-compatible high-*k* dielectrics include hBN, as well as Bi_2SeO_5, Bi_2TeO_5, GdOCl, and LaOCl. Topological insulators include Bi_2Se_3, Bi_2Te_3, SnTe, $(Bi, Sb)_2Se_3$, WTe_2, and $MoTe_2$. (b) A trivial insulator. It features a distinct bandgap and lacks surface states. (c) A topological insulator. It has the property that the valence and conduction bands are split by a clear bandgap, yet surface states bridge this gap, allowing electricity to flow freely along the material's surface.

CMOS Compatibility: The extensive, multibillion-dollar infrastructure and well-documented processes for Si manufacturing enable the incorporation of new nanodevices into existing Si circuits, thereby lowering both cost and complexity.

Nanoscale Performance: At the nanoscale, Si demonstrates distinctive quantum characteristics (such as tunable bandgaps in quantum dots (QDs) and versatility in nanomembranes) that are being actively explored for future applications in areas including optoelectronics, sensors, and brain-inspired (neuromorphic) computing.

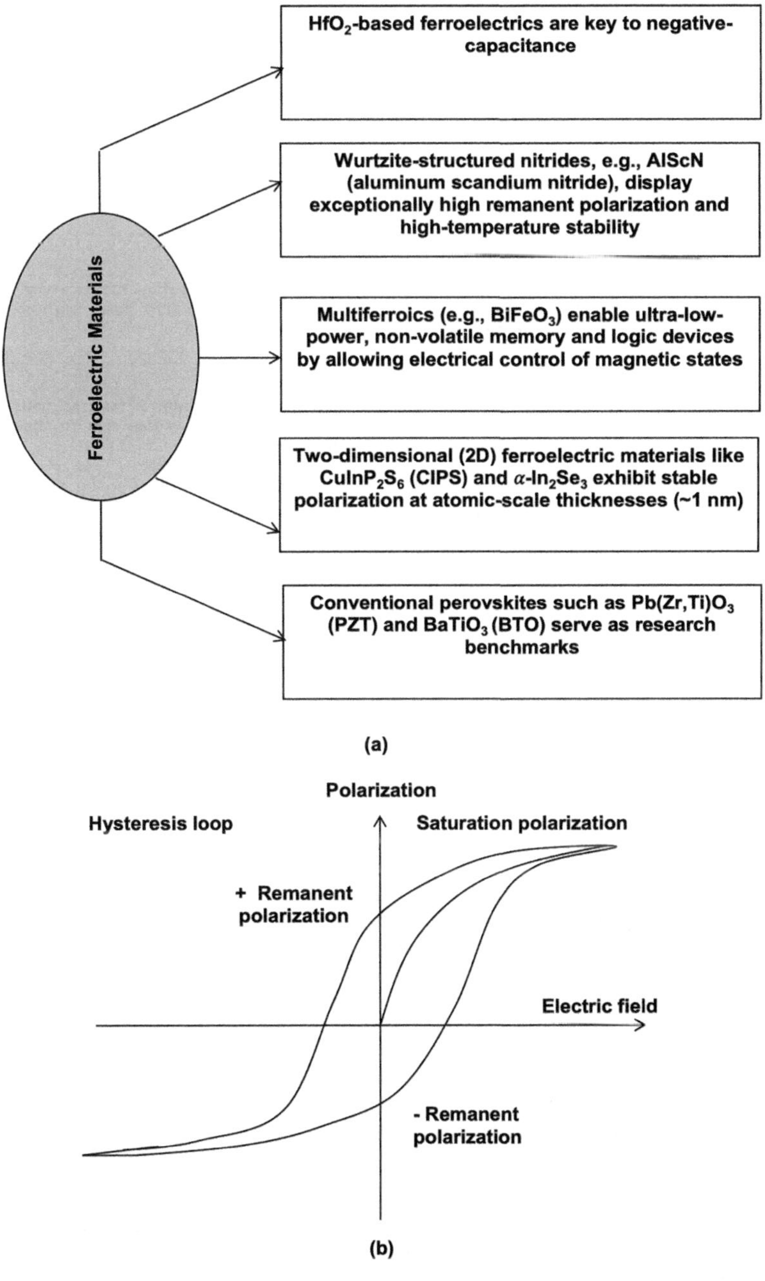

FIGURE 3.4 (a) Ferroelectric materials used in CMOS fabrication. These are HfO_2-based ferroelectrics; wurtzite-structured nitrides, e.g., AlScN (aluminum scandium nitride); multiferroics, e.g., $BiFeO_3$. 2D ferroelectric materials are $CuInP_2S_6$ and α-In_2Se_3. Perovskites are Pb(Zr, Ti)O_3 (PZT) and $BaTiO_3$ (BTO). (b) Ferroelectric materials are characterized by reversible spontaneous electric polarization. This polarization depends on both the current field and its history, resulting in a hysteresis loop when the polarization is plotted against the electric field. Saturation and remanent polarization points are marked on the loop.

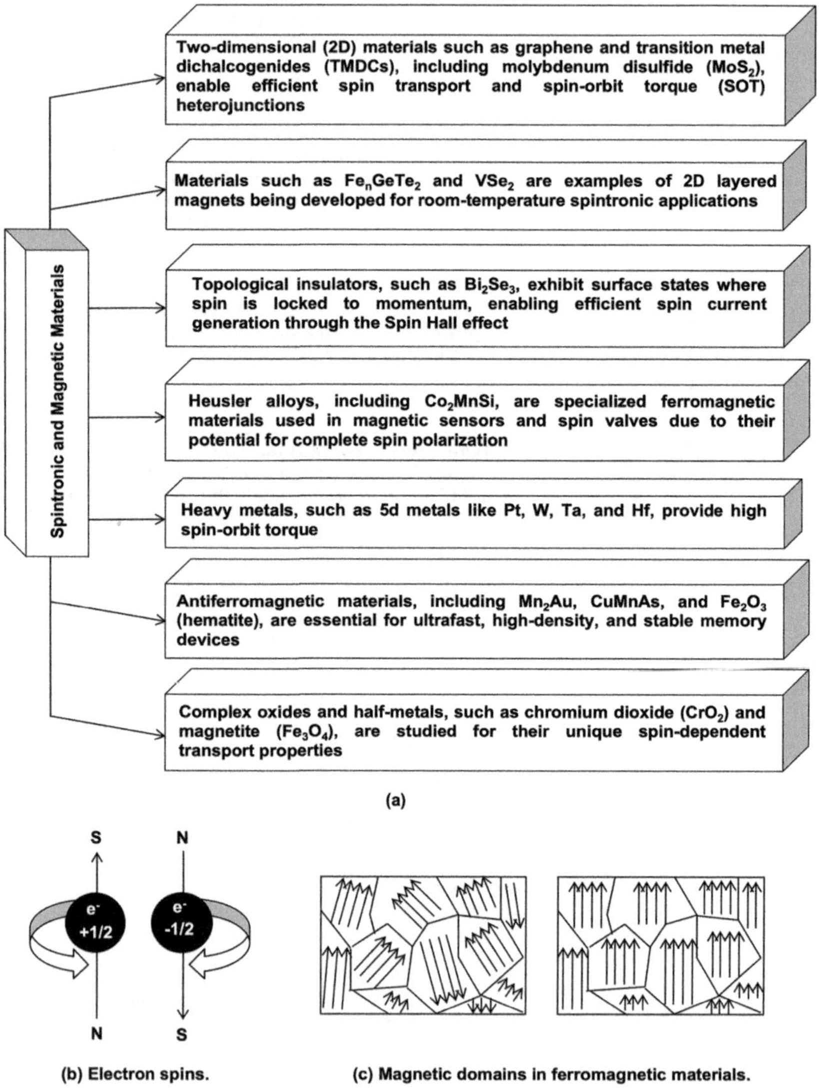

FIGURE 3.5 (a) Spintronic and magnetic materials for CMOS. These are 2D materials such as graphene and TMDCs, including MoS_2; materials such as Fe_nGeTe_2 and VSe_2 (2D layered magnets); topological insulators, such as Bi_2Se_3; ferromagnetic materials, e.g., Heusler alloys, including Co_2MnSi; heavy metals, such as 5d metals like Pt, W, Ta, and Hf; antiferromagnetic materials, including Mn_2Au, CuMnAs, and Fe_2O_3 (hematite); complex oxides and half-metals, such as chromium dioxide (CrO_2) and magnetite (Fe_3O_4). (b) In spintronics, electron spin is often illustrated by representing spin-up (+½) and spin-down (−½) states as magnetic arrows pointing between the north and south poles. (c) Inside ferromagnetic materials, tiny magnetic domains act like teams of atoms whose magnetic fields line up together. When an external magnetic field is applied, these domains align, creating a powerful magnetic field that persists even after the field is removed. This persistent magnetism makes ferromagnets valuable for data storage. The diagram illustrates how these domains look before and after exposure to a strong magnetic field.

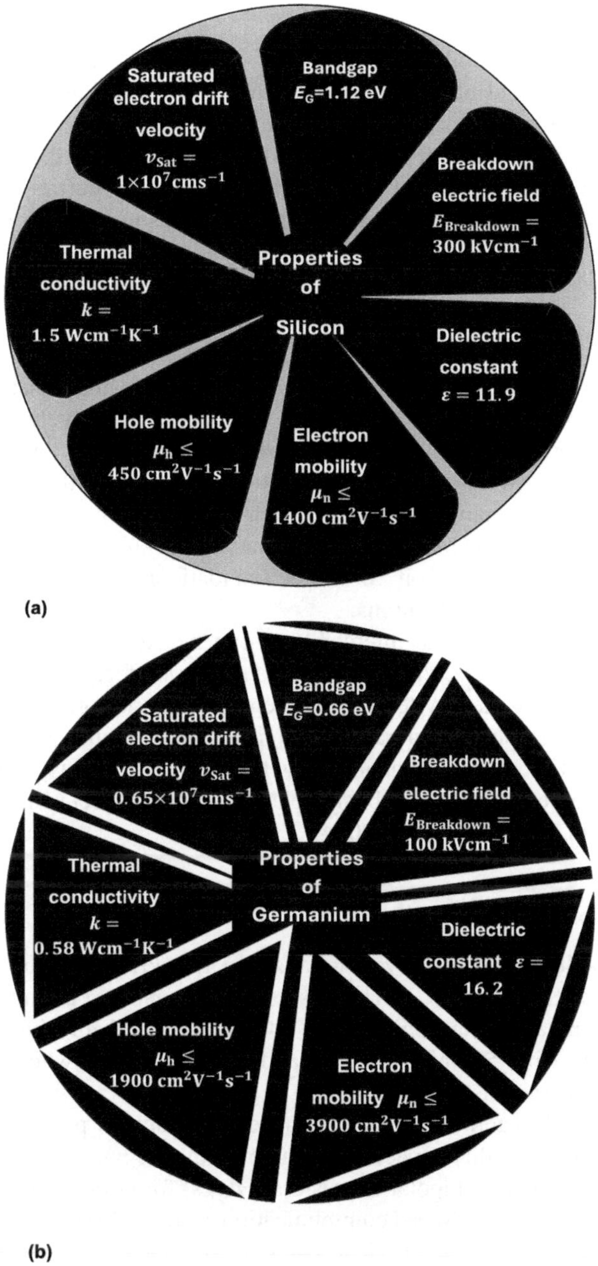

FIGURE 3.6 Properties of elemental semiconductors: (a) silicon and (b) germanium. In part (a), the properties of silicon are given as: bandgap $E_G(\mathrm{Si}) = 1.12$ eV, breakdown electric field $E_{\mathrm{Breakdown}}(\mathrm{Si}) = 300$ kV/cm, dielectric constant $\epsilon(\mathrm{Si}) = 11.9$, electron mobility $\mu_n(\mathrm{Si}) \leq 1400$ cm^2/V s, hole mobility $\mu_h(\mathrm{Si}) \leq 450$ cm^2/V s, thermal conductivity $k(\mathrm{Si}) = 1.5$ W/cmK, and saturated electron drift velocity $v_{\mathrm{Sat}}(\mathrm{Si}) = 1 \times 10^7$cm/s. In part (b), the properties of germanium are shown as bandgap $E_G(\mathrm{Ge}) = 0.66$ eV, breakdown electric field $E_{\mathrm{Breakdown}}$ (Ge) = 100 kV/cm, dielectric constant $\epsilon(\mathrm{Ge}) = 16.2$, electron mobility $\mu_n(\mathrm{Ge}) \leq 3900$ cm^2/V s, hole mobility $\mu_h(\mathrm{Ge}) \leq 1900$ cm^2/V s, thermal conductivity k (Ge) = 0.58 W/cm K, and saturated electrondrift velocity $v_{\mathrm{Sat}}(\mathrm{Ge}) = 0.65 \times 10^7$ cm/s.

3.2.2 Germanium

The excellent carrier mobility positions it as an essential material for applications that demand speed and efficiency at the nanoscale, including contemporary high-frequency and low-power transistors. Additionally, it is used in optoelectronics due to its transparency to infrared light. However, its limited thermal stability and high expense restrict its widespread use. Its electrical properties are illustrated in Figure 3.6(b) (ioffe-Ge 2001). Ge can be processed using existing silicon (CMOS) manufacturing techniques, facilitating its integration on Si wafers.

3.2.3 Germanium-on-Silicon (Ge-on-Si)

The combination of optoelectronics and electronics into a single platform is a highly sought-after capability. In photodetectors, germanium effectively absorbs infrared light, making germanium layers on Si substrates well-suited for optical receivers that combine optics with electronics. In optical modulators, the electro-absorption characteristics of germanium facilitate high-speed, low-power optical switches, addressing the limitations of Si-based modulators. When it comes to QDs, forming organized germanium QDs on Si offers foundational components for quantum computing and sensing applications.

Ge quantum-well photonic interconnects can be created with minimal loss on Si wafers. Ge-rich $Si_{1-x}Ge_x$ virtual substrates serve as effective, high-quality optical waveguides, enabling the low-temperature, epitaxial fabrication of Ge quantum-well devices. As a demonstration, a passive $Si_{0.16}Ge_{0.84}$ waveguide is successfully integrated with two Ge/SiGe multi-quantum-well active components—an optical modulator and a photodetector—to establish a photonic interconnect using a single step of epitaxial growth (Chaisakul et al. 2014).

3.2.4 Silicon-Germanium Alloy

Silicon-Germanium (SiGe) alloys capitalize on the advantages of both materials. The selection of Si versus SiGe materials usually depends on the intended use: Si is used for widespread digital logic, while SiGe is preferred for high-frequency telecommunications (like 5G infrastructure) and optoelectronics, where speed and particular optical characteristics are essential.

SiGe is utilized to address the performance constraints of pure Si, particularly by enhancing carrier mobility (speed) and facilitating integrated optoelectronics, frequently in heterojunction bipolar transistors (HBTs) for critical applications such as 5G infrastructure and high-speed communication systems. Essentially, Ge serves as a "performance enhancer" and "optical gateway," broadening Si's potential in high-speed electronics and integrated photonics by means of precise material engineering.

In heterojunction bipolar transistors (HBTs), incorporating Ge into the base region improves mobility and speed, which are vital for high-frequency electronics. During strain engineering, the addition of Ge alters the lattice spacing, inducing strain that boosts electron and hole mobility, which is essential for advanced MOSFETs.

In optoelectronics, SiGe alloys with specially designed crystal structures (hexagonal) can function as light emitters, facilitating Si-based lasers. SiGe alloys provide improved performance and are under investigation for applications in solar panels and components for quantum computing.

3.2.4.1 Epitaxial Growth

Germanium is combined with Si nanoelectronics mainly through heteroepitaxy to create SiGe alloys and structures. The 4.2% lattice mismatch between Ge and Si results in substantial stress development. Engineers deliberately utilize this strain to alter the band structures of the materials, enhancing carrier (electron and hole) mobility and allowing for particular optical characteristics in optoelectronics.

The key deposition methods for the Ge-on-Si system are solid source MBE and chemical vapor deposition (CVD) (Ye and Yu 2014). The MBE apparatus can achieve ultrahigh vacuum (UHV), allow for precise control of the deposition rate, and facilitate in situ monitoring of the surface during growth. In the Si/Ge epitaxy MBE system, the atomic or molecular flux of the growth materials, such as Si, Ge, and other dopants, is generated by Knudsen effusion cells (for low-melting-point substances) or electron-beam evaporators (for materials with high melting points). Shutters within the UHV chamber can easily turn the atomic flux on and off. Reflection high-energy electron diffraction (RHEED) is frequently utilized with the MBE setup to assess the surface of crystalline substances.

The CVD methods employ gaseous precursors such as GeH_4, SiH_4, Si_2H_6, or $SiCl_2H_2$, which are pyrolyzed at elevated temperatures. Most CVD techniques can achieve exceptional precision and reproducibility in the deposition rates for pseudomorphic SiGe layers. Unlike the MBE technique, in CVD epitaxy, the Ge content of a film will not only depend on the flow rates of the gases but will also be influenced by the substrate's temperature and the pressure within the chamber.

3.2.4.2 Selective Area Growth (SEG)

This sophisticated method involves the targeted growth of Ge in specific, designated areas on a Si substrate by employing a dielectric mask (such as Si dioxide). This technique enables the accurate positioning of Ge devices on a chip while ensuring planarity for subsequent fabrication processes.

3.2.4.3 Layer Transfer and Wafer Bonding

High-quality, thin layers of single-crystal Ge can be transferred to different substrates (including Si, glass, or plastic) through direct bonding and epitaxial lift-off methods.

3.2.4.4 Challenges

The primary obstacle in integration is the 4.2% lattice mismatch between Si and germanium, which can result in defects (threading dislocations) that impair device performance. Additional challenges consist of surface passivation problems for Ge and the management of increased leakage currents stemming from its narrower bandgap in comparison to Si.

To minimize defects, specifically threading dislocations (TDs) and surface roughness, during the growth of Ge on a Si substrate, various advanced epitaxial and post-growth processing methods are utilized by researchers and engineers:

3.2.4.5 Growth Techniques

Two-Step Growth Method: This technique is commonly employed to reduce the occurrence of 3D islands during the initial growth phase and to ensure a smooth surface.

Low-Temperature (LT) Buffer Layer: An initial, ultra-thin (tens of nanometers) Ge "seed" layer is deposited at low temperatures (between 300°C and 400°C) to restrict the movement of Ge atoms and encourage two-dimensional, layer-by-layer growth.

High-Temperature (HT) Growth: A thicker Ge layer is then grown at higher temperatures (approximately 600°C–700°C) on top of the LT buffer layer to enhance material quality and accelerate growth rates.

Compositionally Graded SiGe Buffer Layers: A buffer layer with a gradually increasing concentration of Ge is created between the Si substrate and the final pure Ge layer. This method helps distribute the relaxation of lattice mismatch over a broader area, resulting in the formation of a network of misfit dislocations near the interface and the bending of threading dislocations so they can either annihilate each other or remain confined to the buffer layer, rather than extending to the surface.

3.2.4.6 Post-Growth Techniques

Thermal Cyclic Annealing (TCA): Following the growth phase, the Ge film undergoes multiple rapid heating cycles (e.g., from 600°C to 850°C) in a hydrogen environment. This treatment enhances the mobility of dislocations, allowing them to move, interact, and annihilate, which significantly reduces the threading dislocation density (TDD).

Chemical Mechanical Polishing (CMP): CMP is applied to smooth out rough surfaces that result from the growth of the graded buffer layer or island coalescence, which is essential for the subsequent standard planar CMOS fabrication steps.

3.2.5 Germanium-Tin

Germanium-tin (GeSn) is a highly promising material for applications in optoelectronics, electronics, and spintronics due to its properties of having a direct adjustable bandgap, low effective masses of carriers, and significant spin-orbit coupling. The bandgap of GeSn can be tuned into the near- to mid-infrared spectrum exceeding 2 μm, and it turns into a direct bandgap semiconductor at Sn concentrations usually exceeding 8–10 at.%. These features allow for effective optoelectronics and high-performing transistors that are difficult to achieve with Ge by itself. GeSn can enable effective light emission (lasing) and detection on a Si platform, which is unachievable with indirect bandgap Si or pure germanium.

Remarkably high electron mobilities of up to 6200 and 1500 $cm^2/V\ s$ are obtained in the strain-relaxed $Ge_{0.88}Sn_{0.12}$ film at temperatures of 50 and 300 K, respectively, because of the enhanced electron population in the direct Γ-valley (Tien et al. 2025). In tensile-strained $Ge_{0.96}Sn_{0.04}$ NMOSFET, $\mu_n(\text{GeSn}) = 698\ \text{cm/V s}$ (Chuang et al. 2021). The hole mobility is $\mu_h(\text{GeSn}) = 540\ \text{cm/V s}$ (Moto et al. 2019). The high electron mobility results in high-performance, low-power transistors with reduced sizes for next-generation microchips.

Like Ge, GeSn is compatible with current silicon (CMOS) manufacturing methods, making monolithic integration on Si chips easier. The elevated electron and hole mobilities in GeSn alloys are beneficially exploited by the incorporation of GeSn in source/drain regions or as the channel material. This enhances the efficiency of both P-channel and N-channel MOSFETs as well as FinFETs (Fin field-effect transistors). The small, direct bandgap of GeSn increases the rate of band-to-band tunneling, making it a promising candidate for high-performance TFETs, which are essential for low-power applications and potential quantum computing elements. The successful demonstration of operational GeSn CMOS inverters indicates the possibility of creating fully integrated electronic-photonic circuits on a single platform.

3.2.6 Gallium Arsenide

GaAs's exceptional electron mobility enables much faster devices and higher frequency performance, which is essential for specialized uses such as 5G communication systems. The direct bandgap of GaAs allows for highly efficient conversion of electrical energy to light (in lasers and LEDs) and from light to electrical energy (in solar cells and photodetectors). But the high expense and production difficulties associated with GaAs make it impractical for the mass-market, general-purpose electronics that are primarily dominated by Si. Figure 3.7 presents the properties of III–V compound semiconductors. In Figure 3.7(a), properties of GaAs are displayed (ioffe-GaAs 1998).

The integration of GaAs with large-scale Si in a monolithic manner marks a significant progress in Si photonics. However, attaining high-quality epitaxial growth of GaAs on a Si substrate remains a challenging task. The substantial lattice mismatch of 4.2% between GaAs and Si, along with over 50% mismatch in their thermal expansion coefficients and differing crystal structures, complicates the defect-free growth of GaAs on Si. GaAs adopts a zinc blende crystal structure, whereas Si features a diamond cubic configuration.

Various methods have been implemented to reduce defect formation in GaAs layers grown on Si, which is caused by strain and lattice mismatch. These methods include utilizing SiGe buffer layers as virtual substrates, applying selective area growth (SAG), employing epitaxial lateral overgrowth (ELO), and utilizing the aspect ratio trapping (ART) technique (Rodrigues et al. 2025).

Selective Area Growth (SAG) and Nanoheteroepitaxy (NHE): GaAs is selectively deposited on Si substrates that have been pre-patterned using silicon dioxide (SiO_2) masks. This process employs the aspect ratio trapping (ART) technique to restrict defects to specific small areas, resulting in high-quality, single-crystalline GaAs nanowires or nanoislands.

Nanowire Growth: Utilizing self-catalyzed molecular beam epitaxy (MBE) to grow GaAs/AlGaAs core-shell nanowires on Si wafers presents a promising

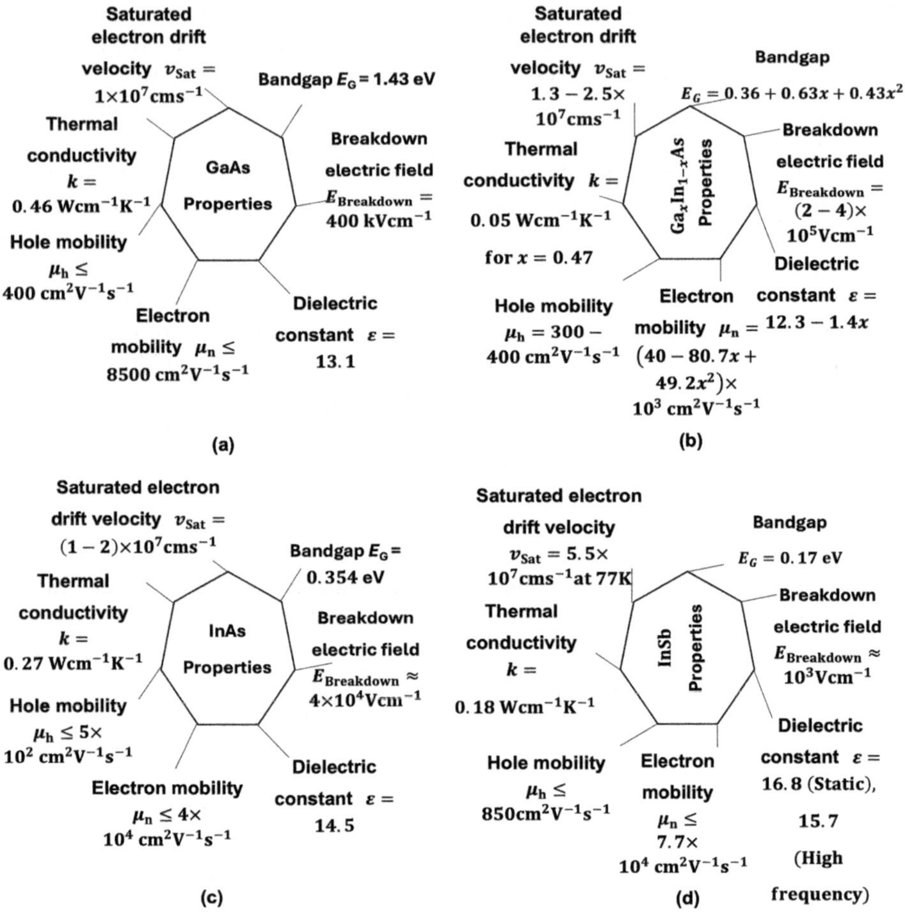

FIGURE 3.7 Main properties of III–V compound semiconductors: (a) GaAs properties are: bandgap $E_G(\text{GaAs}) = 1.43$ eV, breakdown electric field $E_{\text{Breakdown}}(\text{GaAs}) = 400$ kV/cm, dielectric constant $\epsilon(\text{GaAs}) = 13.1$, electron mobility $\mu_n(\text{GaAs}) \leq 8500$ cm/V s, hole mobility $\mu_h(\text{GaAs}) \leq 400$ cm/V s, thermal conductivity k (GaAs) $= 0.46$ W/cm K, and saturated electron drift velocity $v_{\text{Sat}}(\text{GaAs}) = 1 \times 10^7$cm/s. (b) $Ga_xIn_{1-x}As$ properties are: bandgap E_G $(Ga_xIn_{1-x}As) = 0.36 + 0.63x + 0.43x^2$eV, breakdown electric field $E_{\text{Breakdown}}(Ga_xIn_{1-x}As) = (2-4) \times 10^5$ V/cm, dielectric constant $\epsilon(Ga_xIn_{1-x}As) = 12.3 - 1.4x$, electron mobility μ_n $(Ga_xIn_{1-x}As) = (40 - 80.7x + 49.2x^2) \times 10^3$ cm/V s, hole mobility $\mu_h(Ga_xIn_{1-x}As) = 300 - 400$ cm/V s, thermal conductivity k $(Ga_xIn_{1-x}As) = 0.05$ W/cm K for $x = 0.47, 1 - x = 0.53$, and saturated electron drift velocity $v_{\text{Sat}}(Ga_xIn_{1-x}As) = (1.3 - 2.5) \times 10^7$cm/s. (c) InAs properties are: bandgap $E_G(\text{InAs}) = 0.354$ eV, breakdown electric field $E_{\text{Breakdown}}(\text{InAs}) \approx 4 \times 10^4$ V/cm, dielectric constant $\epsilon(\text{InAs}) = 14.5$, electron mobility $\mu_n(\text{InAs}) \leq 4 \times 10^4$ cm/V s, hole mobility $\mu_h(\text{InAs}) \leq 5 \times 10^2$ cm/V s, thermal conductivity k (InAs) $= 0.27$ W/cm K, and saturated electron drift velocity $v_{\text{Sat}}(\text{InAs}) = 1 - 2 \times 10^7$ cm/s. (d) InSb properties are: bandgap $E_G(\text{InSb}) = 0.17$ eV, breakdown electric field $E_{\text{Breakdown}}(\text{InSb}) \approx 10^3$ V/cm, dielectric constant ϵ (InSb) $=$ 16.8 (static), 15.7 (high frequency), electron mobility $\mu_n(\text{InSb}) \leq 7.7 \times 10^4$ cm/V s, hole mobility $\mu_h(\text{InSb}) \leq 850$ cm/V s, thermal conductivity k (InSb) $= 0.18$ W/cm K, and saturated electron drift velocity $v_{\text{Sat}}(\text{InSb}) = 5.5 \times 10^7$ cm/s at 77 K.

strategy for wafer-scale integration without the need for extensive pre-treatment. This approach facilitates the development of hybrid Si/III–V nanowires with interfaces that exhibit low defect densities, making them suitable for quantum information technologies.

Wafer Bonding and Layer Transfer: This "heterogeneous integration" method consists of separately producing GaAs layers and subsequently bonding them to a Si or silicon-on-insulator (SOI) wafer with the help of an adhesive polymer. This circumvents the complications associated with direct epitaxial growth while still enabling the functional merging of materials.

Nano-Ridge Engineering: This approach entails growing GaAs on specially designed Si substrates to manage defect density, allowing for the production of high-performance electrically pumped lasers directly on 300mm Si wafers.

3.2.7 Indium Gallium Arsenide

The superior electron mobility in InGaAs enables devices to function at significantly higher frequencies and speeds than their Si equivalents. This characteristic makes InGaAs the preferred material for advanced wireless communication and data processing applications. Thanks to its direct bandgap properties, InGaAs can effectively emit and absorb light, particularly in the infrared spectrum utilized for fiber-optic communications. Figure 3.7(b) shows the properties of $Ga_xIn_{1-x}As$ (ioffe-GaInAs 1999).

In the field of nanoelectronics, researchers are increasingly investigating hybrid methods to merge the high performance of InGaAs with the cost-effectiveness and established infrastructure of Si platforms, especially for integrated Si photonics.

3.2.8 Indium Arsenide

Electrical properties of indium arsenide are depicted in Figure 3.7(c) (ioffe-InAs 2001). Various innovative methods have been devised to integrate InAs with Si, overcoming the significant lattice mismatch and differing thermal properties of the two materials:

Nanowire (NW) Epitaxial Growth: This is a key technique where InAs nanowires are directly grown on patterned Si substrates utilizing methods such as metal-organic vapor-phase epitaxy (MOVPE) or molecular beam epitaxy (MBE) (Cirlin et al. 2001, Ilkiv et al. 2024). The mechanism of growth of nanowires better accommodates the lattice mismatch compared to planar films, often resulting in the trapping of dislocations at the interface. Numerous processes are designed to be compatible with CMOS technology, functioning at temperatures below the 450°C back-end-of-line thermal limit, and are frequently "self-catalyzed" (without gold) to prevent contamination of the Si substrate.

Layer Transfer and Wafer Bonding: Techniques such as the Smart-cut® process (which involves hydrogen ion implantation and wafer bonding) are employed to transfer thin, high-quality InAs layers onto a silicon-on-insulator (SOI) platform. This enables the creation of electrically isolated InAs nanostructures on Si substrates.

Ion Implantation and Annealing: Indium and arsenic ions are implanted into a Si wafer to form InAs nanocrystals within the Si matrix, followed by rapid thermal annealing. This method is highly compatible with existing large-scale Si technologies.

Template-Assisted Selective Epitaxy (TASE): This technique employs dielectric templates (for instance, Si_3N_4) to designate specific regions for InAs growth, resulting in high-quality, single-crystalline nanostructures that are seamlessly integrated into the Si platform.

Example 3.1 A nanoelectronic device channel composed of indium arsenide InAs is subjected to an electric field. The properties of InAs are as follows: Electron mobility μ_n is 30,000 cm^2/V s, and the applied electric field E measures 100 V/cm. Determine the drift velocity, v_d, of the electrons within the InAs channel.

Solution: The equation that relates drift velocity (v_d), electron mobility (μ_n), and the applied electric field (E) is expressed as:

$$v_d = \mu_n E \tag{3.1}$$

By substituting the provided values into the equation: $\mu_n = 30{,}000\ cm^2/V\ s$, and $E = 100\ V/cm$, we have

$$v_d = 30{,}000\ cm^2/V\ s \times 100\ V/cm = 3 \times 10^6\ cm/s = 3 \times 10^6 \times 10^{-2}\ m/s = 3 \times 10^4\ m/s \tag{3.2}$$

The drift velocity of the electrons in the InAs channel is 3×10^6 cm/s (or 30,000 m/s). This elevated velocity provides a significant advantage for III–V materials over Si in high-speed applications.

3.2.9 Indium Antimonide

Its principal properties are shown in Figure 3.7(d) (ioffe-InSb 2001, Dargys et al. 1978). The effective combination of InSb with Si nanoelectronics allows for various advanced applications:

Mid-Infrared Photodetectors: Due to InSb's narrow bandgap, it is well-suited for mid-wavelength infrared (MWIR) photodetectors (3–8 μm range), which are utilized in night vision, thermal imaging, and atmospheric monitoring. Monolithic integration on Si CMOS read-out integrated circuits significantly lowers costs and enhances scalability.

High-Speed, Low-Power Logic: InSb quantum-well transistors demonstrate exceptionally high electron mobility (exceeding 12,000 $cm^2/V{\cdot}s$ at room temperature), positioning them as promising candidates for ultra-high-speed and very low-power digital logic applications that surpass conventional Si transistors.

Quantum Computing and Photonics: The distinctive electronic and spin characteristics of InSb nanostructures are also of interest for future topological quantum devices and Si photonics.

The main challenge is the considerable 19% lattice mismatch and differences in crystal symmetry between InSb and Si, which generally results in a high defect density with standard growth methods. Researchers are creating innovative techniques to address this issue:

Rapid Melt Growth (RMG): This CMOS-compatible method entails depositing amorphous InSb onto an insulator-covered Si substrate with minimal openings (seed areas). A quick thermal anneal melts the InSb, which is then recrystallized from the Si seed, limiting defects to the seed interface and generating single-crystalline InSb nanostructures with minimal thermal impact on the underlying Si circuitry.

Buffer Layers: Another strategy includes growing InSb on intermediary buffer layers such as GaAs/Ge or AlSb/GaSb, which progressively accommodate the lattice mismatch. The implementation of an interfacial misfit (IMF) array at the buffer/Si interface can effectively regulate dislocation formation.

Nanowires (NWs) and Nanostructures: Cultivating InSb in the form of nanowires or other nanostructures on Si can help alleviate strain due to the high surface-to-volume ratio, providing an avenue for integration. InAs is often utilized as a seed layer for the catalyst-free growth of InSb nanowires.

Electrodeposition: This economical method is being investigated to fill predefined templates on Si substrates with polycrystalline InSb, followed by a recrystallization process to obtain single-crystalline material.

3.3 WIDE-BANDGAP (WBG) SEMICONDUCTORS

3.3.1 Silicon Carbide

Figure 3.8 presents the properties of wide-bandgap semiconductors. In Figure 3.8(a), properties of 4H–SiC are given (ioffe-SiC 2001).

Merging Silicon Carbide (SiC) with Si nanoelectronics leverages the established CMOS technology of Si alongside the enhanced power, thermal, and optical characteristics of SiC, resulting in hybrid platforms for sophisticated electronics, particularly in power devices, quantum sensing, and photonics. The resulting hybrid SiC/Si platforms are employed across a range of state-of-the-art nanoelectronic and nanophotonic applications:

Extreme Environment Sensors: The remarkable thermal resilience and chemical stability of SiC enable the creation of micro-sensors and flexible electronics capable of functioning at temperatures exceeding 300°C in extreme conditions (for instance, aerospace, automotive, or combustion applications).

Biomedical Devices: Transparent bioelectronic systems based on SiC can be incorporated at the tips of medical devices such as endoscopes for enhanced diagnostics and treatments, like tissue ablation, owing to SiC's outstanding biocompatibility and transparency.

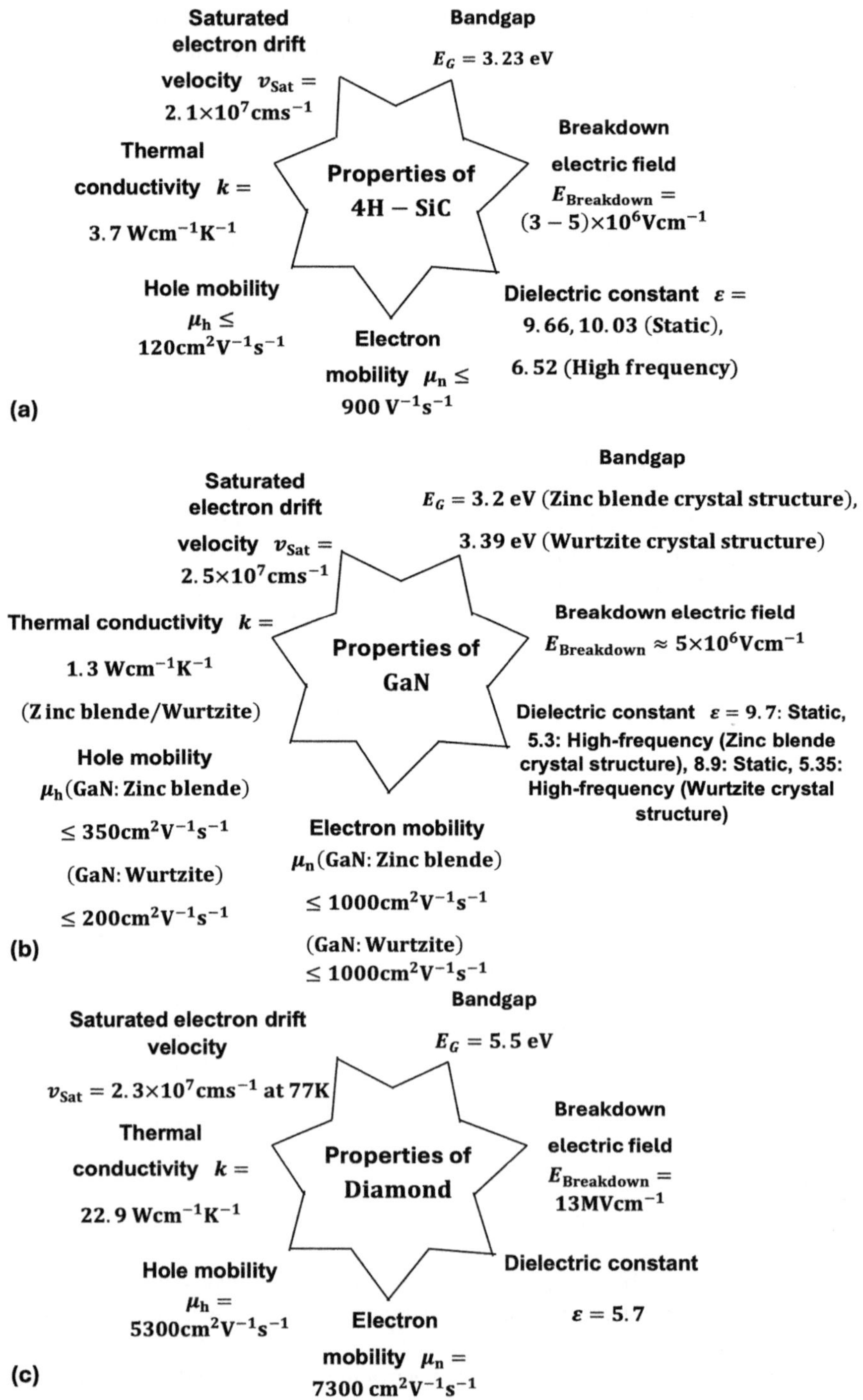

FIGURE 3.8 Main properties of wide-bandgap semiconductors: Properties of 4H–SiC are:

(Continued)

FIGURE 3.8 (Continued) bandgap E_G(4H – SiC)=3.23 eV, breakdown electric field $E_{Breakdown}$ (4H – SiC)≈(3–5)×10^6 V/cm, dielectric constant ϵ(4H – SiC) = 9.66, 10.03 (static), 6.52 (high frequency), electron mobility μ_n (4H – SiC) ≤ 900 cm/Vs, hole mobility μ_h (4H – SiC) ≤ 120 cm/Vs, thermal conductivity k (4H – SiC) = 3.7 W/cm K, and saturated electron drift velocity v_{Sat} (4H – SiC) = 2.1 ×10^7cm/s. (b) Properties of GaN are: bandgap E_G (GaN) = 3.2 eV (zinc blende crystal structure), 3.39 ev (wurtzite crystal structure), breakdown electric field $E_{Breakdown}$ (GaN) ≈ 5 × 10^6 V/cm (zinc blende/wurtzite crystal structure), dielectric constant ϵ(GaN) = 9.7: static, 5.3: high –frequency (zinc blende crystal structure), 8.9: static, 5.35 : high–frequency (wurtzite crystal structure), electron mobility μ_n(GaN : zinc blende) ≤ 1000 cm/V s, electron mobility μ_n (GaN : wurtzite) ≤ 1000 cm/V s, hole mobility μ_n (GaN: zinc blende) ≤ 350 cm/V s, hole mobility μ_n (GaN : Wurtzite) ≤ 200 cm/V s, thermal conductivity k (GaN) = 1.3 W/cm K (zinc blende/wurtzite), and saturated electron drift velocity v_{Sat} (GaN) = 2.5 × 10^7cm/s. (c) Properties of diamond are: bandgap E_G (diamond) = 5.5 eV, breakdown $E_{Breakdown}$ (diamond) = 13 MV/cm, dielectric constant ϵ (diamond)=5.7, electron mobility μ_n (diamond)=7300 cm²/V s, hole mobility μ_n (diamond) = 5300 cm²/V s, thermal conductivity k (diamond)=22.9 W/cm K, and saturated electron drift velocity v_{Sat} (diamond) = 2.3×10^7cm/sat 77 K.

Quantum Sensing and Photonics: SiC features color centers with quantum characteristics that are utilized for extremely sensitive detection of magnetic fields and temperatures. Merging these sensors with nanophotonics and nanoelectronics on a single chip is vital for improving signal collection and achieving scalable quantum systems.

High-Power/High-Frequency Electronics: SiC devices are particularly effective in high-voltage, high-frequency, and high-power settings, rendering them superior to conventional Si in applications like electric vehicles and industrial drives. Integrated thermal management systems (on-chip cooling) are being developed to handle the heat produced by high-efficiency SiC power nanoelectronics.

Combining SiC and Si presents challenges due to the substantial lattice mismatch (approximately 20%) and thermal expansion differences, which can lead to crystal imperfections. Researchers employ various techniques to address these challenges:

SiC-on-Insulator (SiCOI): This approach, which aligns with the well-established silicon-on-insulator technology, utilizes wafer bonding methods to transfer a high-quality, epitaxially grown SiC layer onto a silicon dioxide-on-silicon substrate. This scalable technique is compatible with current industrial practices.

Heteroepitaxial Growth: The direct growth of 3C–SiC on a Si substrate is a cost-effective strategy, often employing buffer layers like porous Si to enhance adhesion and reduce defects.

Nanowire (NW) Synthesis: Si–SiC core–shell nanowires are produced by carburizing Si nanowires through methods such as chemical vapor deposition (CVD) or plasma-assisted bonding. This process facilitates the integration of the excellent electrical conductivity of the Si core with the chemical stability and biocompatibility of the SiC shell.

3.3.2 Gallium Nitride

Its main properties are presented in Figure 3.8(b) (ioffe-GaN 2001). Combining Gallium Nitride (GaN) with Si nanoelectronics merges GaN's exceptional power and frequency capabilities with Si's intricate circuit functionalities, facilitating efficient power conversion, RF, and optoelectronic devices (such as LEDs and lasers) on unified chips. The effective combination of GaN with Si nanoelectronics paves the way for enhanced performance and novel functionalities:

High-Efficiency Power Electronics: GaN devices present higher breakdown voltages, quicker switching speeds, and decreased conduction losses in comparison to Si. This makes them well-suited for power applications like rapid chargers, power supplies, and solar inverters.

RF and Wireless Communications: The elevated electron mobility of GaN supports operation at gigahertz (GHz) frequencies, positioning it as a vital material for 5G and upcoming 6G base stations and mobile technology.

Optoelectronics and Displays: Heterogeneous integration facilitates the combination of GaN light-emitting diodes (LEDs) with amorphous silicon (a-Si:H) thin-film transistor (TFT) circuits, enabling the production of flexible, high-brightness displays.

Monolithic Integrated Circuits (ICs): The ultimate aim is to achieve a completely monolithic system-on-a-chip (SoC) where drivers, power devices, sensors, and digital control functionalities are all merged onto a single, compact, cost-efficient GaN-on-Si IC.

3.3.2.1 Challenges

The direct deposition of GaN-on-Si substrates poses considerable material science obstacles:

Lattice Mismatch: A significant lattice mismatch (~17%) between the crystal structures of GaN and Si results in elevated defect densities at their interface.

Thermal Mismatch: A considerable disparity in thermal expansion rates (~54%) induces stress that can result in cracking and reliability concerns within the GaN layer during high-temperature fabrication procedures.

Etching Effects: GaN cannot be directly deposited on Si due to back-etching impacts, which may lead to the formation of polycrystalline layers.

3.3.2.2 Key Integration Approaches

Organizations and researchers such as imec and Intel have devised innovative solutions to address these challenges and facilitate functional GaN-on-Si integration:

Buffer Layers and Superlattices: This is a primary technique for MOCVD (metal-organic chemical vapor deposition) epitaxy. AlN/GaN superlattice and AlGaN buffer layers are positioned between the Si substrate and the GaN epitaxial layer to manage strain and diminish threading dislocation density.

Wafer-Level Heterogeneous Integration (Bonding): This method entails fabricating GaN devices on one wafer while producing standard Si CMOS circuits on another and then bonding them together at the wafer level. This allows for the independent optimization of each material.

Nanostructured Substrates: Utilizing patterned or nanostructured substrates, such as silicon nanoporous pillar arrays (Si-NPA), provides a three-dimensional stress-release mechanism, which significantly alleviates strain and enhances quality.

3D Integration: Advanced 3D hetero-integration techniques, including die-to-wafer bonding, amalgamate the technologies onto a single platform to develop hybrid structures such as SOI-GaAs/Ge/GaN/Si.

3.3.3 Diamond

Its main properties are given in Figure 3.8(c) (Evolve Diamonds 2025). Combining diamond with Si nanoelectronics presents significant enhancements in performance, particularly for heat management and power devices, through approaches such as Silicon-on-Diamond (SOD) substrates, which create hybrid systems that utilize the established processing capabilities of Si alongside the superior thermal conductivity of diamond, while addressing the challenges of diamond's limited substrate size and high defect density. The amalgamation of diamond and Si nanoelectronics is a vibrant area of research focused on revolutionizing a variety of technologies:

High-Power and RF Electronics: Development of diamond-based transistors and RF components is underway for use in electric vehicles, power grids, and wireless communication systems, where effective heat dissipation and high- power capacity are essential.

Integrated Photonics and Quantum Systems: The potential to merge diamond's distinctive optical and mechanical characteristics with traditional Si photonics allows for the creation of advanced integrated nanophotonic and optomechanical circuits suited for quantum networking and ultra-sensitive sensing.

3.3.3.1 Key Elements of Integration

Thermal Management: The main motivation for integration is to utilize the diamond layer as an exceptional heat spreader, transferring heat away from "hot spots" in high-power Si electronics, which serves as a major limitation in conventional CMOS technology. This leads to the development of quicker, smaller, and more efficient electronic circuits.

Improved Performance: Diamond's superior traits (high carrier mobility, high critical electric field) can potentially enable devices to function at higher frequencies and voltages than those achievable using Si on its own.

Quantum Applications: In the domain of quantum computing, integrated diamond membranes containing nitrogen-vacancy (NV) color centers are essential for the development of hybrid quantum systems, sensors, and nanophotonic circuits that can be integrated with existing electronic control systems.

3.3.3.2 Integration Methods and Challenges

A primary challenge in integration stems from the substantial lattice mismatch and varying material properties between diamond and Si, which may result in elevated thermal boundary resistance and interfacial defects. Various techniques have been devised to address this:

Chemical Vapor Deposition (CVD): Polycrystalline or single-crystal diamond films are grown directly on Si substrates using methods such as microwave plasma CVD (MPCVD). A significant concern with direct growth is the creation of intermediate layers like silicon carbide ($\beta - SiC$) or amorphous carbon at the interface, which can influence the quality of the subsequent diamond layer and the overall device efficacy. Managing the growth conditions, such as the methane concentration in the plasma, is crucial in controlling the competitive formation of these interfacial layers and achieving desired film characteristics.

Direct Bonding: Innovative techniques involve the direct bonding of pre-fabricated, high-quality, ultra-thin single-crystal diamond membranes onto various substrates (including Si, fused silica, and lithium niobate) without using an intermediary adhesive. This "gentle" bonding technique preserves the diamond's high crystalline quality and facilitates independent optimization of the photonic/quantum diamond layer and the underlying Si nanoelectronics.

3.4 CARBON-BASED AND TWO-DIMENSIONAL NANOMATERIALS

3.4.1 Carbon Nanotubes

CNTs exhibit quasi-ballistic transport characteristics with an extended carrier mean free path. Transport properties are utilized to determine the field-effect mobility (79,000 $cm^2/V\ s$) and to estimate the intrinsic mobility (>100,000 $cm^2/V\ s$) in semiconducting CNTs at room temperature (Dürkop et al. 2004).

In Si, carrier transport is restricted by three-dimensional scattering, resulting in lower carrier mobility compared to CNTs, particularly at ultra-low voltages. CNTs offer a faster and more efficient gate delay (such as being up to five times quicker at a 120 nm gate length) and reduced power consumption (potentially up to 100 times lower in certain cases). They also demonstrate enhanced immunity to short channel effects (SCEs) and exhibit improved performance in dimensions under 3 nm.

By incorporating CNTs into microsystems, the nanostructures can successfully bridge the gap between the nanoscale and the macroscopic world through a microscale framework (Albrecht et al. 2003, Haugen et al. 2013). Successful integration has resulted in the creation of various hybrid Si/CNT nanoelectronic devices:

High-Performance Transistors: CNTFETs have demonstrated superior performance compared to Si transistors regarding speed and energy efficiency, offering prospects for quicker and more power-efficient electronic devices.

Sensors: The extraordinarily high surface-to-volume ratio of CNTs renders them ideal for extremely sensitive chemical and gas sensors that can be directly integrated onto a single Si chip alongside processing electronics.

Memory and Switches: Non-volatile switches and memory cells leveraging CNT networks in contact with Si nanocrystals have been showcased.

Integrated Circuits (ICs): Various research teams have successfully created functional microprocessors utilizing CNT-based circuits that align with industry-standard Si design practices, indicating a promising direction for electronic systems beyond Si.

Integration Methods: The amalgamation of CNT processing with conventional Si nanoelectronics is mainly accomplished through wafer-level techniques that are compatible with CMOS, such as localized chemical vapor deposition (CVD) and methods for controlled assembly and transfer. The primary techniques for achieving this integration include direct synthesis and post-synthesis assembly/transfer, each presenting its own unique engineering challenges along with corresponding solutions.

Direct Synthesis (In situ): CNTs are directly cultivated on Si or Si microsystems through approaches like chemical vapor deposition (CVD).

A major concern is that the growth of CNTs necessitates elevated temperatures (800°C–1000°C), which could potentially damage or alter the characteristics of pre-existing Si components and CMOS circuitry. For solving this issue, researchers employ localized synthesis techniques, wherein only designated regions are heated using on-chip microheaters, allowing the adjacent Si to remain at room temperature. Electric fields are also utilized to direct the growth and alignment of the CNTs between specific electrodes.

Post-Synthesis Assembly (Ex-situ/Transfer): CNTs are fabricated separately, purified, organized, and subsequently transferred onto the intended Si substrate using methods such as stamping or inkjet printing. This approach offers the advantage of compatibility with current wafer-level fabrication methods and permits the analysis and sorting of CNTs (e.g., by chirality) prior to integration.

Difficulties faced with CNTs: The lower bandgap (0.4–0.7 eV) of CNTs may pose difficulties concerning off-state power consumption. Additionally, CNTs encounter substantial challenges related to large-scale and uniform production, including issues with alignment and contact resistance.

Example 3.2 A perfect, single-walled carbon nanotube (SWCNT) serves as a ballistic quantum wire within a nanoelectronic device. Determine its peak electrical conductance at low temperatures. The constants provided include the elementary charge, $e \sim 1.602 \times 10^{-19}$ Coulombs (C), and Planck's constant, $h \sim 6.626 \times 10^{-34}$ Joule – seconds $(\mathrm{J \cdot s})$.

Solution: The peak electrical conductance (G) of a ballistic single-walled carbon nanotube is quantized and is expressed by the equation:

$$G = 2G_0 \tag{3.3}$$

where G_0 denotes the conductance quantum, characterized as:

$$G_0 = \frac{e^2}{h} \tag{3.4}$$

First, we compute the conductance quantum (G_0) using:

$$G_0 = \frac{e^2}{h} = \frac{\left(1.602 \times 10^{-19}\right)^2 \text{C}^2}{6.626 \times 10^{-34}\ \text{Js}} = \frac{2.566404 \times 10^{-38}}{6.626 \times 10^{-34}}\ \text{S} \tag{3.5}$$

This simplifies to

$$\begin{aligned} G_0 &= \frac{2.566404 \times 10^{-38+34}}{6.626} = 0.3873 \times 10^{-4} = 3.873 \times 10^{-5}\ \text{S} \\ &= 3.873 \times 10^{-5} \times 10^{6}\ \mu\text{S} \end{aligned} \tag{3.6}$$

This is roughly 38.7 μS.

Next, we calculate the maximum conductance of the SWCNT using:

$$\begin{aligned} G = 2G_0 &= 2 \times 3.873 \times 10^{-5}\ \text{S} = 7.746 \times 10^{-5}\ \text{S} = 7.746 \times 10^{-5} \times 10^{6}\ \mu\text{S} \\ &= 77.46\ \mu\text{S} \approx 77.5\ \mu\text{S} \end{aligned} \tag{3.7}$$

The highest electrical conductance of the ballistic SWCNT is approximately 77.5 μS.

3.4.2 Graphene

Mobility values exceeding 200,000 cm^2/V s have been reached at electron densities around ~$2 \times 10^{11}\ cm^{-2}$ by utilizing single-layer graphene suspended approximately 150 nm above a Si/SiO_2 gate electrode (Bolotin et al. 2008).

Graphene's remarkable carrier mobility and thermal conductivity could result in processors for future devices that are significantly faster and more energy-efficient, provided that manufacturing hurdles are overcome. Its outstanding thermal conductivity presents benefits in heat management for densely packed nanoelectronic devices. The main challenge for using graphene in traditional digital logic lies in its absence of an intrinsic bandgap, which prevents it from being easily switched "off" like silicon transistors (FETs). Researchers are investigating approaches such as the creation of graphene nanoribbons or the integration of graphene into hybrid structures to address this limitation. Current studies frequently concentrate on graphene-on-silicon hybrid devices, capitalizing on the strengths of both materials (for instance, using graphene for rapid photodetection on a Si chip) to develop high-performance devices that are compatible with existing manufacturing techniques.

Integrating graphene with Si nanoelectronics requires addressing issues related to scaling and preserving graphene's distinctive properties, utilizing techniques such as Chemical Vapor Deposition (CVD) for large-scale growth, transfer methods (e.g., PMMA-assisted wet transfer), and post-CMOS processing to construct heterostructures, which facilitate advanced uses like transparent electronics, high-frequency devices, and flexible sensors. Researchers utilize a variety of techniques to integrate graphene with Si platforms:

Direct Growth on Si or Germanium: A straightforward thermal annealing method with a copper catalyst and carbon source enables graphene to be grown directly on a Si substrate. Additionally, it can be grown on Si wafers with epitaxially grown germanium (Ge).

Transfer Techniques: One prevalent approach involves synthesizing graphene on a separate metal catalyst (such as copper foil) via chemical vapor deposition (CVD) and subsequently transferring the graphene layer onto the chosen Si/SiO_2 substrate. Wafer-to-wafer transfer techniques are being improved for enhanced scalability and better quality control.

Epitaxial Growth on SiC: Graphene can be grown epitaxially on SiC substrates, which can then be affixed to a top silicon layer using a method known as "smart-cut."

Electrochemical Reduction: Graphene oxide (GO) can be applied to a hydrogen-terminated Si surface and then electrochemically reduced to create a graphene/Si interface, providing a rapid and scalable technique that protects the Si surface from oxidation.

Process Compatibility and Scalability: Traditional CMOS processes involve high temperatures and various chemicals that risk damaging graphene's fragile 2D structure or introducing contamination.

Solution: To address this, low-temperature growth and transfer techniques are being developed, or graphene can be integrated after the majority of high-temperature CMOS processes are completed (post-CMOS integration).

Challenges—Material Quality and Uniformity: Achieving consistent, high-quality, large-area graphene on wafers poses a challenge, as defects, residues from transfers, and edge disorder can significantly impair its exceptional electronic properties.

Solution: Enhancing metrology at the wafer scale, utilizing cleaner transfer methods, and encapsulating the graphene with materials such as hexagonal boron nitride (hBN) can help protect it during subsequent processing stages.

3.4.3 Molybdenum Disulfide

As a two-dimensional material, MoS_2 is particularly well-suited for significant downscaling. MoS_2 can be reduced to a single atomic layer, allowing for extreme miniaturization of devices beyond silicon's limitations, where quantum tunneling becomes a significant challenge. Room-temperature-processed MoS_2 thin films displayed an average electron mobility of 50 $cm^2/V{\cdot}s$, which can be enhanced to 68 $cm^2/V{\cdot}s$ with processing at 200°C (Xue et al. 2025).

Transistors made from MoS_2 demonstrate a substantially higher on/off current ratio and reduced off-state leakage current, which leads to a marked decrease in power consumption during standby, crucial for modern portable electronics. Moreover, thinner transistors use less energy.

Monolayer MoS_2 possesses a direct bandgap, making it ideal for efficient switching with high on/off ratios and applications in optoelectronics, since it enables quicker generation and recombination of electron-hole pairs. The layered nature of MoS_2 means its surfaces do not have the "dangling bonds" that are common in silicon, which create performance-limiting interface traps. This leads to cleaner interfaces with dielectric layers and lower leakage currents. Additionally, MoS_2's mechanical flexibility presents opportunities for the development of flexible and wearable electronic devices, an area where rigid silicon falls short.

3.4.3.1 Applications

The combination of MoS_2 and silicon is being investigated for various cutting-edge nanoelectronic and nanophotonic applications:

High-Performance Transistors: Due to MoS_2's direct bandgap and impressive charge mobility, it shows promise as a material for energy-efficient FETs, serving as an alternative to silicon at very small scales.

Integrated Photonics: Thanks to the material's strong interactions with light and a bandgap that varies with thickness, it enables the development of on-chip optical components, including integrated photodetectors, optical polarizers, and phase modulators integrated with silicon photonic waveguides.

Logic Circuits: Researchers have successfully created integrated logic circuits (such as inverters, NAND, and NOR gates) on a wafer scale utilizing MoS_2 FETs with buried gates in the silicon substrate.

Novel Heterostructures: The combination of MoS_2 with silicon nanowires (SiNWs) leads to heterostructures that are applied in areas like high-capacity battery technology and photocathodes for water splitting.

Incorporating Molybdenum Disulfide (MoS_2) with silicon nanoelectronics requires the direct growth of high-quality MoS_2 on silicon nanostructures (such as nanowires or waveguides) using methods like Chemical Vapor Deposition (CVD) or Atomic Layer Deposition (ALD) for enhanced optoelectronic (photodetectors, modulators, sensors) and electronic applications, while addressing issues related to charge transport at the 2D/3D interface, frequently utilizing buffer layers like hBN or Al_2O_3 to maintain the properties of MoS_2.

3.4.3.2 Key Integration Strategies

Direct Growth on Silicon: This approach focuses on depositing MoS_2 films directly onto silicon or silicon-on-insulator (SOI) substrates, which streamlines the process by removing intricate transfer steps and lowering costs.

Chemical Vapor Deposition (CVD) and ALD: These vapor-phase techniques are regarded as highly effective due to their alignment with semiconductor industry standards and their capability for wafer-scale manufacturing (e.g., 200mm and 300mm wafers). ALD processes can precisely manage the number of MoS_2 layers while ensuring uniformity.

Challenges: Direct deposition onto bare Si or SiO_2 may compromise the quality of MoS_2, resulting in distorted morphology and diminished optical characteristics due to interactions with the substrate.

Buffer Layers: To safeguard the fundamental properties of MoS_2 during direct growth, ultra-thin buffer layers, such as hexagonal boron nitride (hBN) or alumina (Al_2O_3), can be placed between the MoS_2 and the silicon substrate. These layers contribute to restoring sharp crystallinity, minimizing surface roughness, and protecting the 2D material from harmful substrate influences.

Heterogeneous Integration and Transfer: In certain situations, MoS_2 produced in other locations is transferred onto pre-fabricated silicon nanostructures or photonic waveguides. This method enables independent optimization of the two materials but presents challenges related to mechanical transfer.

Example 3.3 A field-effect transistor (FET) made from monolayer molybdenum disulfide MoS_2 shows an electron mobility μ of 100 cm^2/V s at room temperature. The carrier concentration n within the channel is found to be 1×10^{13} cm^{-2}. Determine the two-dimensional (2D) conductivity σ_{2D} of the material.

Solution: We use the relationship connecting two-dimensional (2D) conductivity σ_{2D}, carrier concentration n, mobility μ and the elementary charge e:

$$\sigma_{2D} = ne\mu \tag{3.8}$$

Now, let's insert the provided values into this formula:

$$\sigma_{2D} = 1 \times 10^{13}\ \mathrm{cm}^{-2} \times 1.602 \times 10^{-19}\ \mathrm{C} \times 100\ \mathrm{cm}^2/\mathrm{V\ s} \tag{3.9}$$

Performing the multiplication yields:

$$\begin{aligned}\sigma_{2D} &= 1.602 \times 10^{13-19+2}\ \mathrm{cm}^{-2} \times \mathrm{C/V\ s} = 1.602 \times 10^{-4}\ \mathrm{S} = 1.602 \times 10^{-4} \times 10^{6}\ \mu\mathrm{S} \\ &= 160.2\ \mu\mathrm{S}\end{aligned} \tag{3.10}$$

The two-dimensional conductivity of the MoS_2 channel is 160.2 μS.

Note: The standard unit for electrical conductivity is the Siemens per meter (S/m). Depending on the context, conductivity may also be represented using an alternative unit. For instance, in the examination of 2D materials, conductivity can be presented in Siemens (S) or micro-Siemens (μS).

3.4.4 Tungsten Diselenide

It presents distinct advantages for next-generation nanoelectronics, particularly due to its capability to operate as an atomically thin (2D) semiconductor that maintains stable electrical properties and exhibits high carrier mobility, even at monolayer thickness. A peak hole mobility of about 200 $cm^2/V{\cdot}s$ was attained in WSe_2 due to a decrease in scattering sources (Wu et al. 2016).

Its direct bandgap in its monolayer form positions it as a strong contender for integrated photonics and optoelectronics applications. The main obstacle for WSe_2 lies in the absence of a large-scale, high-yield manufacturing process that matches the established methods used for silicon.

The integration enables numerous advanced applications:

Enhanced Optoelectronics: WSe_2 serves as a highly effective atomic-scale light source or photodetector on a silicon platform. By linking the WSe_2 layer with silicon grating-waveguide structures, researchers have shown marked improvements in directed light emission, which is vital for on-chip optical communication and data transmission.

CMOS Logic Circuits: WSe_2 plays a crucial role in developing high-performance P-type field-effect transistors (P-FETs), a task that can be quite difficult with some other 2D materials. The combination of P-type WSe_2 with N-type materials such as MoS_2 on a silicon substrate enables the realization of fully integrated CMOS circuits.

Enhanced Doping and Contact: Scientists employ techniques like ion implantation, a conventional process in silicon technology, to generate precise doping profiles within WSe_2. The fabrication methods are also refined to achieve low-resistance ohmic contacts between metal electrodes (such as gold or titanium) and the WSe_2 layer, which is essential for optimal device performance.

Edge Passivation: A significant hurdle to scaling WSe_2 devices is addressing edge defects. Integration techniques include passivating the edges of WSe_2 nanoribbons with amorphous WO_xSe_y to enhance carrier mobility and overall device performance.

The main approaches for integrating WSe_2 with silicon platforms include:

Dry Transfer Method: High-quality, exfoliated WSe_2 monolayers or multilayers are accurately placed onto pre-formed silicon or silicon-on-insulator (SOI) structures (such as waveguides or nanobeam cavities) utilizing a polydimethylsiloxane (PDMS) or polycarbonate stamp. This technique permits meticulous alignment while minimizing damage to the fragile 2D material, although it might affect the quality (Q-factor) of high-performing silicon cavities.

Chemical Vapor Deposition (CVD) and Growth: WSe_2 can be directly synthesized on a silicon substrate through methods like modified reverse-flow thermal CVD. This technique focuses on achieving wafer-scale growth, which aligns better with traditional silicon fabrication techniques than mechanical exfoliation.

Solution Processing/Thin-Film Deposition: This method involves applying uniform WSe_2 films (often in the form of nanocrystals) onto the silicon substrate from a solution, providing a scalable approach for creating heterojunctions.

3.4.5 Black Phosphorus

It possesses exceptional characteristics for high-speed, low-power nanoelectronics because of its excellent mobility and adjustable direct bandgap. At room temperature, the carrier mobility of black phosphorous (BP) films can reach 1000 cm^2/V s, while the on/off ratio exceeds 10^5 (Zhong 2022).

When examined at the nanoscale, the performance of black phosphorus can exceed that of silicon in certain uses, such as the efficiency of 100 nm thick solar cells. The adjustable direct bandgap of black phosphorus renders it particularly suitable for optoelectronic applications and high-frequency devices, an area where silicon falls short due to its indirect bandgap. The primary obstacle to the commercialization of black phosphorus is its inadequate stability in environmental conditions and the necessity for dependable, large-scale production techniques.

Integrated platforms provide solutions for advanced nanoelectronic and optoelectronic devices, particularly in the mid-infrared (mid-IR) range, where silicon lacks transparency.

Photodetectors made from BP feature a narrow, direct, and tunable bandgap that spans from visible light to mid-IR, making it an excellent choice for room-temperature on-chip photodetectors. Integrated detectors have demonstrated high responsivity and quick response times.

Optical Modulators and Light Emitters: Hybrid systems combining silicon and BP have been utilized to develop on-chip mid-IR LEDs and highly effective optical modulators, which are crucial components for fully integrated photonic circuits.

High-Performance Transistors: Black phosphorus nanoribbon (BPNR) FETs show impressive carrier mobility and significant on/off ratios, positioning them as attractive channel materials for cutting-edge electronic platforms.

Sensing Applications: The capability to integrate both BP photodetectors and modulators on a single chip offers potential for on-chip biosensing, night vision, and various advanced sensing applications.

Researchers have established multiple techniques for integrating BP into silicon-based devices:

Van der Waals (vdW) Heterostructures: The two-dimensional, layered structure of BP, which is held together by weak van der Waals forces, enables its transfer and integration onto silicon substrates without needing lattice matching.

Mechanical Exfoliation and Transfer: High-quality BP flakes are mechanically peeled off and placed onto silicon-on-insulator (SOI) waveguides and other nanostructures. This technique is essential for laboratory-scale research and prototyping.

Solution-Based Processing: Sonochemical exfoliation techniques can yield BPNRs with specified size and structure, making them promising for large-scale integration into future CMOS and heterogeneous electronic platforms.

Three-Dimensional (3D) Hybrid Integration: This method merges low-loss silicon waveguides with high-field confinement plasmonic nanostructures and BP photodetectors in a 3D configuration, achieving high responsivity in the near-infrared spectrum.

Despite encouraging outcomes, challenges persist, especially concerning the environmental stability of BP (which is vulnerable to oxygen and moisture) and the demand for scalable manufacturing methods suited for mass production.

3.4.6 Hexagonal Boron Nitride

Incorporating hexagonal boron nitride (hBN) processing with silicon nanoelectronics markedly enhances device functionality by utilizing hBN's excellent electrical insulation properties, high thermal conductivity, and atomically smooth surface as a protective and high-quality dielectric layer (Xie et al. 2025a).

3.4.6.1 Key Advantages

Outstanding Dielectric Properties: hBN serves as a superb electrical insulator possessing a significant bandgap (~5.8 eV for bulk) and a breakdown voltage much higher than that of silicon dioxide (SiO_2). This leads to smaller device dimensions, better performance, and increased capacitance in components such as field-effect transistors (FETs) and modulators.

Improved Carrier Mobility: The atomically flat characteristics of hBN, which have minimal dangling bonds and trap states, greatly decrease carrier scattering when utilized as a substrate or encapsulating layer. This results in a notable enhancement in carrier mobility within adjacent 2D materials (like graphene) and silicon nanowires.

Efficient Thermal Management: Due to its high in-plane thermal conductivity, hBN is ideal for regulating heat in next-generation, miniaturized electronics, mitigating localized heating and increasing the overall reliability of devices.

Integration of Quantum Emitters: hBN can accommodate stable, bright single-photon emitters that align with silicon photonic frameworks. This facilitates the creation of integrated quantum nanophotonic circuits and on-chip optical communication systems.

Chemical and Mechanical Resilience: hBN is both chemically and mechanically solid and inert, functioning as an efficient passivation and encapsulation layer that safeguards delicate nanostructures from environmental deterioration.

3.4.6.2 Uses in Silicon Nanoelectronics

2D Transistors and Tunneling Barriers: The atomically thin structure of hBN offers ideal barriers for low-power transistors.

Quantum Photonics: Development of scalable quantum circuits featuring on-chip single-photon sources, essential for quantum computing and communication.

Hybrid Devices: The mixed-dimensional (2D–3D) integration method enables innovative device designs, including reconfigurable FETs and high-speed electro-absorption modulators, which feature improved performance and smaller footprints. Merging 1D silicon nanowires with 2D hBN for reconfigurable transistors (RFETs) or combining hBN memristors with CMOS technology

3.4.6.3 Processing and Integration Techniques

Integration frequently requires addressing the issue of hBN's elevated synthesis temperatures, which can surpass the thermal limits of traditional CMOS techniques. Popular methods include:

Mechanical Exfoliation and Transfer: High-quality hBN flakes are mechanically exfoliated and placed onto silicon or silicon dioxide substrates. This technique is common in research but lacks scalability.

Wafer-Scale Growth (CVD/ALD): Researchers are working on scalable approaches such as metal-organic chemical vapor deposition (MOCVD) and ALD to directly grow hBN films on large silicon wafers at lower, CMOS-compatible temperatures (<380°C), thereby removing the requirement for transfer methods.

Van der Waals Assembly: This method entails stacking layers of various 2D materials (such as hBN and graphene) onto silicon waveguides with careful alignment to form intricate heterostructures.

3.5 DIELECTRICS AND TOPOLOGICAL INSULATORS

Dielectrics and topological insulators (TIs) play unique roles in Si nanoelectronics: dielectrics serve as crucial insulators and materials for charge storage, while TIs represent a newer category of materials for innovative electronic, spintronic, and optoelectronic applications that address the challenges of traditional silicon-based devices at the nanoscale. Dielectrics are insulating materials that are essential for blocking unwanted electrical currents and storing electrical energy. As device dimensions decrease in Si nanoelectronics, traditional materials such as silicon dioxide (SiO_2) become excessively thin, leading to significant current leakage through quantum tunneling. To mitigate this issue, high-*k* (high dielectric constant) and low-*k* (low dielectric constant) materials are utilized.

The reasons for their use include:

Electrical Insulation: It is used to separate components (such as transistors and interconnects) to prevent short circuits.

Enhanced Capacitance: In transistors, high-*k* materials serve as gate dielectrics to boost gate capacitance, enabling smaller, more densely packed transistors to maintain performance while reducing leakage currents.

Reduced Power Consumption and Improved Speed: Low-*k* dielectrics are used in interconnect layers to reduce parasitic capacitance, thereby reducing signal delays and power loss, facilitating high-speed operations.

Reliability: Dielectrics offer high breakdown strength, enabling devices to withstand high electric fields without failing. They are also essential components in capacitors and memory devices (such as DRAM, ReRAM, and FeRAM) for efficient charge storage.

The dielectrics used include (King 2018):

Silicon Dioxide (SiO_2): The traditional and most widespread gate dielectric and insulator, though it has limitations at tiny scales.

High-*k* Dielectrics: Hafnium Oxide (HfO_2) is a leading choice for modern CMOS transistors at nodes of 65 nm and below because of its high dielectric constant (~25). Zirconium Oxide (ZrO_2) is another high-*k* material frequently utilized in gate dielectrics. Aluminum Oxide (Al_2O_3) is used in thin-film transistors and in non-volatile memory applications.

Low-*k* Dielectrics—Porous Silica and Polymer-based Nanocomposites: These are designed with increased porosity to lower the dielectric constant.

2D Materials: New materials, such as hexagonal boron nitride (hBN), are being researched for applications in flexible and ultra-thin electronics.

Example 3.4 A ferroelectric memory device (FeRAM) utilizes a thin layer of Hafnium Oxide (HfO_2) as its ferroelectric component, incorporated within a conventional CMOS fabrication process. The primary characteristic utilized for data retention is the remanent polarization (P_r), which represents the enduring polarization that persists after the external electric field is eliminated. Given: Material: Doped (HfO_2), Remanent Polarization (P_r): 15 μC/cm^2, Device Area (A): 0.1 μm^2, Inquiry: Determine the overall switchable charge (ΔQ) stored within this nanoscale memory cell.

Solution: The switchable charge is generally calculated as twice the remanent polarization ($2 \times P_r$) per device area, since the polarization can alternate between a positive ($+P_r$) and a negative ($-P_r$) state, representing binary values (0 or 1). Converting units to standard SI (Coulombs and meters): $P_r = 15\ \mu\mathrm{C/cm^2} = 15 \times 10^{-6}\ \mathrm{C/cm^2}$. Converting cm^2 to m^2: 1 cm = 0.01 m, thus $1\ \mathrm{cm^2} = (0.01)^2\ \mathrm{m^2} = 1 \times 10^{-4}\ \mathrm{m^2}$. $P_r = 15 \times 10^{-6}\ \mathrm{C}/\left(1 \times 10^{-4}\ \mathrm{m^2}\right) = 15 \times 10^{-2}\ \mathrm{C/m^2} = 0.15$ $\mathrm{C/m^2}$. $A = 0.1\ \mu\mathrm{m^2} = 0.1 \times \left(10^{-6}\right)^2\ \mathrm{m^2} = 0.1 \times 10^{-12}\ \mathrm{m^2} = 1 \times 10^{-13}\ \mathrm{m^2}$

Calculating the total switchable charge (ΔQ): The connection between polarization (P), indicating charge per unit area) and the total charge (Q) is defined as $P = Q/A$. The total switchable charge is represented by

$$\Delta Q = 2P_rA = 2 \times 0.15\ \mathrm{C/m^2} \times 1 \times 10^{-13}\ \mathrm{m^2} = 3 \times 10^{-14}\ \mathrm{C} \quad (3.11)$$

The overall switchable charge stored in the memory cell amounts to 3×10^{-14} C (equivalent to 30 femtocoulombs, or 30 fC. This minute charge difference is essential for the non-volatile storage of information at the nanoscale.

3.5.1 Integration Processes for Dielectrics

A significant challenge in integrating dielectrics with topological insulators (TIs) or other advanced materials, such as 2D materials, is ensuring a high-quality interface that prevents damage or unintentional doping.

Atomic Layer Deposition: The ALD method is favored for depositing thin, uniform high-*k* dielectrics, thanks to its ability to achieve conformality and precise thickness control. However, a significant obstacle is the chemical inertness of certain novel materials, such as graphene and TIs, which complicates direct deposition.

Dry Transfer Method: An emerging technique involves fabricating the dielectric layer (for instance, Al_2O_3 or HfO_2) on a distinct substrate and subsequently transferring it dry onto the surface of the target material, such as a TI or a 2D material. This method avoids direct deposition processes that may harm the delicate TI surface and helps ensure a clean interface.

Surface Modification and Passivation: To enhance interface quality during direct deposition, techniques such as functionalization, plasma treatment, or the application of ultra-thin buffer layers are used.

Topological insulators (TIs) represent a more recent advancement in the field of condensed matter physics. TIs represent a category of materials characterized by an insulating time-reversal invariant (TRI) band structure (Gilbert 2021). In these materials, robust spin–orbit interactions cause an inversion of the bandgap at an odd number of time-reversed points within the Brillouin zone.

TIs exhibit distinctive properties such as the spin–momentum locking of surface states, the magneto-electric effect, and an absence of backscattering. They function as electrical insulators in their bulk form but possess highly conductive, spin-polarized metallic states at their surfaces. The reasons for their use include:

Dissipationless Transport: The surface states are "topologically protected" against backscattering from non-magnetic impurities and defects, enabling current to flow with minimal dissipation (little energy loss and heat generation). This offers a significant advantage in addressing heat dissipation and power consumption challenges in compact nanoelectronics.

Spintronics Applications: The distinctive spin-momentum locking characteristic of TI surfaces allows for the generation of spin-polarized currents through an electric field. This holds promise for new memory and logic devices (spintronics) that use electron spin to store and process information, potentially enabling faster, more energy-efficient computing.

Integration with Silicon: Ongoing research aims to incorporate TI nanostructures, such as nanowires and thin films, into existing Si-based platforms to develop novel hybrid devices, including high-performance field-effect transistors (FETs) and photodetectors.

Robustness: The properties of conducting surface states are resilient to external disturbances, such as surface imperfections and material distortions, making them dependable for nanoscale applications where conventional materials struggle with defects.

Optoelectronics: TIs exhibit unique optical properties (such as a high refractive index and broad light absorption), making them suitable for advanced optoelectronic devices, including ultra-broadband photodetectors and silicon-integrated solar cells.

The topological insulators employed include HgTe/CdTe quantum wells and transition-metal dichalcogenides like Bi_2Te_3 and Bi_2Se_3:

Bismuth Selenide (Bi_2Se_3): It is a well-studied 3D TI, recognized for its robust surface states at room temperature, and utilized in experimental FETs and photodetectors.

Bismuth Telluride (Bi_2Te_3): It is also extensively studied, especially for its outstanding thermoelectric properties and application in hybrid structures. The topological characteristics of a 100 nm Bi_2Te_3 thin film, created using a co-deposition method with a specially designed electron-beam evaporator, were investigated (Kander et al. 2024). The presence of features associated with topological surface states (TSS) was observed in the Bi_2Te_3 film.

Bismuth-Antimony alloys ($Bi_{1-x}Sb_x$): The first 3D TI discovered experimentally, used to investigate fundamental properties and potential device applications.

Tin Telluride (SnTe) is a topological crystalline insulator currently under investigation for high-performance silicon-based photodetectors.

Example 3.5 A 2D topological insulator nanoribbon has been observed experimentally to exhibit a total electrical conductance (G) of around $2e^2/h$ at lower temperatures. Calculate this conductance's numerical value in Siemens (S). Additionally, determine the anticipated resistance (R) of this nanoribbon in Ohms (Ω). Given Constants: Elementary charge, $e \sim 1.602 \times 10^{-19}$ C; Planck's constant, $h \sim 6.626 \times 10^{-34}$ Js.

Solution: A 2D topological insulator (also referred to as a quantum spin Hall insulator or QSHI) is a type of atomically thin substance that acts as an electrical insulator in its bulk while having stable, conducting states along its one-dimensional edges. To find the numerical value of the conductance, the conductance is expressed as

$$\begin{aligned} G &= \frac{2e^2}{h} = \frac{2\left(1.602 \times 10^{-19}\right)^2 \text{C}^2}{6.626 \times 10^{-34}\text{ Js}} = \frac{2 \times 2.566404 \times 10^{-38}}{6.626 \times 10^{-34}}\text{ S} \\ &= 0.774646 \times 10^{-4}\text{ S} \approx 0.774646 \times 10^{-4} \times 10^{6}\ \mu\text{S} \approx 77.5\ \mu\text{S} \end{aligned} \tag{3.12}$$

Thus, the conductance value is roughly 77.5 μS. Note that

$$\frac{\text{C}^2}{\text{Js}} = \frac{\text{C}}{\text{s}} \times \frac{\text{C}}{\text{J}} = \text{A} \times \frac{1}{\frac{\text{J}}{\text{C}}} = \frac{\text{A}}{\text{V}} = \text{S} \tag{3.13}$$

To calculate the expected resistance, we note that the resistance (R) is the reciprocal of the conductance (G), given by

$$R = \frac{1}{G} = \frac{1}{0.774646 \times 10^{-4}\text{ S}} = 12909.122\ \Omega = \frac{12909.122\text{ k}\Omega}{1 \times 10^3} \approx 12.9\text{ k}\Omega \tag{3.14}$$

The expected resistance value is approximately 12.9 kΩ.

3.5.2 Integration Processes for Topological Insulators

The primary objective of integrating TIs is to exploit their distinct conducting surface states while preserving their bulk insulating characteristics, often necessitating a high surface-area-to-volume ratio that makes nanostructures appealing.

Molecular Beam Epitaxy (MBE): MBE is an essential technique for the growth of high-quality TI thin films (e.g., Bi_2Te_3, Sb_2Te_3) directly onto substrates like GaAs or Ge-buffered silicon wafers. MBE provides precise control over film thickness and crystalline quality at relatively low temperatures, which is a significant advantage for integration with existing silicon processing. The resulting films often display distinct layered crystal structures.

Chemical Vapor Deposition (CVD): This method is recognized for its scalability and adaptability. CVD and similar techniques, including metal-organic CVD (MOCVD), are employed to create large-area TI films and nanostructures, making them suitable for industrial use.

Physical Vapor Deposition (PVD): Enhanced PVD techniques are being developed to optimize the uniform, large-area growth of TI materials, which is vital for scalable device manufacturing.

Mechanical Exfoliation: For research purposes, high-quality TI nanostructures can be obtained from bulk layered crystals through mechanical exfoliation, although this approach is not feasible for large-scale production.

Chemical Methods: Solvothermal synthesis can yield TI nanoplates (such as Bi_2Te_3), which can then be converted into large-area films using techniques like spin-coating, compatible with some existing fabrication processes.

3.6 FERROELECTRIC MATERIALS

Ferroelectric materials are utilized in Si nanoelectronics for low-power memory, energy-efficient logic, and innovative AI hardware due to their capacity for non-volatile memory (maintaining data without power), sharp subthreshold swing (enhanced switching), and in-memory computing (integrating storage and processing) by utilizing their switchable electric polarization, particularly with compatible substances like Hafnium Oxide (HfO_2). They facilitate the development of advanced transistors (FeFETs, FeSFETs) and photonic devices, which are essential for computing advancements beyond Moore's Law and AI chip technology.

3.6.1 Hafnium Oxide (HfO_2)

The primary use of hafnium oxide is as a gate insulator in field-effect transistors (FETs) and dynamic random-access memory (DRAM) capacitors. HfO_2 possesses a dielectric constant of roughly 20 (varying with deposition technique and structure), which is four to six times higher than that of SiO_2 ($\approx$ 3.9). As transistor sizes decrease, the SiO_2 gate oxide layer becomes exceedingly thin, leading to substantial power leakage from quantum tunneling. The elevated dielectric constant of HfO_2 permits a physically thicker layer to maintain equivalent electrical performance (capacitance), effectively minimizing leakage current and power usage. Further, HfO_2 remains thermally and chemically stable when in contact with the silicon substrate at elevated processing temperatures, making it essential for compatibility with standard CMOS fabrication processes.

In addition to its role as a conventional gate dielectric, the distinct characteristics of HfO_2 are being investigated for next-generation nanoelectronic devices. Doped or undoped HfO_2 can demonstrate ferroelectricity (stable polarization) when the film thickness is minimal (approximately 6 nm). This positions it as a potential "game-changer" for the development of high-speed, low-power, non-volatile memory solutions, such as ferroelectric random-access memory (FeRAM) and ferroelectric field-effect transistors (FeFETs), which could connect existing memory technologies like SRAM, DRAM, and Flash.

By utilizing the existing materials infrastructure of the semiconductor sector, hafnia-based ferroelectrics have been integrated into multiple memory systems, such as ferroelectric random-access memory (FeRAM), ferroelectric tunnel junctions (FTJs), ferroelectric field-effect transistors (FeFETs), and ferroelectric memcapacitors (FeCAPs) (Hwang et al. 2025).

Using HfO_2 in silicon nanoelectronics typically employs ALD for uniform and accurate film growth, often using an interface engineering layer (such as ultra-thin SiO_2) or doping (with Si, Al, or Zr) to control phase (amorphous/crystalline) and prevent silicon oxidation. This is followed by post-deposition annealing (at low/high temperatures for a short duration) to enhance crystallinity, stoichiometry, and electrical characteristics for MOSFETs or ferroelectric memory applications. Alternative solutions, such as sol-gel methods, are gaining attention.

3.6.2 Lead Zirconate Titanate (PZT) Pb (Zr, Ti)O_3

PZT is used for its piezoelectric and ferroelectric properties, enabling the development of advanced pressure and vibration sensors, micro-actuators for precise motion, energy harvesters that convert vibrations into electrical power, ultrasonic transducers, and non-volatile memory devices (FeRAM) (Bretos et al. 2016). The remnant polarization property of PZT enables the production of non-volatile memory that retains information even without a continuous power supply, positioning it as a strong candidate for high-speed, low-power memory solutions integrated with silicon.

Incorporating PZT into silicon nanoelectronics entails depositing thin films (using techniques such as sputtering, sol-gel, and CVD), followed by patterning (including photolithography and etching), and post-deposition annealing (Rapid Thermal Annealing) to promote perovskite crystallization. Techniques such as wafer bonding or transfer printing are utilized to ensure compatibility with CMOS, employing buffer layers (like $LaNiO_3$ or ZTO) for improved adhesion and as barriers, while paying close attention to thermal management to align with silicon's lower thermal budget. Critical processes emphasize the formation of high-quality, oriented films, stress management, and the design of electrodes for constructing sensors, actuators, or modulators.

3.6.3 Strontium Bismuth Tantalate

Strontium bismuth tantalate (SBT) is primarily used in silicon nanoelectronics to create non-volatile ferroelectric random-access memories (FeRAMs or FRAMs) (Wouters et al. 2006). One of the most notable benefits of SBT is its exceptional fatigue resistance. It shows minimal degradation, even after 10^9–10^{12} cycles of polarization switching, which is considerably superior to that of other ferroelectric materials such as PZT (lead zirconate titanate). This remarkable endurance is vital for memory devices that require a long lifespan. SBT is also used to produce ferroelectric-gate field-effect transistors (FeFETs), which are essential components of high-performance FeRAMs. SBT films have been effectively incorporated into metal/ferroelectric/insulator/semiconductor (MFIS) stacked-gate structures on Si substrates, often utilizing buffer layers to prevent interfacial diffusion and enhance interface quality.

The main obstacle to merging SBT with silicon is the high-temperature annealing, typically around 750°C or higher, which is essential for crystallizing SBT and inducing ferroelectricity but can harm the functionality of pre-existing Si transistors. To address this issue, specialized fabrication techniques have been devised:

Replacement-Gate Process (Gate-Last Approach): This innovative method is crucial for area-scalable FeFETs (Ferroelectric Field-Effect Transistors) and involves creating standard silicon transistor patterns using a temporary "dummy-gate" material, then removing the dummy material to reveal precisely formed grooves (which are then filled with the SBT precursor), placing gate electrodes (e.g., Iridium, Ir) on top of the SBT, and executing a final annealing process for the entire stack to facilitate SBT crystallization.

This strategy safeguards the sensitive transistor sidewalls from harmful etches or high temperatures during the critical stages of SBT formation.

Silicon-on-Insulator (SOI) Integration: SBT capacitors can be effectively integrated onto SOI substrates, using the underlying insulating layer to shield the active silicon layer and adjacent transistors from the specific processing requirements of SBT, while maintaining the integrity of SBT properties during SOI processing.

Deposition Techniques: SBT thin films are often deposited utilizing several key methods:

Sputtering: This PVD technique uses high-energy ions to knock atoms from an SBT target, which then condense onto the silicon substrate, forming a thin film.

Spin Coating/Sol-Gel Method: This chemical approach involves preparing a solution of metal-organic precursors (such as bismuth isopropoxide and strontium tantalum isopropoxide in methoxyethanol) and applying it to the substrate via spin coating. The film is subsequently dried and annealed to yield the final crystalline SBT.

Metal-Organic Chemical Vapor Deposition (MOCVD): This method involves introducing volatile metal-organic precursors into a reactor, where they decompose on a heated substrate to form the SBT film. MOCVD provides outstanding control over film thickness and composition and is well-suited to industrial applications.

3.6.4 Copper Indium Phosphide Sulfide

Copper indium phosphide sulfide (CIPS) is primarily used in silicon nanoelectronics to produce high-efficiency, non-toxic optoelectronic devices, particularly as a material for light-emitting diodes (LEDs) and solar cells in the form of QDs (Bangolla et al. 2023). This material is being investigated as a safer substitute for cadmium- or lead-based QDs.

The primary methods for integrating CIPS with silicon substrates include:

Mechanical Exfoliation and Transfer: It is a widely used approach in research to develop proof-of-concept devices. The synthesis of CIPS single crystals begins with growing high-quality bulk crystals, typically via chemical vapor transport. Exfoliation involves obtaining thin, few-layer flakes by repeatedly peeling the bulk crystal with adhesive tape, such as Scotch tape. The resulting flakes are then placed onto the desired Si substrate, often utilizing an intermediate layer of PDMS for accurate positioning. This process creates a van der Waals (vdW) interface with the Si or another material grown on the Si (such as SiO_2) or an additional 2D material, which helps to mitigate the interface challenges (dangling bonds, thermodynamic instability) that are typically present with conventional ferroelectrics on Si.

For scalability and cost-effectiveness in manufacturing, solution-based deposition methods are being researched for similar materials (like $CuInS_2$). These methods usually consist of nanoparticle precursor synthesis, which involves generating CIPS nanoparticles in a liquid solution; film deposition, where techniques like spin coating, drop casting, or inkjet printing are used to apply a layer of these nanoparticles onto a silicon wafer; and annealing/conversion, a subsequent thermal treatment that is necessary to transform the nanoparticle film into a dense, high-quality semiconductor layer.

3.6.5 Poly (Vinylidene Fluoride-Co-Trifluoroethylene) Copolymer P(VDF-TrFE)

Poly (Vinylidene Fluoride-Co-Trifluoroethylene) Copolymer [P(VDF-TrFE)] is used in silicon nanoelectronics due to its ferroelectric, piezoelectric, and pyroelectric properties, serving as a flexible alternative to inorganic materials in applications such as non-volatile memory (FeRAM), ferroelectric-gate transistors, various sensors (pressure, temperature), and energy-harvesting systems (Lee et al. 2015). It integrates with silicon through interface engineering, promoting high-performance, flexible, and self-powered electronic devices. Thus, P(VDF-TrFE) connects rigid silicon electronics with flexible organic materials, facilitating advanced functionalities in the future of flexible and wearable electronics.

The integration of (VDF-TrFE) into silicon nanoelectronics is mainly achieved via solution-based techniques such as spin coating and Langmuir-Blodgett deposition, complemented by post-processing steps such as thermal annealing and electrical poling to enhance the material's ferroelectric properties.

The standard fabrication procedure includes several essential stages to guarantee compatibility with silicon substrates and achieve the target functional properties.

Solution Preparation: P(VDF-TrFE) powder is dissolved in a solvent, such as methyl ethyl ketone (MEK) or a mixture of *N,N*-dimethylformamide and acetone, to formulate a casting solution. The solution's concentration is crucial for regulating the thickness of the resulting film.

Film Deposition: The spin-coating method is widely used because it is straightforward and cost-effective, enabling the deposition of uniform films ranging in thickness from tens of nanometers to micrometers on silicon wafers. The film's thickness can be adjusted by varying the spin speed and solution concentration. The Langmuir-Blodgett (LB) deposition technique is advantageous for producing ultra-thin films, including molecular monolayers, down to just a few nanometers. Inkjet/spray coating, screen printing, and other scalable methods compatible with roll-to-roll processes are also investigated for the mass production and integration into flexible electronics.

Solvent Evaporation and Drying: After deposition, the wet film undergoes solvent evaporation and drying to remove the solvent. The parameters of this step, including temperature and evaporation rate, are vital for managing the film's porosity and surface roughness, which directly influence electrical shorting conditions.

Thermal Annealing: The dried films are subjected to thermal annealing after deposition, usually at temperatures between 140°C and 150°C (below the melting point), to enhance crystallinity and facilitate the transition to the desired ferroelectric β-phase (or orthorhombic phase). This step significantly improves the piezoelectric response.

Electrode Integration and Patterning: Bottom electrodes (e.g., platinum or aluminum) are generally sputtered onto the silicon substrate (or an intermediate oxide layer) before the polymer is deposited. Top electrodes are added following the processing of the P(VDF-TrFE) film.

The patterning of the polymer layer is commonly performed using specific techniques, such as dry etching with oxygen plasma, as traditional photolithography developers can damage the polymer. Nano-imprinting lithography is also employed for patterning.

Electrical Poling: To activate the ferroelectric attributes, a high electric field is applied across the film's thickness, aligning the molecular dipoles to ensure a significant remnant polarization. This can be achieved through electrode or corona poling techniques.

3.7 MAGNETIC MATERIALS

Sort (2025) offers an overview of the utilization of magnetic materials beyond traditional memory applications, examining their role in neuromorphic and brain-inspired computing through ideas such as magnetic synapses and neurons.

3.7.1 Ferromagnetic Materials

Ferromagnetic (FM) materials play a vital role in spintronics, enabling non-volatile memory (MRAM) and logic by manipulating electron spin rather than charge. These materials offer a fresh way to cut down the energy demands of memory and logic devices by harnessing the power of spin (Xie et al. 2025b). They thus allow the creation of low-power, high-density devices, such as magnetic tunnel junctions (MTJs), that utilize multiferroics for voltage control and 2D magnets for advanced functionalities. FM metals and alloys, such as Cobalt (Co), nickel (Ni), and iron (Fe), and half-metals (exhibiting complete spin polarization for effective spin injection and detection), along with alloys like NiFe and CoFe, are utilized in synthetic antiferromagnets (SyAF) and ferro-spintronic heterostructures (such as $BiFeO_3$/Co), to propel advancements in magneto-electric logic and nano-magnetics. Heusler alloys, intermetallic compounds of transition metals (Ni, Mn, and Ti), show promise owing to their ability to achieve high spin injection efficiency and facilitate low-loss spin-wave propagation. Dilute magnetic semiconductors (DMS) are compounds like manganese (Mn)-doped germanium (Ge) or silicon (Si) developed to enhance compatibility with current Si-based semiconductor systems. Multiferroic and complex oxides, including materials that display both ferroelectric and FM properties (such as bismuth ferrite), are being tailored for low-energy, multi-state memory applications. The exceptional characteristics of 2D nanomaterials like graphene and hexagonal boron nitride (hBN) are being investigated to develop spintronic devices.

Integrating FM materials with silicon for nanoelectronics involves various techniques, such as ion implantation (which embeds FM nanoparticles within Si), template-assisted growth (utilizing AAO templates for FM nanotube formation), vapor deposition (CVD, MOCVD for depositing FM layers or shells), and directed assembly. This aims to merge magnetic features with CMOS compatibility for spin-based applications (spintronics). Core processes emphasize careful management of nanoparticle dimensions, positions, and crystal quality, frequently employing in situ treatments or protective layers (like SiO_x) to enhance stability. These techniques enable the development of spintronic devices such as spin-polarized contacts, magnetic memory, and spin-qubit elements integrated with conventional silicon technology, paving the way for spin-charge conversion and manipulation. The primary integration techniques are:

Ion Implantation and Annealing: The approach is to introduce magnetic ions (such as Ni) into Si, often accompanied by hydrogen co-implantation to create extensive nanocavities. Subsequent annealing after implantation causes the metal ions to settle into these cavities, forming uniform FM nanoparticles (e.g., Ni in Si). Implantation produces well-defined FM nanostructures within defect-free Si, ideal for spin-based applications.

Template-Assisted Synthesis: It uses porous templates, such as anodized aluminum oxide (AAO), to facilitate the growth of FM materials (e.g., Ni, Co, Fe) via electrochemical deposition, CVD, or ALD, creating nanowires or nanotubes. This process generates consistent one-dimensional nanostructures with customized magnetic properties and compatibility with Si.

Vapor Deposition (CVD/MOCVD/MBE): Chemical vapor deposition (CVD) or metal-organic CVD (MOCVD) permits the epitaxial growth of thin FM layers or core/shell nanostructures (e.g., Co/Si, Fe/Si). It enables precise layer-by-layer deposition for superior-quality interfaces and heterostructures, which are vital for quantum devices.

Surface Functionalization/Modification: It involves altering the surfaces of FM nanoparticles (like Fe_3O_4) with polymers or inorganic compounds to prevent clumping and oxidation, ensuring compatibility with wet-chemical processing and Si integration.

Directed Assembly: It employs electric, magnetic, or fluidic forces to guide pre-synthesized FM nanomaterials (nanoparticles, nanowires) to assemble selectively on patterned Si surfaces. The method offers scalable production of functional nanodevices by arranging nanomaterials into specific designs.

3.7.2 Antiferromagnetic Materials

Antiferromagnetic materials, with their almost invisible stray fields, lightning-fast spin precession, and remarkable resistance to outside magnetic forces, are capturing the spotlight and emerging as essential players in the future of spintronic technology (Liu et al. 2025). The materials suitable for silicon Si nanoelectronics mainly consist of specific metallic alloys based on Manganese (Mn), including PtMn, IrMn, and FeMn, as well as certain oxides such as NiO and Fe_2O_3. They are utilized for advanced spintronics as they allow for quicker, more compact, and energy-efficient devices with enhanced security; they provide THz-speed switching, are free from stray magnetic fields, exhibit resilience to external influences, and enable adequate spin-orbit torque (SOT) manipulation for computing and memory applications, addressing the shortcomings of conventional magnets. Their introduction into Si nanoelectronics generally requires sophisticated deposition and fabrication methods:

Sputter Deposition: This widely used technique is used to create heterostructures featuring metallic antiferromagnets such as PtMn, FeMn, and IrMn on silicon substrates. It aligns well with established silicon manufacturing processes.

Nanoparticle Synthesis: This method incorporates antiferromagnetic nanoparticles (such as Fe_3O_4 or Co/CoO) into a silicon matrix, which can be achieved via ion-beam synthesis or "plug and play" methods compatible with planar Si technology.

Layered Structures: Antiferromagnetic layers are often combined with heavy-metal (HM) layers (such as Pt and W) to facilitate electrical control via SOTs, a crucial mechanism for the development of next-generation devices.

3.7.3 Spin-Tunnel Barrier Materials

While adding a tunnel barrier has unlocked spin injection from FM metals into GaAs, the quest to achieve the same in silicon continues to be just out of reach (Min et al. 2006). In the field of silicon (Si) nanoelectronics, Al_2O_3, MgO, and graphene are being explored as materials for spin-tunnel barriers (STBs) to address the conductivity mismatch between FM metals and semiconductors, which is essential for effective spin injection and detection. In contrast, SrO has been less frequently examined in this area than the others.

Al_2O_3 (Aluminum Oxide): It is a standard, widely used tunnel barrier material, recognized for forming a stable, uniform, high-quality insulating layer, typically deposited by ALD or sputtering. It exhibits dependable performance in spintronic applications, with quality heavily influenced by the deposition technique; ALD provides excellent control over both thickness and surface roughness. Al_2O_3 barriers align well with spin absorption theories in graphene spin valves and are suitable for high-volume manufacturing. However, they produce cluster-like films on graphene surfaces via conventional evaporation, requiring a minimum thickness of about 3 nm to prevent pinhole formation.

MgO (Magnesium Oxide): This is another standard tunnel barrier material, particularly recognized for facilitating high tunnel magnetoresistance (TMR) ratios in MTJs due to its preferred crystal orientation (notably (100) texture). It delivers excellent performance in flexible spintronic devices, with TMR ratios reaching approximately 190%. But the sputtering process for MgO can significantly damage or amorphize the underlying graphene flakes (up to four layers thick), making evaporation or alternative, gentler deposition methods more appropriate for devices based on graphene.

Graphene: It is employed as an atomically thin (single-layer) tunnel barrier itself. Graphene is chemically stable, thermally robust, and nearly defect-free in its plane, which helps inhibit the formation of interface defects typical in oxide barriers. It effectively resolves the traditional conductivity mismatch between metals and semiconductors, presenting very low resistance-area (RA) products that are crucial for viable Si spintronic devices, and it enables clean magnetic switching behavior in nanowire devices. While it exhibits high in-plane conductivity, it shows poor out-of-plane conductivity, making it an effective vertical tunnel barrier. Further, it shows lower spin polarization than optimized Co/MgO or Co/Al_2O_3 interfaces, and achieving actual tunneling may necessitate functionalization or layer stacking in specific configurations.

The integration of STB materials, such as MgO, Al_2O_3, and graphene, into Si nanoelectronics primarily relies on advanced thin-film deposition methods and precise interface engineering to ensure epitaxial quality and compatibility with semiconductor manufacturing processes. The main techniques for incorporating these materials include PVD methods like sputtering and molecular beam epitaxy (MBE), as well as chemical vapor deposition (CVD) methods like ALD.

3.7.4 2D Magnetic Materials

Two-dimensional magnetic materials shrink channel thickness down to the atomic scale, unlocking far greater electrostatic control over the device (Houssa et al. 2016). Materials such as $Cr_2Ge_2Te_6$ (CGT), Fe_3GeTe_2 (FeGeTe), and CrI_3 are used in silicon nanoelectronics to create energy-efficient spintronic devices, including non-volatile memory and advanced logic systems (Houssa et al. 2016). CGT displays significant ferromagnetism at its edges, which can be nano-patterned to fabricate magnetic nanowires and possibly artificial spin-ice structures. Fe_3GeTe_2 (FGT) is recognized for its metallic characteristics and strong perpendicular magnetic anisotropy. It serves as an essential material for injecting spin-polarized carriers into other 2D materials (such as WSe_2) within vertical heterostructures, which is vital for spintronic applications. CrI_3 exhibits switchable magnetic and electronic behavior (e.g., the transition between antiferromagnetic and ferromagnetic states in its bilayer form) that can be manipulated by gate voltage or strain, offering avenues for voltage-controlled magnetism.

Combining 2D magnetic materials with silicon nanoelectronics mainly employs heterogeneous integration approaches, particularly mechanical exfoliation and transfer or direct synthesis techniques (like chemical vapor deposition/transport). The objective is to create van der Waals (vdW) heterostructures that eliminate the challenges of lattice mismatch and temperature incompatibility arising from conventional epitaxial growth on silicon.

3.7.5 Nickelates, Titanates, and Manganites

Nickelates, titanates, and manganites are complex oxides. They are remarkable for their intense electron-electron interactions, unlocking fascinating phenomena like high-temperature superconductivity, dramatic metal-insulator transitions, and colossal magnetoresistance (CMR) (Galatsis et al. 2015). They are all distinct categories of materials that typically adopt the perovskite crystal structure. These provide exclusive properties not available in traditional semiconductors. Nickelates exhibit remarkable functional properties, including metal-insulator transitions and the potential for high-temperature superconductivity in thin-film form. Titanates (e.g., $SrTiO_3$ and $BaTiO_3$) are primarily used for their high-k dielectric properties and ferroelectric behavior. Strontium titanate (STO) serves as an essential buffer layer that can be directly deposited on silicon, maintaining an atomically sharp interface, and acts as a template for other perovskite oxides. Manganites (such as $LaMnO_3$) are known for their CMR and fascinating resistive switching behaviors, making them suitable for applications in non-volatile memory and neuromorphic computing.

Incorporating functional complex oxides, such as nickelates, titanates, and manganites, into Si nanoelectronics requires specialized methods, primarily advanced thin-film deposition techniques and interface engineering via buffer layers. Direct growth poses difficulties due to fundamental crystal chemistry and mechanical mismatches, such as lattice mismatch and the formation of unwanted silicides. The essential integration methodologies include:

PVD, which enables precise, kinetically controlled, layer-by-layer growth, is vital for maintaining the interface at the atomic scale. Molecular beam epitaxy (MBE) is a prevalent technique for producing high-quality epitaxial oxide films on Si, enabling accurate control over material fluxes, temperature, and oxygen partial pressure (often employing ozone-assisted MBE for nickelates) to avoid creating an amorphous native SiO_2 layer or silicide inclusion that could diminish the desired electronic characteristics. Pulsed laser deposition (PLD) is another significant method for depositing complex oxide thin films and superlattices, capable of forming single-crystal layers on silicon by laser ablation of the target material in a vacuum, followed by specific annealing procedures.

Chemical solution deposition (CSD) methods offer a potentially cost-effective and scalable approach to integrating functional oxides, often involving spin-coating precursor solutions. Precursor deposition involves applying metal-organic precursors to the Si substrate via spin coating. The next step, annealing, consists of heating the deposited layer to high temperatures to form the crystalline oxide film. This method can be combined with physical techniques to construct complex nanostructures.

Interface engineering is a pivotal component of these methods, as it manages the interaction between the oxide and silicon to surmount fundamental incompatibilities. Buffer layers are often employed, with intermediate layers or templates commonly used. Strontium titanate ($SrTiO_3$, or STO), yttrium-stabilized zirconia (YSZ), and CeO_2 are widely used as buffer layers, epitaxially grown on the Si substrate to support the subsequent growth of the complex oxide. Template techniques involve specific preparations of the buffer layer, such as applying a half-monolayer of Sr on a clean Si surface before STO growth, which aid surface passivation and control the epitaxial orientation of the growing oxide film. Epitaxial layer transfer is an innovative approach that enables the room-temperature transfer of single-crystal functional oxide layers onto a silicon substrate, thereby circumventing the challenges of high-temperature growth on Si. These integrated processes allow for the investigation of novel functionalities (e.g., CMR, metal-insulator transitions, ferroelectricity) that these complex oxides bring to the well-established Si nanoelectronics framework.

3.7.6 Permanent Magnetic Metals and Heavy Platinum-Group Metals

Combining permanent magnetic metals and heavy platinum-group metals (PGMs) with silicon Si nanoelectronics enables miniaturization, boosts energy efficiency, increases data storage capacity, and encourages the development of advanced spintronic devices (Ryan 2010). Their inclusion utilizes specialized deposition and nanofabrication techniques, primarily PVD and chemical vapor deposition (CVD), to maintain compatibility with current silicon processing and to avoid material interdiffusion.

3.8 DISCUSSION AND CONCLUSIONS

Next-generation nanoelectronics beyond CMOS need new materials to address the fundamental limitations of scaling and power consumption inherent in traditional silicon. These materials facilitate novel device concepts that exploit various physical phenomena (such as spintronics and quantum mechanics) to enhance performance and energy efficiency. Necessary emergent materials consist of narrow- and wide-bandgap semiconductors along with CNTs and 2D materials. Ferroelectric materials, dielectrics, topological insulators, and magnetic materials are essential to the progress of CMOS nanoelectronics, primarily by enabling new functionalities such as non-volatile memory, enhanced energy efficiency, and spintronic applications, thereby overstepping the boundaries of conventional silicon-based devices.

As the field of nanoelectronics advances and new materials emerge, research and engineering are increasingly focused on developing logic devices that leverage these materials' distinctive characteristics for computational applications. These nanoscale devices serve as essential components in upcoming electronic circuits and systems that utilize these materials to perform computing functions. This development requires tackling the constraints of conventional silicon technology and investigating innovative device architectures to facilitate ongoing improvements in computing performance. In the forthcoming chapter, we will examine electronic logic devices anticipated to handle information processing in nanoelectronics.

REFERENCES

Albrecht P. M., R. M. Farrell, W. Ye and J. W. Lyding 2003 Direct integration of single-walled carbon nanotubes with silicon, *2003 Third IEEE Conference on Nanotechnology, 2003. IEEE-NANO 2003*, San Francisco, CA, USA, 12–14 August, pp. 327–330, vol.2, https://doi.org/10.1109/NANO.2003.1231784

An S. H., J. Park and M. Kim 2023 Recent advances in single crystal narrow band-gap semiconductor nanomembranes and their flexible optoelectronic device applications: Ge, GeSn, InGaAs, and 2D materials, *Journal of Materials Chemistry C*, 11: 2430–2448, https://doi.org/10.1039/D2TC05041B

Bangolla H. K., R. R. Nallapureddy and S. Kumar 2023 Cu-rich copper indium sulfide thin films deposited by co-evaporation for photovoltaic applications, *Journal of Materials Science: Materials in Electronics*, 34: 341, https://doi.org/10.1007/s10854-022-09441-w

Bolotin K.I., K.J. Sikes, Z. Jiang, M. Klima, G. Fudenberg, J. Hone, P. Kim and H.L. Stormer 2008 Ultrahigh electron mobility in suspended graphene, *Solid State Communications*, 146(9–10): 351–355.

Bretos I., R. Jiménez, M. Tomczyk, E. Rodriguez-Castellón, P. M. Vilarinho and M. L. Calzada 2016 Active layers of high-performance lead zirconate titanate at temperatures compatible with silicon nano- and microelectronic devices, *Scientific Reports*, 6: 20143, https://doi.org/10.1038/srep20143

Chaisakul P., D. Marris-Morini, J. Frigerio, D. Chrastina, M.-S. Rouifed, S. Cecchi, P. Crozat, G. Isella and L. Vivien 2014 Integrated germanium optical interconnects on silicon substrates. *Nature Photonics*, 8: 482–488, https://doi.org/10.1038/nphoton.2014.73

Chuang Y., C.-Y. Liu, G.-L. Luo and J.-Y. Li 2021 Electron mobility enhancement in GeSn n-channel MOSFETs by tensile strain, *IEEE Electron Device Letters*, 42(1): 10–13, https://doi.org/10.1109/LED.2020.3041051

Cirlin G. E., N. K. Polyakov, V. N. Petrov, V. A. Egorov, D. V. Denisov, B. V. Volovik, V. M. Ustinov, Zh. I. Alferov, N. N. Ledentsov, R. Heitz, D. Bimberg, N. D. Zakharov, P. Werner and U. Gösele 2001 Incorporation of InAs nanostructures in a silicon matrix: Growth, structure and optical properties, *Materials Science and Engineering: B*, 80 (1–3): 108–111, https://doi.org/10.1016/S0921-5107(00)00624-3

Dargys A., R. Sedrakyan and J. Pôela 1978 Measurement of the electron drift velocity in InSb up to fields of 800 V/cm in the presence of impact ionization, *Physica Status Solidi (a)*, 45 (2): 387–392.

Dürkop T., S. A. Getty, E. Cobas, and M. S. Fuhrer 2004 Extraordinary mobility in semiconducting carbon nanotubes, *Nano Letters*, 4(1): 35–39, https://doi.org/10.1021/nl034841q

Evolve Diamonds 2025 https://www.evolvediamonds.com/electronic-properties-of-diamond/

Galatsis K., C. Ahn, I. Krivorotov, P. Kim, R. Lake and K. L. Wang 2015 A material framework for beyond-CMOS devices, *IEEE Journal on Exploratory Solid-State Computational Devices and Circuits*, 1: 19–27, https://doi.org/10.1109/JXCDC.2015.2424832

Gilbert M. J. 2021 Topological electronics, *Communications Physics*, 4: 70, https://doi.org/10.1038/s42005-021-00569-5

Haugen T. B., B. Q. Ta, E. Halvorsen, N. Hoivik and K. E. Aasmundtveit 2013 Integration of carbon nanotubes in microsystems: Local growth and electrical properties of contacts, *Materials (Basel)*, 6(8): 3094–3107. https://doi.org/10.3390/ma6083094

Houssa M., A. Dimoulas, and A. Molle (Eds.) 2016 *2D Materials for Nanoelectronics*, CRC Press (Taylor & Francis Group), Boca Raton, FL, pp. 467–483.

Hwang H., S. Youn, and H. Kim 2025 Recent advances in ferroelectric materials, devices, and in-memory computing applications. *Nano Convergence* 12, 55, https://doi.org/10.1186/s40580-025-00520-2

Ilkiv I. V., V. V. Lendyashova, B. B. Borodin, V. G. Talalaev, T. Shugabaev, R. R. Reznik and G. E. Cirlin 2024 Formation of InAs nanoislands on silicon surfaces and heterostructures based on them, *Semiconductors*, 58: 222–226, https://doi.org/10.1134/S1063782624030059

ioffe-GaAs 1998 GaAs-gallium arsenide: Electrical properties, electronic archive: New semiconductor materials, Characteristics and Properties, A.F. Ioffe Physical-Technical Institute, https://www.ioffe.ru/SVA/NSM/Semicond/GaAs/electric.html

ioffe-GaInAs 1999 $Ga_xIn_{1-x}As$: Band structure and carrier concentration, Characteristics and Properties, A.F. Ioffe Physical-Technical Institute, https://www.ioffe.ru/SVA/NSM/Semicond/GaInAs/bandstr.html

ioffe-GaN 2001 GaN: Band structure and carrier concentration, Characteristics and Properties, A.F. Ioffe Physical-Technical Institute, https://www.ioffe.ru/SVA/NSM/Semicond/GaN/bandstr.html

ioffe-Ge 2001 Ge-germanium: Electrical properties, Characteristics and Properties, A.F. Ioffe Physical-Technical Institute, https://www.ioffe.ru/SVA/NSM/Semicond/Ge/electric.html

ioffe-InAs 2001 InAs-indium arsenide: Electrical properties, Characteristics and Properties, A.F. Ioffe Physical-Technical Institute, https://www.ioffe.ru/SVA/NSM/Semicond/InAs/electric.html

ioffe-InSb 2001 InSb-indium antimonide: Electrical properties, Characteristics and Properties, A.F. Ioffe Physical-Technical Institute, https://www.ioffe.ru/SVA/NSM/Semicond/InSb/electric.html

ioffe-Si 2004 Si-silicon: Electrical properties, Characteristics and Properties, A.F. Ioffe Physical-Technical Institute, https://www.ioffe.ru/SVA/NSM/Semicond/Si/electric.html

ioffe-SiC 2001 SiC-silicon carbide: Electrical properties, Characteristics and Properties, A.F. Ioffe Physical-Technical Institute, https://www.ioffe.ru/SVA/NSM/Semicond/SiC/ebasic.html

Kander N. S., B. Gajar, S. Islam, S. Moulick, S. Guchhait and A.K. Das 2024 Evidence of topological surface states in Bi_2Te_3 thin film grown by electron beam evaporator through co-deposition technique, *Materials Chemistry and Physics*, 326, 129790, https://doi.org/10.1016/j.matchemphys.2024.129790

King S. W. 2018 Beyond the highs and lows: A selectively colorful yet chilly perspective on the future of dielectrics in nanoelectronic devices, *ECS Meeting Abstracts*, p. 1181, vol. MA2018-01, https://doi.org/10.1149/MA2018-01/17/1181

Lee Y. T., H. Kwon, J. S. Kim, H.-H. Kim, Y. J. Lee, J. A. Lim, Y.-W. Song, Y. Yi, W.-K. Choi, D. K. Hwang and S. Im 2015 Nonvolatile ferroelectric memory circuit using black phosphorus nanosheet-based field-effect transistors with P(VDF-TrFE) polymer, *ACS Nano*, 9(10): 10394–10401, https://doi.org/10.1021/acsnano.5b04592

Liu J., J. Lu, S. Peng, Z. Liu, Y. Zhang, J. Qiao, S. Wang, W. Li, J. Chen, Z. Wang, R.-W. Li, Y. Zhang and W. Zhao 2025 Antiferromagnetic materials: From fundamentals to applications, *Matter*, 8(11): 102472, https://doi.org/10.1016/j.matt.2025.102472

Min B.-C., K. Motohashi, J. C. Lodder and R. Jansen 2006 Tunable spin-tunnel contacts to silicon using low-work-function ferromagnets, *Nature Materials*, 5(10): 817–822.

Moto K., N. Saitoh, N. Yoshizawa, T. Suemasu and K. Toko 2019 Solid-phase crystallization of densified amorphous GeSn leading to high hole mobility (540 cm^2/V s), *Applied Physics Letters*, 114(11): 112110, https://doi.org/10.1063/1.5088847

Rodrigues A., A. Kamath, H.-S. Illner, N. Kafi, O. Skibitzki, M. Schmidbauer and F. Hatami 2025 Monolithically integrated GaAs nanoislands on CMOS-compatible Si nanotips using GS-MBE, *Nanomaterials*, 15(14): 1083, https://doi.org/10.3390/nano15141083

Ryan M. 2010 PGM highlights: Platinum in next-generation materials for data storage, *Platinum Metals Review*, 54(4): 244–249, https://doi.org/10.1595/147106710X525263

Sort J. 2025 Beyond storage: The expanding role of magnetism in intelligent computing, *APL Materials*, 13(7): 070401–070405, https://doi.org/10.1063/5.0283713

Tien K.-Y., Y.-Y. Chen, C.-Y. Liu, H.-S. Kao and J.-Y. Li 2025 Extremely high electron mobility in GeSn epitaxial films by chemical vapor deposition, *Advanced Electronic Materials*, 11(11): 2400925, https://doi.org/10.1002/aelm.202400925

Wouters D. J., D. Maes, L. Goux, J. G. Lisoni, V. Paraschiv, J. A. Johnson, M. Schwitters, et al. 2006 Integration of SrBi2Ta2O9 thin films for high density ferroelectric random access memory, *Journal of Applied Physics*, 100: 051603, https://doi.org/10.1063/1.2337359

Wu Z., Z. Luo, Y. Shen, W. Zhao, W. Wang, H. Nan, X. Guo, L. Sun, X. Wang, Y. You and Z. Ni 2016 Defects as a factor limiting carrier mobility in WSe_2: A spectroscopic investigation, *Nano Research*, 9: 3622–3631, https://doi.org/10.1007/s12274-016-1232-5

Xie J., A. E. Yekta, F. A. Mamun, K. Zhu, M. Chen, S. Pazos, W. Zheng, X. Zhang, et al. 2025a On-chip direct synthesis of boron nitride memristors, *Nature Nanotechnology*, 20: 1596–1604, https://doi.org/10.1038/s41565-025-01988-z

Xie Y., S.Y. Zhang, Y. Yin, N. Zheng, A. Ali, M. Younis, S. Ruan and Y.-J. Zeng 2025b Emerging ferromagnetic materials for electrical spin injection: Towards semiconductor spintronics. *npj Spintronics*, 3: 10, https://doi.org/10.1038/s41467-025-66523-z

Xue J., T. Xia, T. Li, J. He, S. Wang, W. Li, J. Ge, R. Tan, Y. Dai, et al. 2025. Room-temperature solution processing of high-mobility MoS_2 thin films, *Nature Communications*, 16: 11538, https://doi.org/10.1038/s41467-025-66523-z

Ye H. and J. Yu 2014 Germanium epitaxy on silicon, *Science and Technology of Advanced Materials*, 15(2): 024601, https://doi.org/10.1088/1468-6996/15/2/024601

Zhong Q. 2022 Intrinsic and engineered properties of black phosphorus, *Materials Today Physics*, 28: 100895, https://doi.org/10.1016/j.mtphys.2022.100895

4 CMOS and Upcoming Electronic Logic Devices

4.1 INTRODUCTION

CMOS remains the primary technology, but "beyond CMOS" innovations are emerging to address future challenges in AI, big data, and sustainable computing. Various developing technologies are poised to either enhance or ultimately replace conventional CMOS to satisfy the need for faster processing speeds and reduced energy consumption. Besides fin field-effect transistors (FinFETs), gate-all-around (GAA) and nanosheet transistors are currently used in high-performance chips, providing improved electrical control and enabling further miniaturization by surrounding the channel with the gate on all sides. Alternative devices in this competition include forksheet FETs, complementary field-effect transistors (CFETs), tunneling, negative capacitance, ferroelectric field-effect transistors, FETs made from 2D materials, single-electron transistors (SET), and nanoelectromechanical (NEM) switches. In this chapter, we shall review the progress in these devices in recent years.

4.2 FINFET

A FinFET is a three-dimensional, non-planar type of transistor that features a vertical silicon "fin" as its channel, surrounded by a gate on multiple sides (double or multi-gate), which provides enhanced control over short-channel effects, reduced leakage, and improved performance compared to planar CMOS, serving as a key element in contemporary nanoelectronics leading up to the advent of GAA structures. The design consists of fins carved from the substrate, coated with high-*k* dielectrics and metal gates, with alternatives such as bulk (common substrate) or SOI (isolated fins), facilitating improved scaling and energy efficiency in advanced chip technologies (Maurya and Bhowmick 2022, Karimi et al. 2024).

Figure 4.1 provides an illustration of two types of FETs (planar and FinFET) fabricated on SOI wafers. The FinFET configuration serves as an essential connector between conventional planar CMOS and next-generation "beyond CMOS" designs such as nanosheets and GAA devices. It is characterized by a three-dimensional, non-planar format in which the conducting channel is represented by a thin, vertical "fin" of semiconductor material.

4.2.1 Core Structural Components

The fin is a slender, upright structure, usually composed of doped silicon or more modern materials such as InGaAs or 2D crystals (e.g., MoS_2), for enhanced scaling, encompassing the source, drain, and channel regions. The gate is an electrode that wraps around the fin on three sides (top and both lateral walls), forming a tri-gate

DOI: 10.1201/9781003753469-4

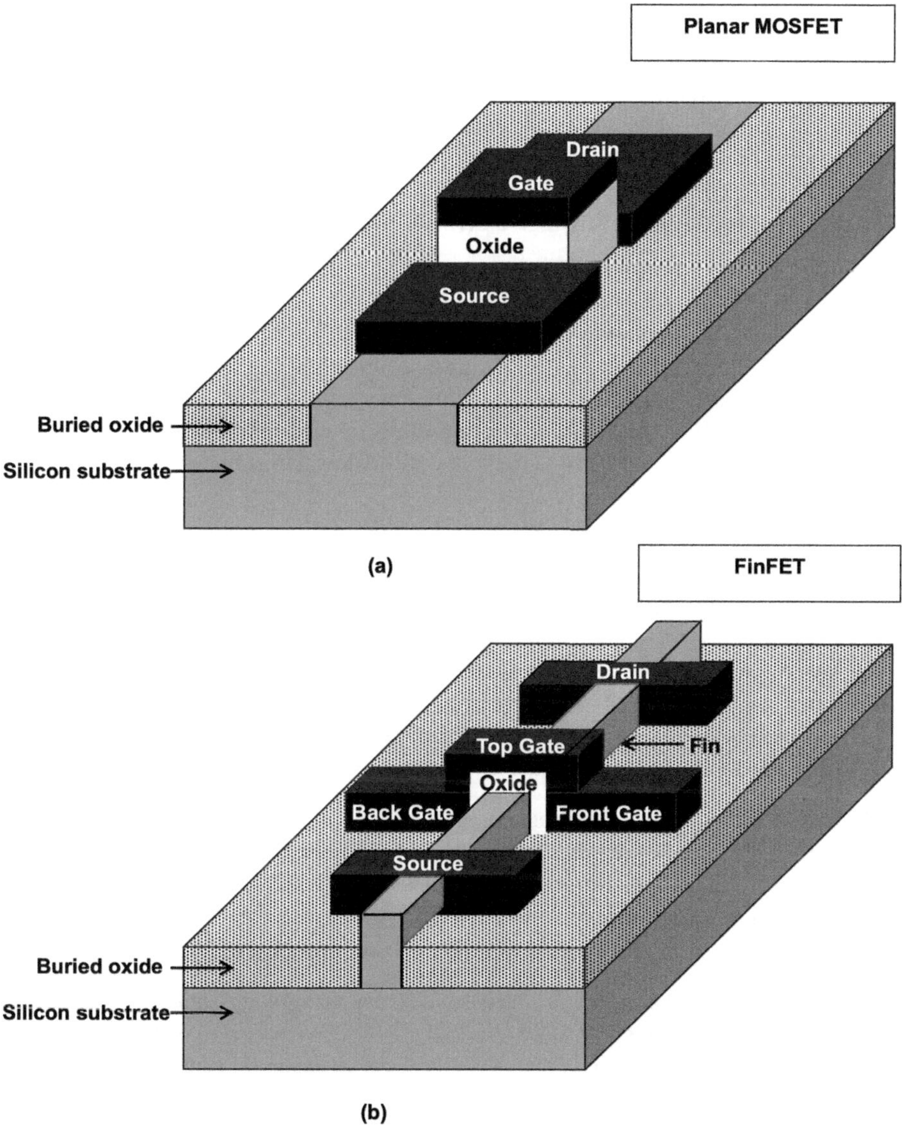

FIGURE 4.1 Two types of FETs made on SOI wafers: (a) Planar MOSFET and (b) FinFET. Both devices are fabricated on an SOI wafer with silicon substrate and buried oxide. For both devices, the source, drain, and gate electrodes are shown in black color with oxide under the gate electrode. In (a), there is only one gate at the top, while in (b) there are three gates: top, left, and right.

structure. This three-dimensional coverage provides better electrostatic control over the channel than planar gates. For the gate insulator, traditionally an oxide layer, contemporary FinFETs utilize high-*k* dielectrics, e.g., Hafnium oxide, combined with metal gates known as *High-k/Metal gate* (HK/MG), to reduce leakage and effectively tackle short-channel effects.

4.2.2 Substrate Variations

Bulk FinFETs are manufactured on standard silicon wafers; this option is more economical but requires complex isolation techniques. SOI FinFETs are created on silicon-on-insulator wafers, which naturally lower leakage and parasitic capacitance, making them suitable for high-frequency operations.

4.2.3 Advanced Structural Variations Beyond Standard CMOS

Ground-plane (GP) FinFETs incorporate ground planes beneath the source/drain to further reduce drain-induced barrier lowering (DIBL). In multi-fin structures, several fins are arranged in parallel to boost drive strength (current), and covered by a single gate. In the independent-gate (IG) FinFETs, both the front and rear gates are kept separate, allowing independent regulation of the threshold voltage V_{Th}. For 2D-material FinFETs, research is concentrating on fins constructed from a single atomic layer of 2D materials to achieve the ultimate scaling limits.

4.2.4 Why Is the FinFET Structure Used?

The shift to this three-dimensional configuration was essential as planar transistors became increasingly susceptible to high leakage currents (short-channel effects) when their sizes decreased below 22 nm. The FinFET structure enables lower power consumption, with a marked decrease in static leakage and higher performance due to accelerated switching speeds resulting from increased drive current per unit area, and better scalability, enabling the semiconductor sector to advance toward and surpass the 7 nm node before moving to gate-all-around field-effect transistors (GAAFETs).

Example 4.1 At room temperature ($T = 300$ K), a FinFET device exhibits a subthreshold swing (SS) of 68 mV/dec. Assume that the ideal factor $m = 1 + C_{Dep}/C_{Ox}$ primarily determines the subthreshold swing, where C_{dep} is the depletion capacitance and C_{ox} represents the gate oxide capacitance per unit area. Calculate the ratio of the depletion capacitance to the gate oxide capacitance $\left(C_{dep}/C_{ox}\right)$ for this device.

Solution: We must understand that the theoretical minimum subthreshold swing at room temperature is around 60 mV/dec or $S_{Ideal} = (kT/q) \times \ln(10) \sim 60$ mV/dec. The subthreshold swing for a device that is not ideal can be expressed using the formula:

$$S = S_{Ideal} \times m = S_{Ideal} \times \left(1 + \frac{C_{Dep}}{C_{Ox}}\right) \tag{4.1}$$

We will identify the known values: $S = 68$ mV/dec and $S_{Ideal} \sim 60$ mV/dec. Next, we will rearrange the formula to isolate the ratio C_{Dep}/C_{Ox}:

$$\frac{S}{S_{Ideal}} = 1 + \frac{C_{Dep}}{C_{Ox}} \tag{4.2}$$

or,

$$\frac{C_{Dep}}{C_{Ox}} = \frac{S}{S_{Ideal}} - 1 \tag{4.3}$$

By substituting the known values, we compute:

$$\frac{C_{\text{Dep}}}{C_{\text{Ox}}} = \frac{68\ \text{mV/dec}}{60\ \text{mV/dec}} - 1 = 1.133 - 1 \tag{4.4}$$

yielding

$$\frac{C_{\text{Dep}}}{C_{\text{Ox}}} \approx 0.133 \tag{4.5}$$

Thus, the ratio of the depletion capacitance to the gate oxide capacitance $C_{\text{Dep}}/C_{\text{Ox}}$ is approximately 0.133. This relatively low ratio suggests that the FinFET structure achieves effective gate control over the channel (the "fin"), reducing the impact of the depletion region and alleviating short-channel effects compared to conventional planar MOSFETs, which tend to have higher ratios.

4.3 GATE-ALL-AROUND FIELD-EFFECT TRANSISTOR

The GAAFET serves as the main successor to the FinFET design (Srivastav and Acharya 2023, Liao et al. 2025). Its essential configuration features a conducting channel that is entirely encircled by the gate electrode from every angle. This 360-degree gating offers enhanced electrostatic performance, significantly reducing leakage current and short-channel effects as devices transition below the 3 nm threshold. Let us explore two popular gate-all-around FET designs: the GAA-NWFET and the GAA-NSFET in Figure 4.2.

4.3.1 Core Structural Components

Geometrically, the channel is created using either horizontally or vertically stacked nanosheets or nanowires (NWs). Nanosheets (NSs) are broad, thin rectangular channels that provide greater drive current due to their increased effective width. NWs are either cylindrical or square channels, offering superior electrostatic control due to their symmetrical design. A gate electrode is a metal gate that fully encloses the channel and is separated from the channel by a high-*k* dielectric layer (e.g., HfO_2). Inner spacers are essential dielectric layers placed between the gate and the source/drain to minimize parasitic capacitance and determine the gate width. Source and drain components are epitaxial regions situated at the ends of the suspended channel.

4.3.2 Structural Orientations

In lateral GAA (LGAA), the current flows parallel to the substrate surface, a technique employed by industry leaders for 2 nm nodes. In vertical GAA (VGAA), the channel is oriented perpendicular to the substrate, enabling a significantly higher packing density because the horizontal area no longer constrains the channel length.

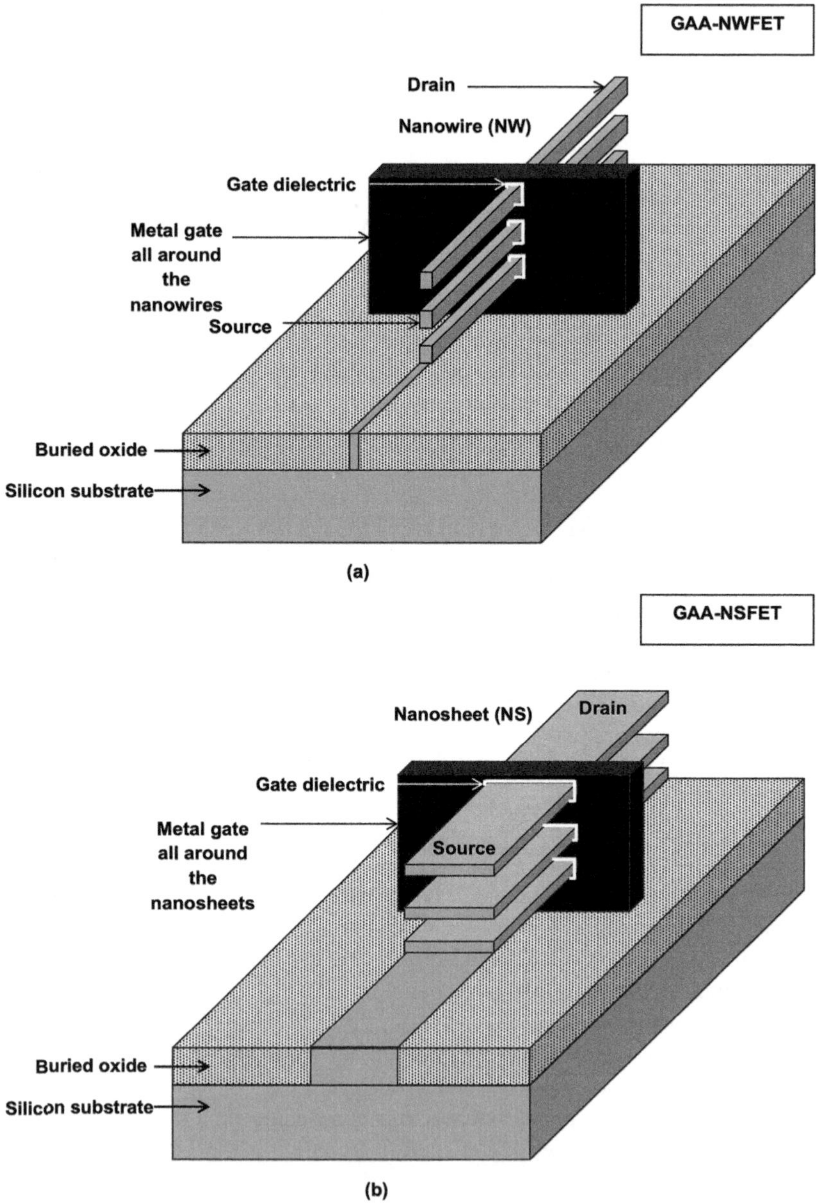

FIGURE 4.2 Common gate-all-around FET configurations: (a) GAA-NWFET and (b) GAA-NSFET. Both devices are fabricated on an SOI wafer. The silicon substrate and buried oxide of the wafer are shown. For both devices, the source and drain terminals are indicated. In (a), these terminals are located at the ends of nanowires passing through the surrounding gate insulator. So, the gate insulator envelops the nanowires from all sides, making electrostatic action of the gate more effective as compared to a simple two-dimensional MOSFET. In (b), these terminals lie at the opposite sides of nanosheets passing across the gate dielectric. In this case, the gate dielectric wraps around the nanosheets from all sides, which enhances the action of the gate considerably, as with nanowires.

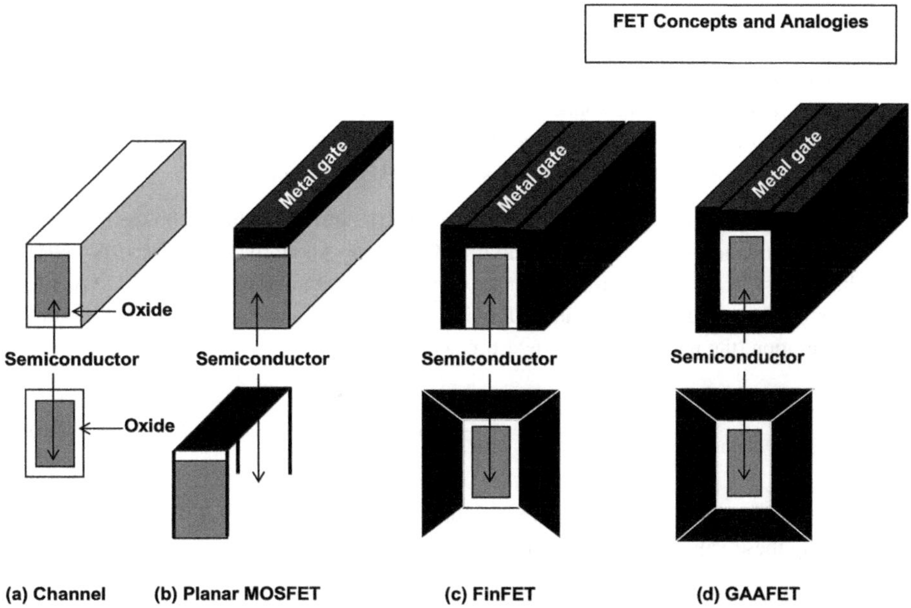

FIGURE 4.3 2D vs. 3D structural analogies of planar MOSFET, FinFET, and GAAFET: (a) Channel, (b) Planar MOSFET (flat roof): Imagine an open tent with a flat ceiling and open sides. As the tent shrinks, controlling the space from just the ceiling (the gate) becomes tricky. (c) FinFET (Walled room): Now, picture a room with tall walls on three sides. The silicon "fin" enables these vertical walls, allowing the gate to wrap around and grip the channel from multiple directions, providing much better control. (d) GAAFET (Fully enclosed secure room): Finally, envision a room completely surrounded on all sides, like a vault. This full enclosure gives the gate total command over the channel, slashing leakage and boosting performance.

4.3.3 Advantages in Beyond-CMOS Scaling

In extremely thin NWs (30–50 nm), the entire channel volume contributes to conduction, enhancing carrier mobility and minimizing scattering. The fully surrounding structure enables accurate regulation of the threshold voltage, ensuring stability within 10 mV, in contrast to over 50 mV in previous designs. Area efficiency is achieved because stacking nanosheets increases logic cell density by 30% compared to conventional FinFETs.

In summary, GAAFETs are next-generation transistors for sub-3 nm nodes, utilizing nanoscale silicon NWs or nanosheets (NSs) as channels, completely enveloped by a gate for optimal electrostatic management, unlike the three-sided enclosure of FinFETs, which facilitates improved performance, reduced power consumption, and scalability through vertical stacking of channels for enhanced current drive within the same chip area. GAAFETs have evolved from the FinFET's 3D fin structure to a genuine 3D channel wrap, employing stacked nanoscale sheets or wires for superior performance essential to future advanced logic applications.

The concepts underlying planar MOSFETs, FinFETs, and GAAFETs can be understood by drawing simple analogies, as in Figure 4.3.

Example 4.2 A silicon-based gate-all-around field-effect transistor (GAAFET) featuring a cylindrical nanowire channel is currently under development for a technology node that goes beyond CMOS. The following parameters are provided: Channel radius r_{Ch} : 3 nm (nanometers), Gate oxide thickness t_{ox} : 0.8 nm, Relative permittivity of silicon ϵ_{Si} : 11.7, and relative permittivity of the gate oxide, ϵ_{ox} : 3.9 (SiO_2). Determine the natural length λ of this GAAFET device.

Solution: The natural length λ is a key characteristic length that indicates how effectively the gate influences the channel potential and mitigates short-channel effects. A lower value of λ indicates improved electrostatic control. The natural length λ of this GAAFET device can be calculated using the following formula: For a cylindrical gate-all-around configuration, the natural length λ is estimated by:

$$\lambda \approx r_{\mathrm{Ch}} \sqrt{\frac{\epsilon_{\mathrm{Si}}}{\epsilon_{\mathrm{Ox}}} \ln\left(1 + \frac{t_{\mathrm{Ox}}}{r_{\mathrm{Ch}}}\right)} \tag{4.6}$$

Specifying the provided values: $r_{\mathrm{Ch}} = 3$ nm, $t_{\mathrm{ox}} = 0.8$ nm, $\epsilon_{\mathrm{Si}} = 11.7$, $\epsilon_{\mathrm{ox}} = 3.9$
Plugging these values into the equation for λ:

$$\begin{aligned} \lambda \approx 3\ \mathrm{nm}\sqrt{\frac{11.7}{3.9} \ln\left(1 + \frac{0.8\ \mathrm{nm}}{3\ \mathrm{nm}}\right)} &= 3\ \mathrm{nm}\sqrt{3 \ln(1 + 0.267)} = 3\ \mathrm{nm}\sqrt{3 \ln(1.267)} \\ &= 3\ \mathrm{nm}\sqrt{3 \times 0.2366} = 3\ \mathrm{nm}\ \sqrt{0.7098} \\ &= 3\ \mathrm{nm} \times 0.8425 = 2.5275\ \mathrm{nm} \end{aligned} \tag{4.7}$$

The natural length λ of this GAAFET device is roughly 2.53 nm. This minimal value demonstrates the gate's outstanding electrostatic control of the channel, a significant benefit of GAAFETs compared to planar or FinFET designs.

4.4 FORKSHEET FET ARCHITECTURE

Figure 4.4 shows the evolution from planar MOSFET to FinFET, GAAFET, and then forksheet FET. The forksheet CFET architecture represents a bold leap forward from the GAA nanosheet transistor, developed by Imec, Leuven, Belgium, as a vital bridge between today's nanosheets and tomorrow's complementary FET (CFET) technology (Weckx et al. 2017, Verschueren et al. 2025). Imagine the journey: nanosheet to forksheet to CFET. Nanosheets, the current champions at 3 and 2 nm nodes, stack silicon layers horizontally and wrap them in a gate on all sides. Forksheet steps in as the astute intermediary, inserting a dielectric wall between NMOS and PMOS devices. This innovation lets transistors huddle closer together, shrinking cell height and footprint beyond what nanosheets alone can achieve. The roadmap's final destination is CFET, where NMOS and PMOS are stacked vertically, nearly erasing lateral spacing and doubling density. Forksheet earns its title as a bridge by offering a smooth, non-disruptive transition: it borrows many fabrication steps from nanosheets while facilitating and expediting the scaling demands that precede the intricate vertical stacking of CFETs, taking center stage.

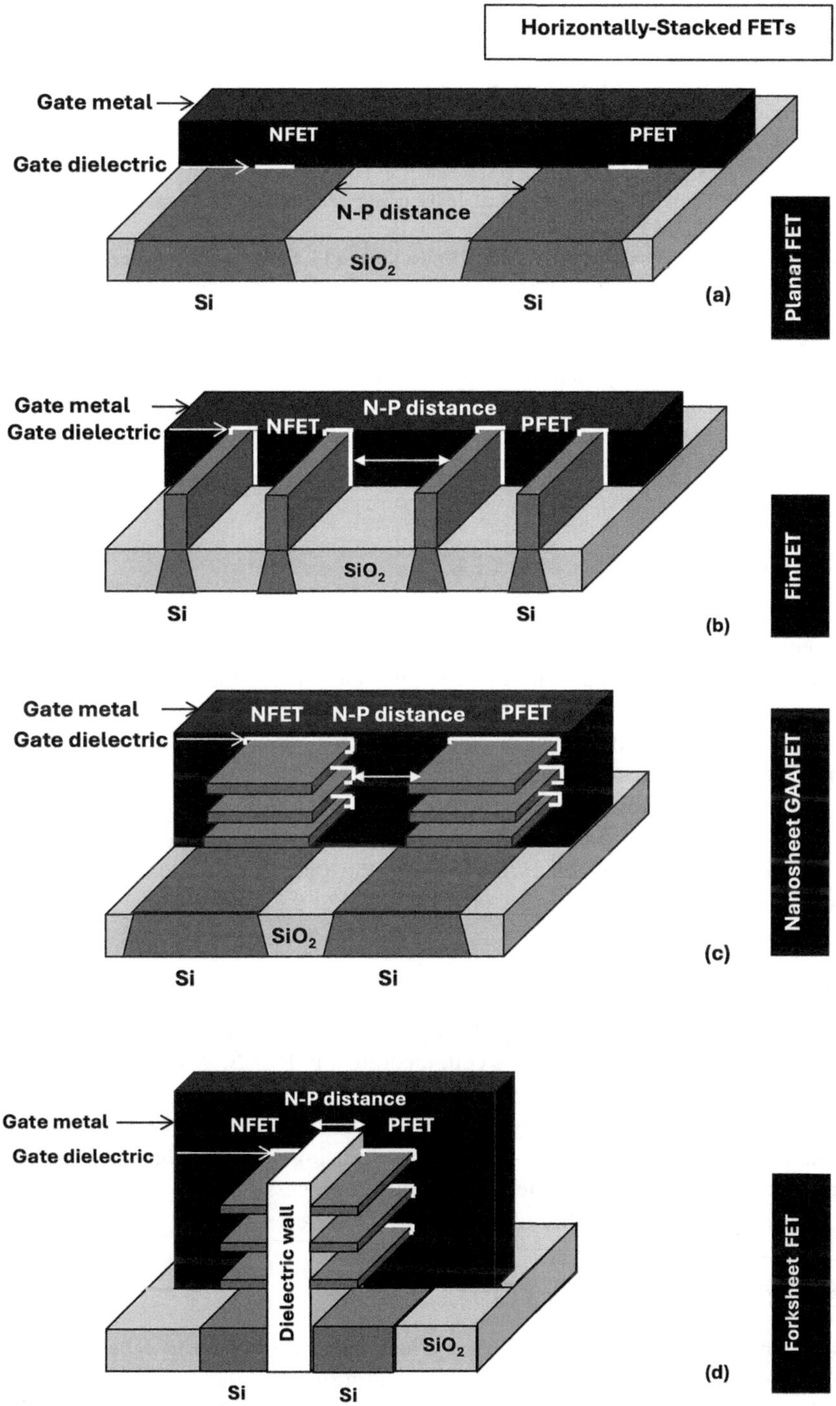

FIGURE 4.4 Diagram illustrating how the physical gap between PFET and NFET devices, known as P–N spacing, decreases across different horizontally organized FET technologies:

(Continued)

FIGURE 4.4 (Continued) (a) planar FET, (b) FinFET, (c) GAAFET, and (d) forksheet FET. In FinFETs, the distance between NFET and PFET devices is typically smaller than in planar FETs, though process constraints still set a lower limit. This happens because planar FETs rely on isolation regions, such as shallow trench isolation, to prevent unwanted interactions, but as these technologies scaled down, the P-to-N spacing became an obdurate obstacle to reducing cell height. The move to 3D FinFET structures enables closer packing, but some limitations remain. GAAFETs take integration further by using vertically stacked nanosheets that naturally require less lateral space, increasing standard cell density and diminishing cell height, especially at advanced nodes like 3 nm. The forksheet FETs go a step beyond. The forksheet FET arranges N-type and P-type transistors side by side, bringing them much closer together than before. A slender central dielectric wall, often made of silicon nitride, separates the two, shrinking the gap between N and P regions to just 17 nm—far tighter than in standard nanosheets. More recently, innovators have introduced the "outer wall" forksheet, which streamlines manufacturing by moving the dielectric wall to the cell's edge instead of between the transistors. This shift not only simplifies the process but also enhances the use of strain engineering to further boost device performance.

4.4.1 Inner-Wall Forksheet FET

The first forksheet device architecture stood out for its use of a dielectric wall located between NMOS and PMOS devices before the gate was patterned (Figure 4.5). Because this wall sat at the center of a logic standard cell, the design became known as the "inner wall" forksheet. By physically separating the P-gate and N-gate trenches, the inner wall structure enabled much tighter N-to-P spacing than what FinFET or nanosheet devices could achieve, setting the scene for smaller cell areas without sacrificing performance. In this setup, the sheets are steered by a tri-gate forked structure, which inspired the device's name. Although placing the dielectric wall inside the logic cell increases density, the inner-wall variant of forked-sheet designs poses a manufacturability challenge due to the tight squeeze.

4.4.2 Outer-Wall Forksheet FET

The outer-wall fork sheet shown in Figure 4.6 refines the inner-wall idea by shifting the wall to the edge of the standard cell, creating a P–P or N–N wall, thereby making fabrication easier while preserving excellent gate control and performance. This adroit shift allows neighboring cells to share a wall, which can now be made thicker—up to about 15 nm—without increasing cell height. The outer-wall approach not only simplifies the manufacturing process but also improves performance and maintains the ability to scale down cell area.

4.5 COMPLEMENTARY FET

Complementary FET (CFET) device architectures are poised to take the spotlight from GAA nanosheet transistors in the evolution of logic technology (Ryckaert et al. 2018, Sarkar 2024, Boukortt and Stroobandt 2026). In CFETs, NMOS and PMOS transistors are stacked directly on top of one another, eliminating the

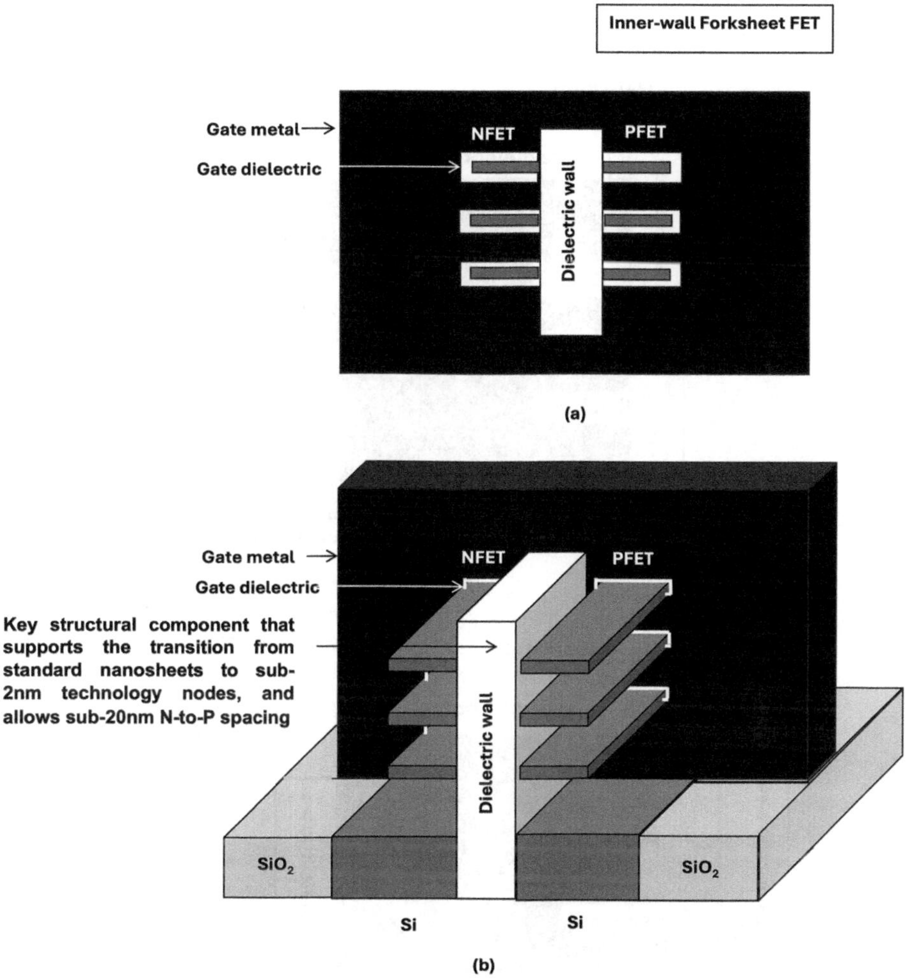

FIGURE 4.5 The inner-wall forksheet device structure: (a) sectional view and (b) perspective view. The inner wall design positions a dielectric barrier right between the P and N channels, pushing scaling to its limits. The inner wall approach raised concerns due to its delicate structure and etching challenges without causing damage. Since the inner wall is built at the very start of the front-end-of-line process, it must withstand a series of intricate etching steps, like the precise removal of sacrificial SiGe for inner spacer formation. Each of these steps risks eroding or thinning the wall, making the choice of materials and the precision of the process absolutely critical.

need for N–P separation at standard cell heights (Figure 4.7). This vertical arrangement allows CFETs to pack twice as many transistors into the same area, reduce interconnect delays, and improve power, performance, and area (PPA) metrics (Zhang et al. 2024).

Figure 4.8 illustrates how the source and drain connections are made to GAAFET and CFET devices. These connections are formed through advanced epitaxial

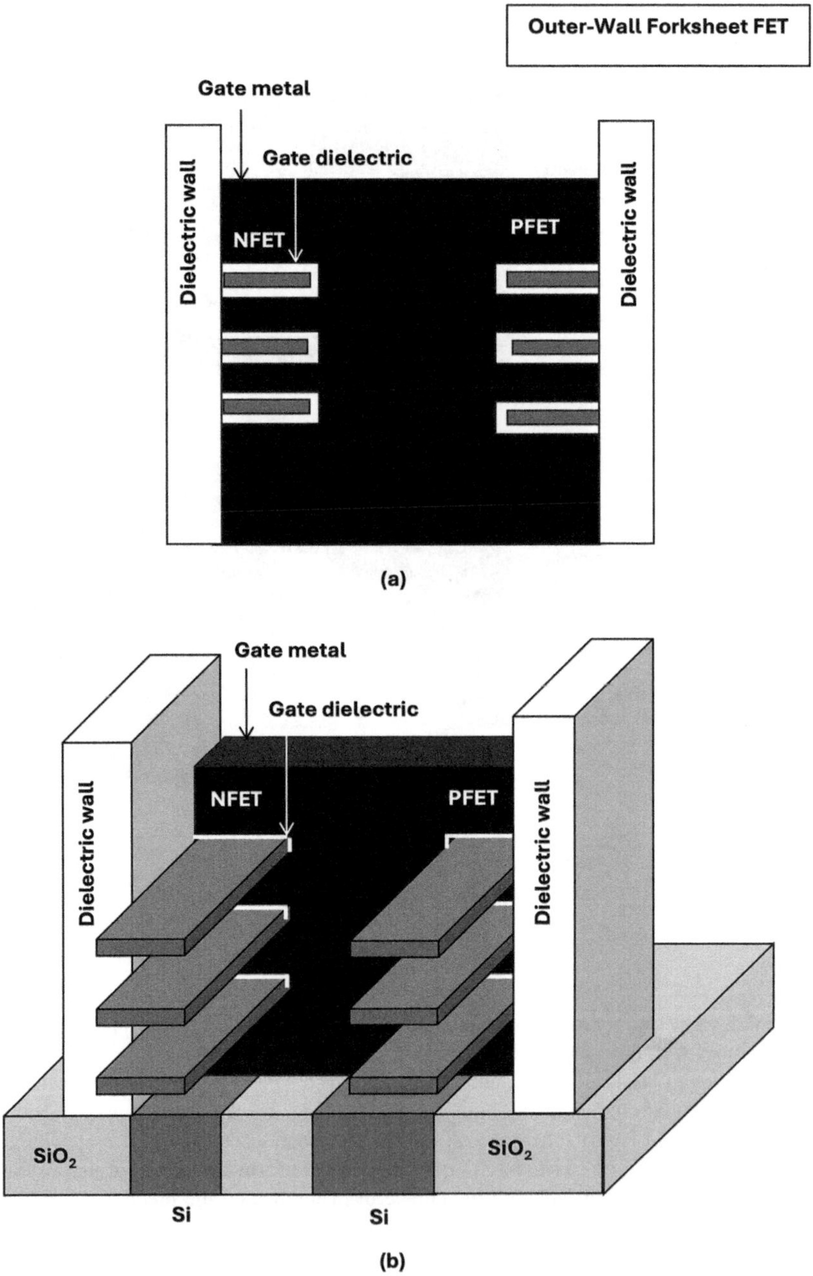

FIGURE 4.6 The outer-wall fork sheet device: (a) representative view, and (b) 3D view. The outer-wall design shifts the dielectric wall to the cell boundary, facilitating manufacturing, providing stronger gate control, and increasing channel stress. There are two walls flanking the device on the two opposite sides. This design emerged as a solution, crafted specifically to overcome the hurdles faced by the inner-wall structure. Since it is built later in the process, after source and drain epitaxy and nanosheet channel release, it can be fabricated using familiar materials such as silicon dioxide and standard fabrication tools.

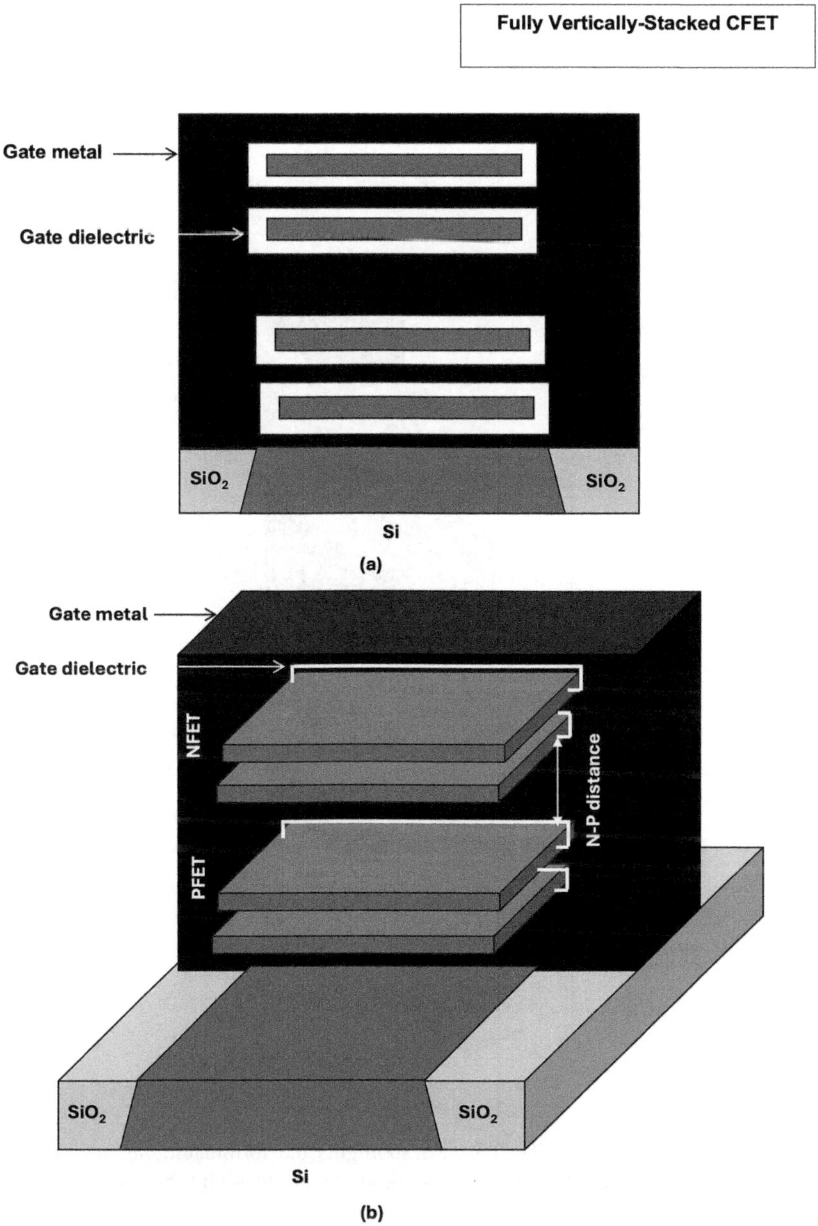

FIGURE 4.7 The complementary field-effect transistor: (a) The sectional view slices through the transistor, unveiling its ingenious vertical stacking that maximizes chip real estate. Instead of placing NFET and PFET side by side, they are stacked directly on top of each other. This ingenious arrangement nearly erases the horizontal gap, achieving even greater density and enabling the smallest standard cell heights yet. (b) The front view shows that both transistors are controlled by a single, vertical or wrap-around gate, orchestrating control from top to bottom.

(Continued)

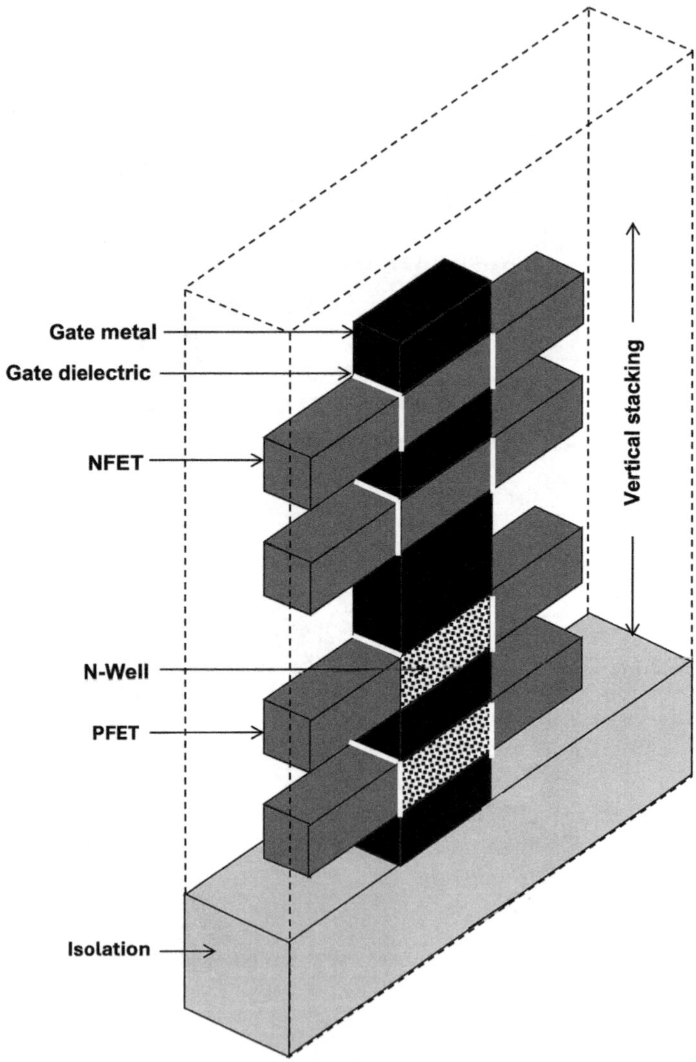

(c)

FIGURE 4.7 (Continued) (c) The 3D view spotlights the formidable engineering hurdles, such as the precision required for high-aspect-ratio etching to sculpt these towering, slender structures.

growth, in which new semiconductor layers sprout from the exposed tips of stacked nanosheets or nanowires. GAAFETs link multiple horizontal channels, whereas CFETs require a more sophisticated vertical stack and precise separation to keep the NFET and PFET pathways apart. Typically, silicon-germanium or silicon layers

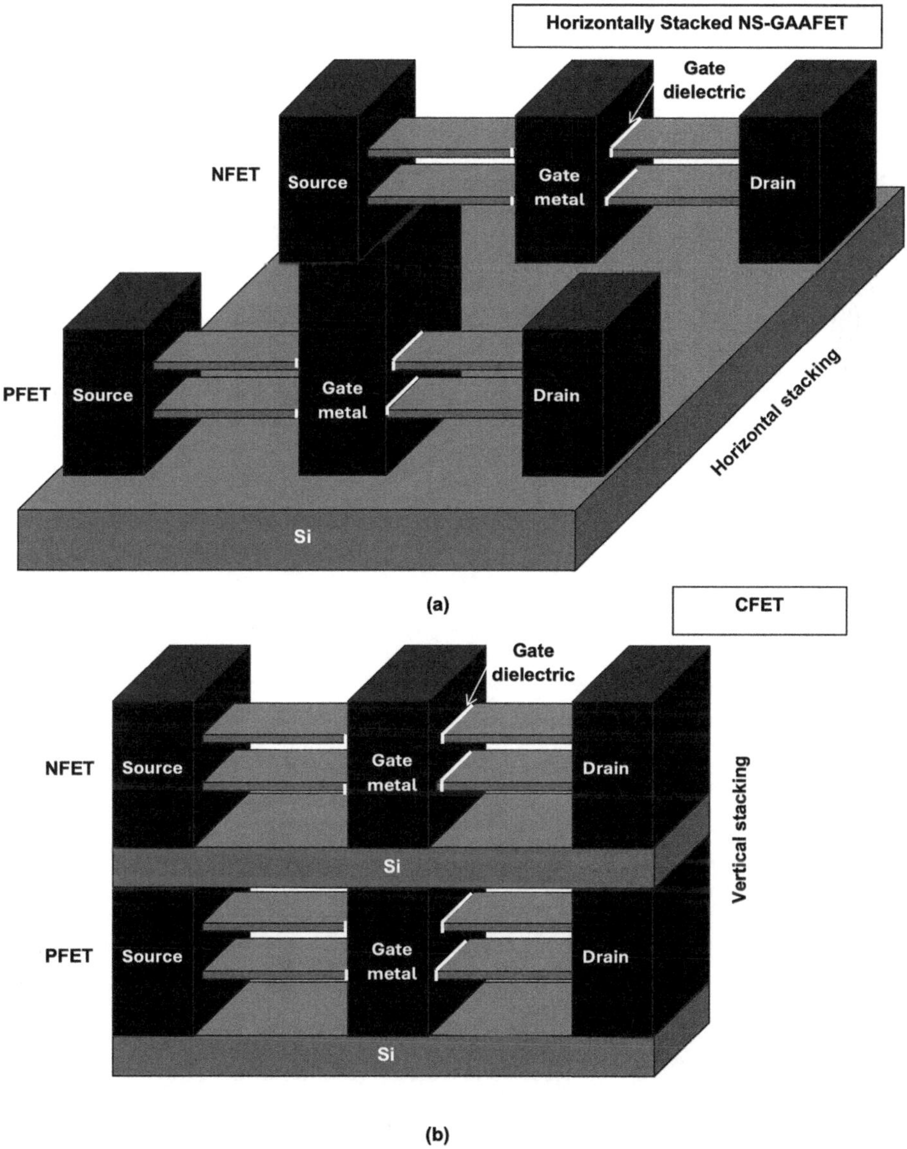

FIGURE 4.8 A side-by-side look at GAAFET and CFET designs: (a) GAAFET and (b) CFET, with their source/drain and gate terminals clearly marked.

are grown at the nanosheet ends, and inner spacers are added to shield the gate from the source and drain, reducing unwanted capacitance.

Two main manufacturing processes have emerged for CFETs: one grows PFET and NFET together in a single, unified process, while the other builds them one after the other (Figure 4.9). The monolithic method promises higher performance and

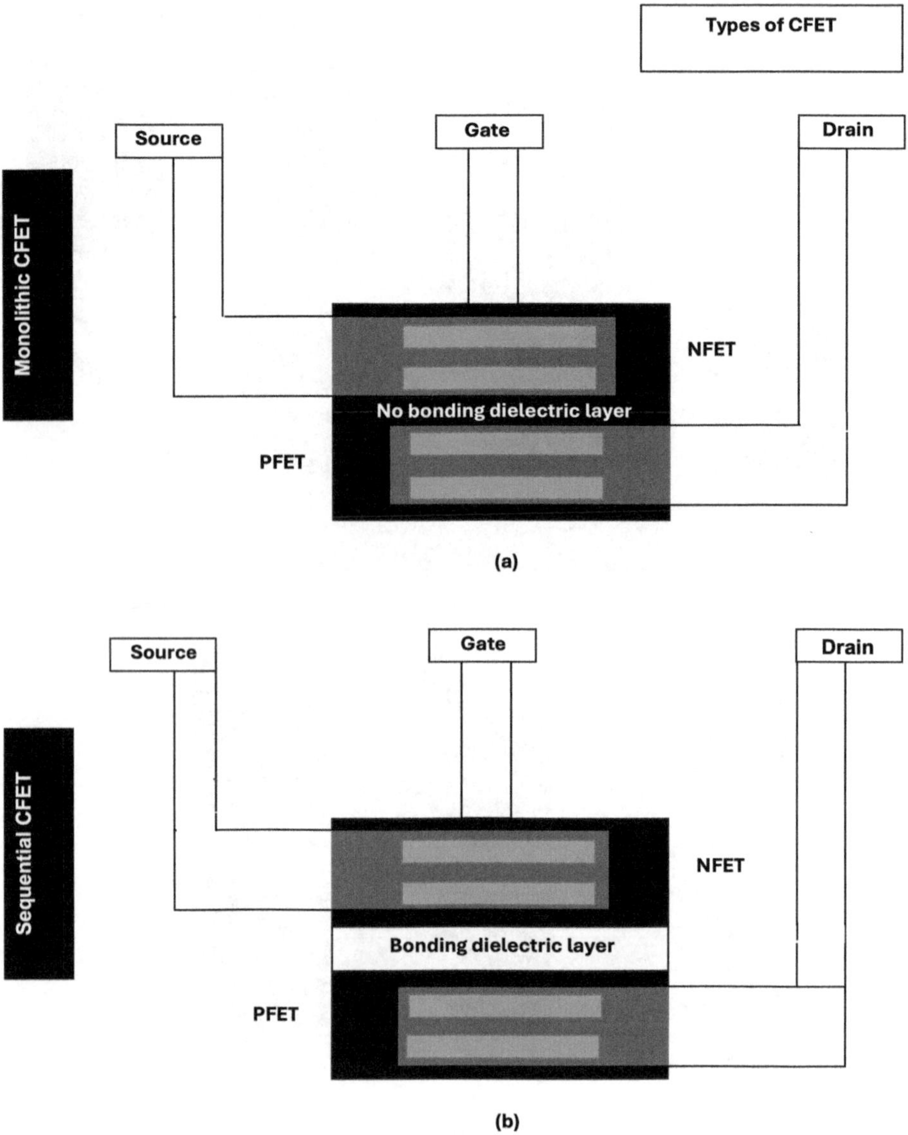

FIGURE 4.9 Diagram comparing (a) monolithic and (b) sequential CFET designs. The main distinction is how they are made: monolithic CFETs are fabricated on a single wafer in a single seamless process, while sequential CFETs require joining two separate wafers. This extra bonding layer not only adds height to the structure but also increases parasitic capacitance, which can hinder device performance.

lower costs but demands extremely high-aspect-ratio etching. The sequential route offers more freedom in choosing channel materials for NFET and PFET but hinges on precise alignment between the two. Both approaches face strict process tolerances, where even minor deviations can lead to unwanted shorts or opens.

4.6 TUNNELING FIELD-EFFECT TRANSISTOR

Tunneling field-effect transistors (TFETs) present a viable option to standard MOSFETs for ultra-low-power logic applications due to their distinct operating principle, which enables lower supply voltages and decreased power usage (Datta et al. 2014, Chen et al. 2025). TFETs use quantum-mechanical tunneling for current regulation, showing promise for low-power performance but encountering issues with low on-currents and hysteresis. Figure 4.10 invites us to examine the fascinating realm of TFETs, focusing on their structure and operation, as we look at their different aspects side-by-side with the familiar MOSFET.

4.6.1 Working Principle

In contrast to traditional MOSFETs that operate via thermionic emission (where carriers overcome a potential barrier), TFETs rely on quantum-mechanical band-to-band tunneling (BTBT). Structurally, a typical TFET features a P–I–N (with a P-type source, intrinsic channel, and N-type drain) or N–I–P configuration. During TFET operation, the gate voltage adjusts the energy-band alignment at the junction between the source and channel. In the OFF state, the energy bands are not aligned, creating a broad barrier for carrier movement, resulting in minimal OFF-state current (I_{OFF}). To being the device into ON state, an adequate gate voltage is applied, whereupon the valence band of the source aligns with the conduction band of the channel, allowing electrons to "tunnel" through this narrow energy barrier from the source to the channel, permitting current flow.

4.6.2 Advantages of TFETs for Logic Applications

TFETs provide a key benefit by achieving a subthreshold swing (SS) below the fundamental limit of 60 mV/dec observed in MOSFETs at room temperature. This feature facilitates reduced power dissipation. The capacity to function at significantly lower supply voltages (V_{DD}) while ensuring a sharp transition between the ON and OFF states considerably lowers both static and dynamic power usage. The sharp subthreshold swing leads to lower operational voltages, allowing for substantial power savings over traditional CMOS logic. Owing to their energy efficiency, TFET-based logic circuits can outperform conventional CMOS designs in energy efficiency, making them suitable for battery-operated devices in the Internet of Things (IoT) and other low-power electronic applications. TFETs offer outstanding scalability for future electronic devices. Because of the band-to-band tunneling process, TFETs display enhanced resistance to short-channel effects as device sizes decrease, which poses a significant challenge for the ongoing scaling of MOSFETs. TFETs also exhibit potential for application in ternary logic circuits, which maintain stable intermediate states, thus opening new avenues beyond conventional binary logic.

4.6.3 Challenges and Research Directions

Despite their promise, several obstacles prevent the widespread commercial use of TFET logic. TFETs often exhibit lower ON-state current (I_{ON}) compared to MOSFETs, potentially slowing circuit performance. Conversely, the OFF-state

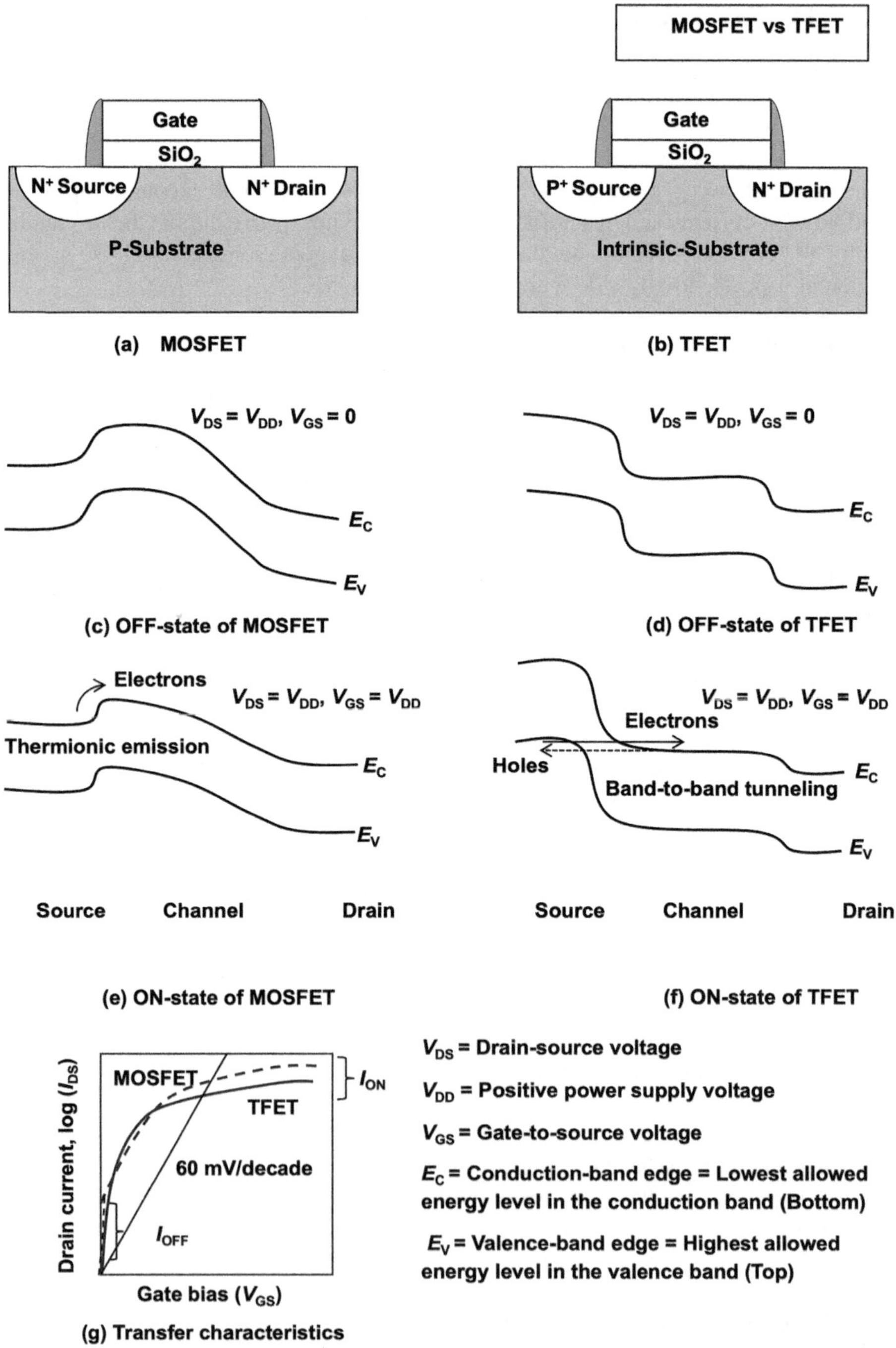

FIGURE 4.10 MOSFET vs. TFET: (a) and (b) Schematic cross-sections of the MOSFET and TFET. (c) and (d) energy-band diagrams of MOSFET and TFET in OFF state, (e) and (f) their

(Continued)

FIGURE 4.10 (Continued) energy-band diagrams in ON state, (g) transfer characteristics of MOSFET and FinFET. (a) and (b) MOSFET has a asymmetric doping (N^+–P–N^+) configuration with an N^+ source and N^+ drain in a P-substrate. Source and drain are of the same type (N^+/N^+) and so interchangeable, leading to bidirectional current direction. TFET has an asymmetric doping configuration (N^+–I–N^+) with P^+ source and N^+ drain in an intrinsic substrate. Source and drain of opposite types (P^+/N^+) and so are fixed, i.e., cannot be interchanged. Hence, the current direction is unidirectional. MOSFET, with its two back-to-back connected P–N junctions, is a three-terminal transistor. TFET is a gated P–I–N diode. Carrier transport in a MOSFET occurs via thermionic emission by drift and diffusion mechanisms, while in a TFET, the carrier transport mechanism is band-to-band tunneling. (c) and (d) In OFF state, $V_{DS} = V_{DD}$ and $V_{GS} = 0$. In a MOSFET, a potential barrier forms when the gate voltage is below threshold, which prevents thermionic emission. In a TFET, the barrier is the forbidden bandgap between the source and the channel. In the OFF state, the source valence band is much lower in energy than the channel conduction band. Tunneling cannot occur because of this misalignment of bands. (e) and (f) In the ON state, $V_{DS} = V_{DD}$ and $V_{GS} = V_{DD}$. In the ON state of a MOSFET, the gate voltage lowers the channel's conduction band, reducing the height of the potential barrier. Current begins to flow when the barrier is low enough for a large number of electrons to flow over it. In the ON-state of a TFET, the gate voltage pulls the channel's conduction band below the source's valence band. This creates a "tunneling window" where the bands overlap vertically, allowing electrons to move horizontally from the valence band to the conduction band. (g) The primary distinction between MOSFET and TFET transfer characteristics is the steepness of the turn-on. MOSFETs are limited to a subthreshold swing of 60 mV/dec at room temperature due to the Boltzmann distribution, while TFETs can achieve sub-60 mV/dec slopes, resulting in a sharper turn-on. MOSFETs typically provide higher ON current because thermionic emission allows for greater carrier flow. In contrast, TFETs generally have lower ON current, as tunneling probability is restricted by barrier width and material properties. However, TFETs offer significantly lower OFF current than MOSFETs, since the tunneling barrier remains wide and effective at low gate voltages.

leakage current remains considerably lower due to reliance on band-to-band tunneling rather than thermionic emission. Furthermore, in certain TFET designs, current can flow in both positive and negative bias conditions, which is known as ambipolarity. The ambipolar conduction in TFETs creates challenges in the design of logic gates and increases power consumption. Another drawback of TFETs is their fabrication complexity. The demanding material and structural requirements, such as the production of sharp doping junctions and high-quality heterostructures (e.g., using III–V or 2D materials), complicate the fabrication process.

Researchers are currently investigating various methods to address these challenges, including innovative device architectures (such as F-shaped TFETs), material engineering with direct bandgap semiconductors (like InAs and Ge), and integrated circuit designs that leverage TFET characteristics for novel logic functions.

Example 4.3 Given Parameters: Supply Voltage (V_{DD}) $= 0.5$ V, OFF-state leakage current (I_{Off}) $= 1$ nA. Assume a complementary TFET inverter configuration, resembling CMOS, where one transistor is in the OFF state, and the other is in the ON state during a stable condition. Determine the static power dissipation (P_{static}) of the TFET inverter when the input is in a steady state (either logic 0 or logic 1).

Solution: In stable CMOS or complementary TFET logic gates, one network, either the pull-up or the pull-down, is active while the other is off, effectively preventing current flow and reducing static power consumption to a minimum. The only current flowing is the minuscule OFF-state leakage current through the series path. We identify the current path. In the static condition (e.g., input low, output high), the current through the circuit is constrained by the transistor's OFF-state leakage current. We use the power calculation formula:

$$P = V \times I \tag{4.8}$$

where V signifies the supply voltage (V_{DD}) and (I) indicates the leakage current (I_{off}). Replacing the variables with the given values:

$$P_{Static} = V_{DD} \times I_{Off} = 0.5\text{ V} \times 1 \times 10^{-9}\text{ A} = 0.5 \times 10^{-9}\text{ W} = 0.5\text{ nW} \tag{4.9}$$

This outcome underscores a significant benefit of TFETs and other devices beyond CMOS: remarkably low static power dissipation compared to traditional scaled MOSFETs, which struggle to maintain low (I_{Off}) at lower supply voltages.

4.7 NEGATIVE CAPACITANCE FIELD-EFFECT TRANSISTOR

Negative Capacitance Field-Effect Transistors (NCFETs) seek to surpass the Boltzmann limit on subthreshold swing by incorporating negative capacitance, thus facilitating more energy-efficient switching (Tu et al. 2018, Rahi et al. 2021). Figure 4.11 presents the structure and equivalent circuits of NCFET, offering a closer look at its inner working principles.

NCFETs logic pertains to digital circuits created with NCFETs, a specific type of transistor that incorporates a ferroelectric material in its gate stack to attain extremely low-power usage and enhanced performance compared to traditional MOSFETs. The fundamental concept of NCFET logic is the utilization of the "negative capacitance" (NC) phenomenon in the ferroelectric layer.

Negative capacitance is an unexpected phenomenon observed in ferroelectric materials, where an increase in stored charge results in a decrease in voltage. In conventional capacitors, charge and voltage rise in proportion to each other; however, ferroelectrics can transition into an unstable state that releases energy, effectively enhancing the internal voltage. The origin of the NC effect is found in the double-well free energy landscape that is typical for ferroelectric materials. Ferroelectric materials inherently possess two stable polarization states (the "wells"). The area between these wells exhibits negative curvature, indicating a state where the energy of the material diminishes as it polarizes. The ferroelectric layer functions as an internal voltage transformer, boosting the gate voltage.

4.7.1 The Impact of NCFET Operation on Logic

The operation of NCFET has manifold effects on logic. An increase in voltage is achieved. When integrated appropriately, the ferroelectric layer provides internal

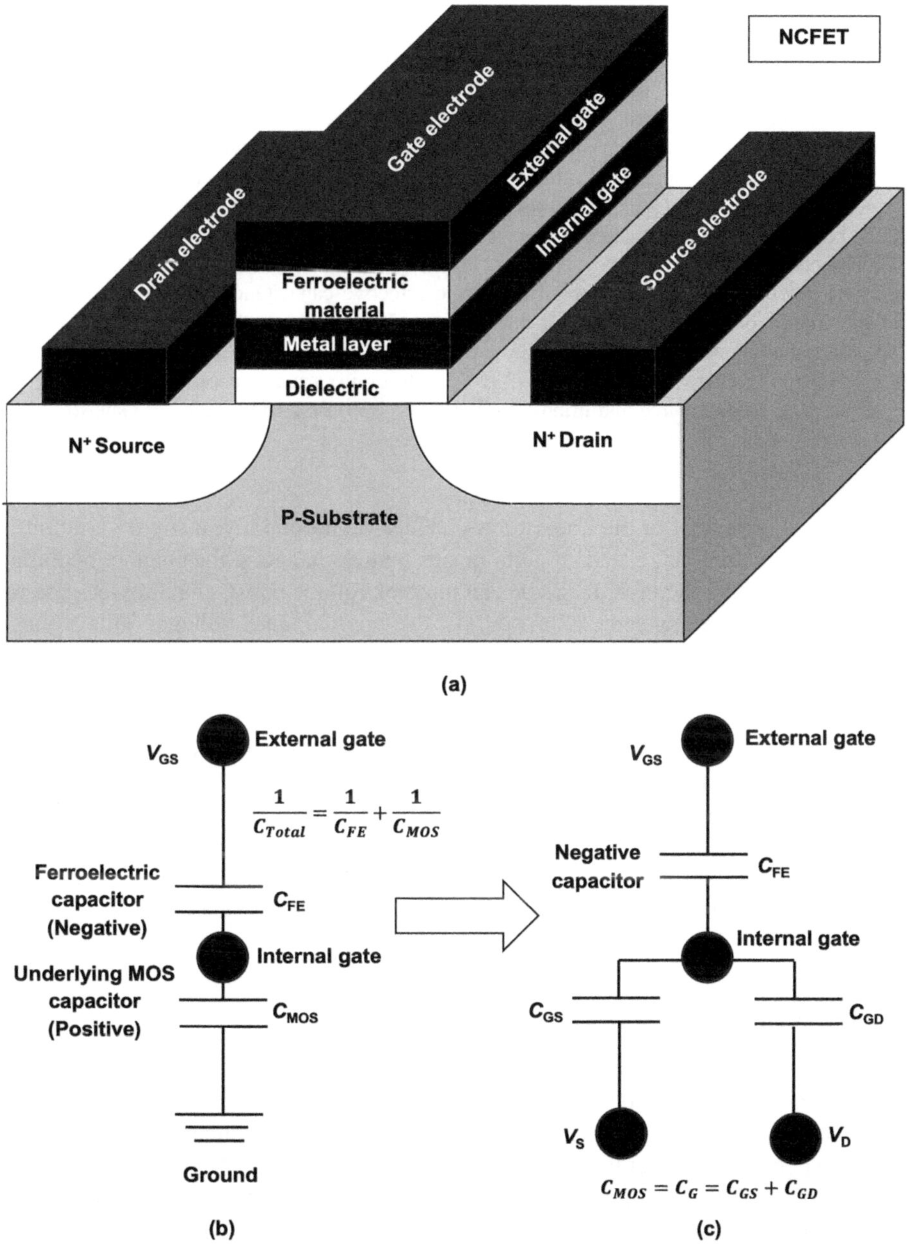

FIGURE 4.11 NCFET and its equivalent circuits: (a) perspective view of NCFET, (b) and (c) equivalent circuit representations for NCFET modeling. (a) Like a MOSFET, the NCFET features N^+ source and N^+ drain regions in a P-substrate. The primary distinction is in the gate stack composition. While a MOSFET uses a metal gate and a conventional thin insulating layer, such as SiO_2 or a high-κ dielectric, an NCFET incorporates a ferroelectric layer to achieve its unique electrical properties. The NCFET structure includes an external metal gate

(Continued)

FIGURE 4.11 (Continued) and an internal floating metal gate positioned between the ferroelectric and dielectric layers, resulting in a metal-ferroelectric-metal-insulator-semiconductor (MFMIS) configuration. (b) An NCFET equivalent circuit models the NCFET device as a series connection of a ferroelectric capacitor (C_{FE}) and a baseline MOSFET capacitor C_{MOS}. The ferroelectric capacitor represents the additional ferroelectric layer in the gate stack and provides the negative capacitance effect, typically described by the Landau-Khalatnikov (L-K) equation. The baseline MOSFET models the underlying transistor, such as a FinFET or GAAFET, with gate capacitance comprising both intrinsic channel and parasitic components. The junction between the two capacitors forms the internal gate node. The external and internal gates are shown. The reciprocal of the effective capacitance of NCFET is the sum of reciprocals of the capacitances of the ferroelectric capacitor and MOSFET capacitor. (c) A modified equivalent circuit of NCFET is drawn by dividing the MOSFET gate capacitance C_{MOS} into C_{GS} (gate-to-source) and C_{GD} (gate-to-drain) capacitances, which connect the internal node to the source and drain terminals. The MOSFET capacitance = Gate-to-source capacitance + Gate-to-drain capacitance.

voltage enhancement to the transistor's channel. Consequently, a slight variation in the external gate voltage leads to a more substantial change in the channel potential. Thus, the ferroelectric layer facilitates an internal voltage boost, enabling the gate-to-channel voltage enhancement to surpass the applied gate voltage. This voltage enhancement enables NCFETs to maintain performance at lower supply voltages than regular MOSFETs, enabling reduced-voltage operation. Additionally, a sharper switching is attained. This voltage amplification allows NCFETs to overcome the inherent "Boltzmann tyranny" limit of 60 mV/dec in the subthreshold swing (SS) observed in standard MOSFETs, thereby minimizing OFF-state leakage current. A lower SS also allows the transistor to transition swiftly between its ON and OFF states at considerably lower operating voltages, resulting in substantial energy savings. The synergy of decreased leakage and lower voltage results in notable reductions in both static and dynamic power usage for logic gates such as inverters, AND, OR, and XOR.

4.7.2 Advantages, Applications, and Challenges

The main benefit of NCFETs is their capability to function at significantly lower supply voltages while retaining performance, leading to a drastic decrease in power dissipation. Besides minimal power usage, NCFETs show enhanced performance through demonstration of increased ON currents and reduced OFF (leakage) currents compared to conventional MOSFETs at equivalent voltages, thereby boosting switching speeds and energy efficiency. Better scaling aids in alleviating short-channel issues that emerge as transistors decrease in size.

NCFET logic is under investigation for a variety of low-power and high-performance use cases. NCFET technology can empower processors to operate at higher frequencies or function at lower voltages without raising power density, which is essential for modern multi-core energy-effective designs. The low-power consumption of NCFETs makes them well-suited for battery-operated and energy-sensitive applications like wearable electronics and Internet of Things (IoT) sensors.

NCFET-based architectures are being utilized in non-volatile memory (for instance, FeFET-based memory cells) and SRAM, providing benefits in retention, read/write speed, and energy efficiency. The remarkable rectification efficiency of NCFETs in low-energy conditions renders them exceptionally effective for applications as RF-DC rectifiers in microwave wireless power transmission systems.

The ferroelectric layer in NCFET may introduce hysteresis, which requires careful management through device design. Research is currently underway to tackle challenges like achieving stable and hysteresis-free operation over the entire voltage span, which is essential for dependable logic circuits. Ongoing studies are concentrating on finding the ideal balance between ferroelectric and MOS capacitor characteristics to realize steep switching without hysteresis. This encompasses the optimization of the thickness and material composition of the ferroelectric and insulating layers.

Example 4.4 A standard MOSFET features a gate control pathway that includes a total effective gate capacitance (C_{ox}). An NCFET alters this configuration by incorporating a ferroelectric (FE) layer, resulting in an effective negative capacitance (CFE) in series with the gate insulator capacitance (C_i). Given values—Gate insulator capacitance (an SiO_2 layer): $C_i = 5\ \mu F/cm^2$, Ferroelectric layer's "negative" capacitance (functioning within the negative capacitance region): $C_{FE} = -10\ \mu F/cm^2$. Assess the total effective gate capacitance (C_{eff}), and evaluate the voltage amplification factor (A_v) of the NCFET structure in comparison to the standard transistor.

Solution: We determine the total effective gate capacitance (C_{eff}). The ferroelectric and insulator capacitances are configured in series within the gate stack (commonly in an MFMIS structure). The equation for series capacitance is as follows:

$$\frac{1}{C_{eff}} = \frac{1}{C_i} + \frac{1}{C_{FE}} \tag{4.10}$$

Inserting the provided values:

$$\frac{1}{C_{eff}} = \frac{1}{5\ \mu F/cm^2} + \frac{1}{-10\ \mu F/cm^2} = \frac{2}{10\ \mu F/cm^2} - \frac{1}{10\ \mu F/cm^2} = \frac{1}{10\ \mu F/cm^2} \tag{4.11}$$

Hence, $C_{eff} = 10\ \mu F/cm^2$. We now assess the voltage amplification factor (A_v). The incorporation of negative capacitance increases the internal voltage ($V_{FE} + V_i$) relative to the voltage across the insulator layer in a traditional FET. The voltage amplification effect can be viewed as the ratio of the effective capacitance to the insulator capacitance. The voltage gain is defined by:

$$A_v = \frac{C_{eff}}{C_i} \tag{4.12}$$

Inserting the calculated C_{eff} and the specified C_i,

$$A_v = \frac{10\ \mu F/cm^2}{5\ \mu F/cm^2} = 2 \tag{4.13}$$

The NCFET design, featuring the aforementioned ferroelectric and insulator layers, achieves a total effective gate capacitance of 10 $\mu F/cm^2$ and delivers a voltage amplification factor of $2\times$. This enhancement in internal voltage facilitates more abrupt transistor activation at lower applied gate voltages, resulting in the desired steep subthreshold swing and greater energy efficiency compared to conventional CMOS devices. The primary advantage of the NCFET lies in its voltage amplification effect, which enables a subthreshold swing (SS) below the classical 60 mV/dec Boltzmann limit.

4.8 FeFET

4.8.1 Operational Mechanisms

Ferroelectric FETs (FeFETs) incorporate a ferroelectric material within the gate stack of a field-effect transistor (Beyer et al. 2020, Yu et al. 2024). The ferroelectric material possesses spontaneous electrical polarization that can be altered by an external electric field and, importantly, maintained after the power is turned off (non-volatile characteristic), allowing for non-volatile memory and logic functions, which leads to reduced power consumption. Therefore, an FeFET acts both as a switch and as a non-volatile storage element, enabling it to execute logic operations while preserving data without the need for power. Figure 4.12 presents the structural diagram and equivalent circuit of an FeFET, and compares its transfer characteristics with those of an NCFET.

FeFET logic represents a cutting-edge technology that leverages the distinct attributes of ferroelectric materials to merge logic and non-volatile memory (NVM) capabilities within a single transistor device.

4.8.2 Differences Between NCFETS and FeFETs

Although they appear physically alike, NCFETs and FeFETs function on different underlying principles and fulfill separate roles. NCFETs are intended for logic functions to reduce power usage in processors, while FeFETs are meant for NVM to retain data without requiring power. Consider NCFETs as efficient logic transistors that switch more proficiently, whereas FeFETs serve as memory devices equipped with a switch.

NCFETs usually incorporate a thinner ferroelectric layer (commonly HZO) and need exact "capacitance matching" with the underlying semiconductor to avoid hysteresis. In contrast, FeFETs typically utilize a thicker ferroelectric layer to ensure stable, non-volatile polarization over time. NCFETs aim for zero or nearly zero hysteresis, employing a "stabilized" negative capacitance to amplify gate voltage, which enables the transistor to switch more rapidly and at lower voltage levels. FeFETs, on the other hand, depend on hysteretic behavior; the "memory window" (the gap within the hysteresis loop) allows differentiation between binary "0" and "1" based on the polarization state of the ferroelectric layer.

NCFETs work toward achieving a Subthreshold Swing (SS) of less than 60 mV/dec, which is a fundamental limit for conventional silicon transistors. In contrast, FeFETs prioritize endurance (the number of times they can be written) and retention (the duration they can maintain data).

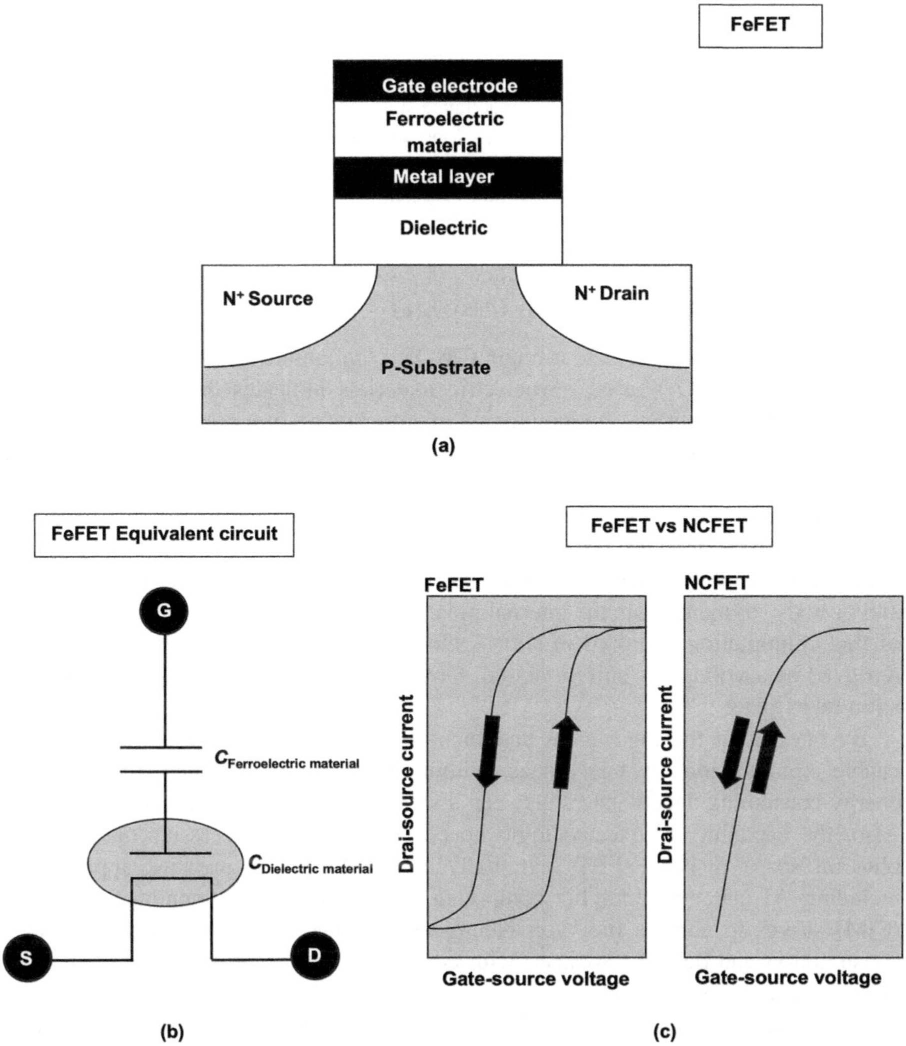

FIGURE 4.12 FeFET and FeFET vs. NCFET. (a) FeFET structure. An FeFET closely resembles a MOSFET in cross-section, sharing the same basic structure of a source, drain, and P-substrate. However, the FeFET includes an additional ferroelectric (FE) layer in the gate stack, usually positioned between the gate metal and the dielectric layer. Hafnium Zirconium Oxide (HZO) is a common ferroelectric material. An extra metal layer is also present. (b) The equivalent circuit of an FeFET. It primarily includes a nonlinear ferroelectric capacitor (C_{FE}) in series with a capacitor $C_{Dielectric}$, representing a MOSFET. C_{FE} models the polarization (P) of the ferroelectric layer. The polarization alters the surface potential and controls the MOSFET threshold voltage, leading to hysteretic $I_{DS} - V_{GS}$ characteristics. The series combination of capacitors is used for modeling memory-in-logic and non-volatile memory applications. (c) Transfer characteristic of FeFET and NCFET. The key difference between FeFET and NCFET transfer characteristics is their hysteresis behavior and subthreshold slope (SS) optimization. FeFETs display a counter-clockwise hysteresis loop, providing a large memory

(Continued)

FIGURE 4.12 (Continued) window for non-volatile memory applications, as the gate polarization persists after voltage removal. They generally follow standard thermionic limits. In contrast, NCFETs are designed for minimal or no hysteresis. Their objective is to stabilize the ferroelectric material in a negative capacitance state, enabling fast, reversible switching for logic applications. NCFETs aim for a steep-slope SS below 60 mV/dec at room temperature, achieved through passive voltage amplification, where the semiconductor surface potential increases more rapidly than the applied gate voltage.

4.8.3 Benefits, Use Cases, and Obstacles

The main benefit of FeFETs is their CMOS compatibility. The advancement of hafnium-oxide (HfO_2)-based ferroelectric materials indicates that FeFETs can be incorporated into current semiconductor manufacturing processes, enhancing their commercial feasibility.

The gate of the FeFET performs a dual function. It contains a ferroelectric material capable of being polarized to toggle the transistor between ON and OFF states while retaining this state even in the absence of power. Logic functions are executed by applying specific voltage pulses to the terminals (gate, source, drain, and/or body) to manipulate the internal polarization state of the FeFET. The outcome of this computation is stored in situ as the device's non-volatile state and can be retrieved as a voltage or current signal. Obviously, a logic-in-memory (LiM) operation takes place.

By integrating the logic gates and memory storage in close proximity, FeFETs enable computation to take place within the memory, thereby reducing the energy-consuming data transfers between distinct processor and memory components. Also, the capability to execute logic operations and store the results in the same physical space makes FeFETs extremely efficient for data-intensive applications, including AI and neuromorphic computing. Hence, LiM or computing-in-memory (CiM) strives to address the "von Neumann bottleneck" by alleviating the power consumption and delays associated with continuous data transfers between separate processing and memory units.

Non-volatility of FeFETs provides inherent data persistence. The polarization direction in the gate material affects the channel conductivity of the transistor, representing a memory state ("0" or "1"). This state is preserved without continuous power, unlike traditional CMOS logic. The data stored remains intact when power is lost, providing a significant benefit for low-power and always-on applications.

The ability to reconfigure FeFETs (reconfigurability) enables one device to carry out various functions, such as memory, logic, and frequency conversion. Certain FeFET configurations can be programmed to execute various logic functions, such as AND, OR, or XOR, using the same hardware by changing read voltages or input programming. The logic function of an individual FeFET-based circuit can be altered in real time by modifying the timing of the input signal or read voltage, enabling hardware reconfiguration without needing to redesign the physical circuit.

The steep subthreshold swing characteristics of FeFETs allow for lower voltage operations, resulting in decreased power consumption compared to standard CMOS circuits at lower voltages. Logic circuits based on FeFETs can achieve enhancements in performance metrics such as propagation delay and dynamic power when compared to other NVM-based logic solutions (e.g., those utilizing magnetic tunnel junctions).

Challenges to FeFETs encompass material-level issues such as variability in film thickness, controlling ferroelectric phases, and addressing the "first switch effect" (trap generation during initial operations), which can influence reliability and durability. The ferroelectric material may display a history effect, whereby its state is contingent on prior operations. This complicates precise programming and represents a focus of ongoing research aimed at mitigation. Current studies, including the exploration of 2D ferroelectric materials, seek to tackle these challenges.

Example 4.5 An N-type FeFET incorporates a ferroelectric film made of hafnium oxide (HfO_2), which has a thickness (t_{FE}) of 8 nm. The material exhibits a known coercive field (E_c) of 1 MV/cm (megavolts per centimeter). Determine the maximum memory window (MW_{Max}) of this FeFET in millivolts (mV).

Solution: We can use the following equation:

$$MW_{Max} = 2 \times E_c \times t_{FE} \tag{4.14}$$

where MW_{Max} signifies the maximum memory window (in Volts, V), E_c represents the coercive field of the ferroelectric material (in Volts per unit length, typically MV/cm or kV/cm), and t_{FE} indicates the thickness of the ferroelectric film (measured in the relevant length units, such as cm, μm, or nm). The calculated maximum memory window for the FeFET is

$$\begin{aligned} MW_{Max} &= 2 \times 1\ \text{MV/cm} \times 8\ \text{nm} = 2 \times 1 \times 10^6 \frac{\text{V}}{10^{-2}\ \text{m}} \times 8 \times 10^{-9}\ \text{m} \\ &= 2 \times 8 \times 10^{6+2-9} = 16 \times 10^{-1}\ \text{V} = 1.6\ \text{V} = 1.6 \times 1000\ \text{mV} = 1600\ \text{mV} \end{aligned} \tag{4.15}$$

This significant window (1600 mV or 1.6 V) facilitates the clear differentiation between the low-voltage state (logic 0 or OFF) and the high-voltage state (logic 1 or ON).

4.9 2D-MATERIAL FETs

Two-dimensional material field-effect transistors utilize two-dimensional materials such as MoS_2 and ReS_2, providing excellent electrostatic control and high carrier mobility, which enables the creation of high-performance, low-power devices (Robinson 2018, Liu et al. 2022, Rajput et al. 2024). 2D-material FETs represent a promising avenue for advancing high-performance, low-power logic circuits, potentially extending the semiconductor sector's roadmap beyond the capabilities of conventional silicon (Si). Their atomic thickness offers enhanced electrostatic control and resistance to short-channel effects, which are significant challenges when scaling down traditional transistors. The working of a 2D-material FET is readily understood from Figure 4.13.

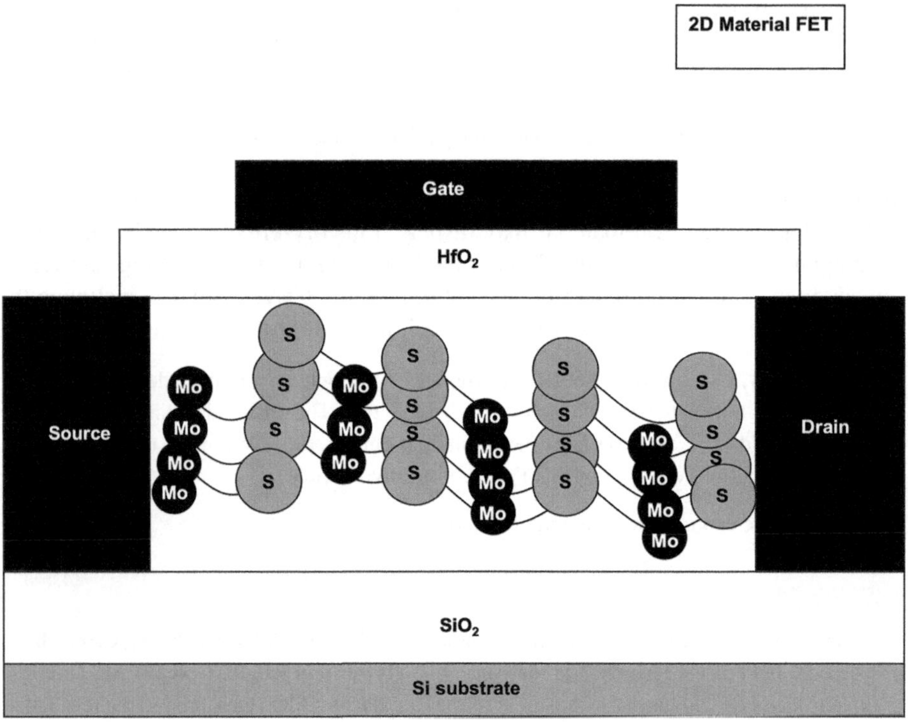

FIGURE 4.13 2D-material FET. A 2D-material field-effect transistor (2D FET) is a three-terminal device that uses an atomically thin layer, typically less than 1 nm, of a 2D semiconductor instead of a traditional bulk silicon channel. This ultra-thin design provides enhanced electrostatic control and reduces short-channel effects. The main components include the 2D channel, which is the active region for charge carrier flow and is often made from transition metal dichalcogenides (TMDs) such as MoS_2 or WSe_2 (the shown device has a MoS_2 semiconductor layer); source and drain electrodes, which are metallic contacts, e.g., gold, palladium, or scandium that inject and collect carriers at the channel ends; a gate electrode, which is a metallic or heavily doped silicon layer that applies voltage to modulate channel conductivity; and a gate dielectric, which is an insulating layer (e.g., SiO_2, HfO_2, or hBN) that separates the gate from the channel. In a back-gated FET, the 2D material is placed on a substrate such as Si/SiO_2, with the silicon serving as a global gate.

4.9.1 Advantages of FET Logic Using 2D Materials

2D-material FETs enable enhanced electrostatics. The atomically thin structure of 2D materials confines charge carriers effectively, enabling the gate electrode to have a stronger impact on the channel and minimizing unwanted OFF-state leakage currents (short-channel effects).

High carrier mobility is a major benefit of 2D materials. Materials like graphene and black phosphorus provide very high carrier mobility, which is vital for achieving the high ON-state currents needed for rapid switching speeds and superior performance in logic applications.

2D-material FETs provide reduced power consumption. The ability to surpass the 60 mV/dec subthreshold swing (SS) limit found in traditional FETs (often referred to as the "Boltzmann Tyranny") allows for a decrease in supply voltage (V_{DD}), resulting in significantly lower static power dissipation. Novel device designs such as Tunnel FETs (TFETs) or Negative Capacitance FETs (NCFETs) often facilitate this reduction.

FETs made from 2D materials can be easily integrated into 3D structures. The van der Waals (vdW) properties of 2D materials make it possible to stack them atop existing silicon circuits without the restrictions of traditional materials concerning lattice matching, paving the way for compact, monolithic 3D integration and 3D integrated circuits (ICs).

4.9.2 Varieties of 2D-Material Logic Devices

Researchers are investigating various device structures and materials to implement 2D logic. Traditional MOSFET-style devices use the conventional thermionic emission method for current flow. Efforts are focused on enhancing materials like MoS_2 and WSe_2 to obtain the required N-type and P-type characteristics for CMOS logic gates. Tunnel FETs (TFETs) leverage quantum-mechanical BTBT for carrier injection, enabling a sub-60 mV/dec SS. Candidates like WTe_2 and black phosphorus, which exhibit small, direct bandgaps and low effective masses, show promise for high-efficiency 2D TFETs. Negative Capacitance FETs (NCFETs) achieve sharp switching behavior beneath the 60 mV/dec threshold while retaining high drive current by adding a ferroelectric layer to the gate structure, NCFETs can achieve an internal voltage amplification effect, resulting Split-Gate FETs (SG-FETs) are based on designs utilizing a single 2D-material channel (for instance, ReS_2 or MoS_2) equipped with multiple gates to perform various logic functions (such as NAND, NOR, AND, OR) within a single transistor, thereby enhancing integration density.

4.9.3 Current Barriers to 2D-Material FETs

Despite noteworthy advancements, several obstacles persist for the commercialization of 2D FET logic. The production of materials and their quality are of utmost importance. Attaining wafer-scale, high-quality 2D materials with consistent properties remains a significant challenge. Apart from the quality of materials, creating low-resistance, stable electrical contacts between 3D metal electrodes and 2D materials without damaging the atomic layers is essential for optimal performance. Moreover, traditional ion implantation techniques can harm 2D materials, making it necessary to refine alternative doping methods, such as charge transfer or electrostatic induction. For integration with silicon, achieving complete compatibility between 2D-material production processes and existing silicon manufacturing methods (CMOS technology) is crucial for large-scale manufacturing.

Current research is actively tackling these challenges, with certain 2D multi-fin FETs already showcasing performance levels that compare favorably with or even surpass Intel's 14 nm node Si FinFETs in specific metrics, underscoring their potential for future logic applications. This technology has the potential to uphold Moore's Law scaling due to the channel material's extraordinary thinness, yet it contends with issues such as contact resistance, short-channel effects, and the integration of both P-type and N-type transistors (CMOS) for high-performing logic circuits. Research initiatives focus on improving the properties of materials, optimizing device design, and refining manufacturing techniques to address these challenges and develop functional, scalable logic chips that utilize 2D materials.

4.10 SET

The SET is a device that uses quantum-mechanical phenomena to control the flow of individual electrons (Sahu and Mazumdar 2024, Nasri et al. 2025). Its design differs markedly from that of a conventional FET, substituting the standard channel for a central nanostructure that functions as an "artificial atom." Figure 4.14 offers a vivid view of the construction of a SET, revealing its intricate design.

4.10.1 Core Structural Components

The fundamental configuration of a SET consists of a central quantum dot or a Coulomb island connected to three terminals: source, drain, and gate. Its functionality is based on quantum-mechanical tunneling and the Coulomb blockade effect, setting it apart from traditional CMOS transistors. Coulomb island (Quantum dot) is a tiny conductive region, typically a metallic nanoparticle, semiconductor nanodot, or a carbon-based structure (such as a nanotube or graphene scroll), with a diameter usually less than 10 nm. This island is sufficiently small that adding even a single electron dramatically raises its electrostatic energy. Tunnel junctions are two extremely thin insulating barriers (commonly oxide layers <10 nm) that separate the island from the source and drain electrodes. These barriers permit electrons to flow only via quantum-mechanical tunneling. Source and drain electrodes are conductive leads that act as the entry and exit points for electrons. The gate electrode is a terminal that is capacitively coupled to the island. It does not allow current to flow but uses an applied voltage to adjust the electrostatic potential of the island, thus facilitating or obstructing electron tunneling.

4.10.2 Advanced "Beyond-CMOS" Implementations

To address the constraints of conventional silicon, several advanced structural variations are being investigated:

Silicon nanowire SETs are made by employing etched silicon nanowires to form a consistent and reliable SET structure that aligns with existing CMOS fabrication techniques. Carbon allotrope islands utilize carbon nanotubes (CNTs), fullerenes (C_{60}), or graphene nanoscrolls as the central island to improve switching speed and decrease the range of the "Coulomb blockade." Vertical nanopillar is a three-dimensional design where a $Si/SiO_2/Si$ nanopillar contains silicon nanodots. This vertical structure enhances integration density for high-performance computing. Hybrid SET-CMOS structures merge SETs with standard CMOS circuits to develop systems that take advantage of the high sensitivity of SETs and the high gain/speed of CMOS.

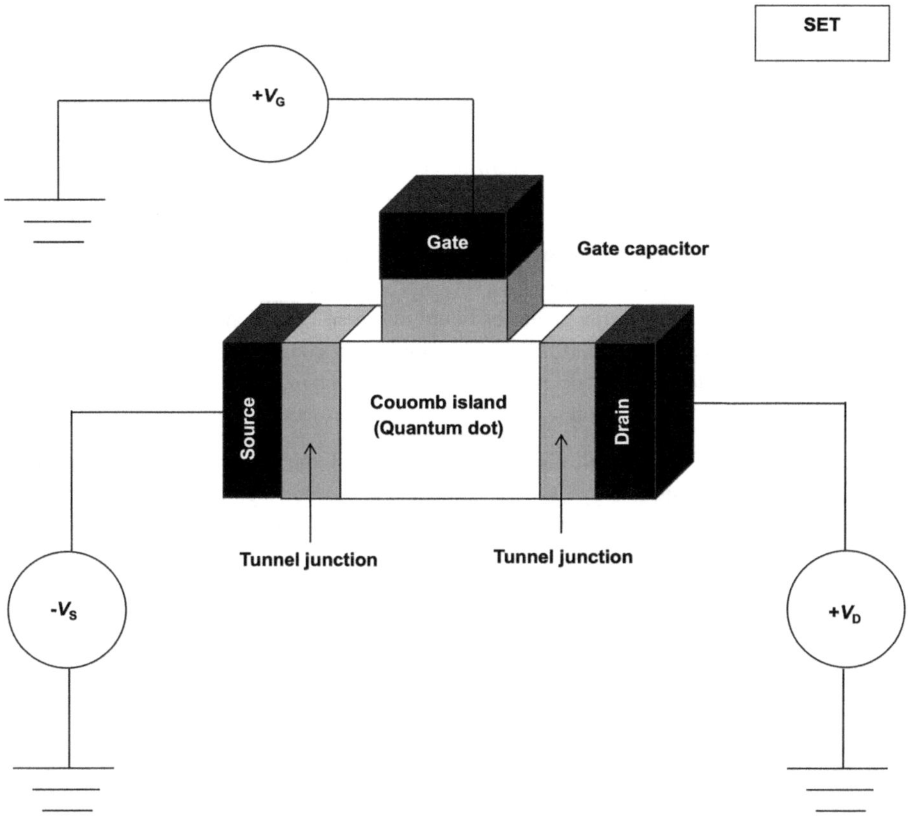

FIGURE 4.14 Architecture of a single-electron transistor. An SET resembles a standard field-effect transistor but operates through quantum tunneling across nanometer-scale junctions. Its core component, the Coulomb island (quantum dot), is a small conducting region, typically a metallic or semiconducting nanodot less than 100 nm, located between the source and drain. This island acts as the channel where electrons are trapped and released individually. The source and drain electrodes serve as the entry and exit points for electrons and are connected to the island by thin insulating barriers. Tunnel junctions, usually less than 10 nm thick layers of oxide, separate the island from the source and drain, allowing electrons to tunnel through by quantum mechanics. The gate electrode, capacitively coupled to the island, controls its electrostatic potential. The gate capacitor in a SET is a key nanoscale component that electrostatically links the gate electrode to the quantum dot. It allows precise tuning of the island's potential to control single-electron tunneling between the source and drain. By adjusting the gate voltage, we can block or permit electron tunneling. Typically, a positive voltage ($+V_G$) is applied to the gate, a negative voltage $(-V_S)$ to the source, and a positive voltage $(+ V_D)$ to the drain.

4.10.3 Operational Principles

The structure is tailored to support two primary phenomena: quantum tunneling in which electrons "leap" through the thin insulating barriers despite lacking the classical energy required to cross them; and Coulomb blockade, which takes place when the island's capacitance is exceedingly small so that the energy necessary to add electron is significantly greater than the thermal energy (kT), effectively preventing current flow until a specific gate voltage is achieved. SETs are regarded as a "beyond CMOS" technology due to their possibilities for ultra-low power usage, high sensitivity, and innovative applications such as quantum computing and densely packed memory devices. However, challenges remain in room-temperature operation and large-scale manufacturing integration. Hybrid circuits that combine CMOS and SETs offer a promising approach to harnessing the advantages of both technologies.

Example 4.6 A single-electron transistor (SET) has a total island capacitance (C_{Total}) of 1 aF (attofarad, equivalent to 1×10^{-18} F). The device is designed to function at a temperature of 77 K (the temperature of liquid nitrogen). The elementary charge of an electron (e) is roughly 1.6×10^{-19} C. The Boltzmann constant (k_{B}) is about 1.38×10^{-23} J/K. Is Coulomb blockade expected to occur effectively at this temperature? The required calculation must be presented to support your response.

Solution: For effective single-electron tunneling and the occurrence of Coulomb blockade, the charging energy (E_{C}) of the island should be significantly greater than the thermal energy ($k_{\mathrm{B}}T$) of the system. This condition is typically expressed as $E_{\mathrm{C}} \gg k_{\mathrm{B}}T$, or in practice $E_{\mathrm{C}} > 10k_{\mathrm{B}}T$.

The charging energy (E_{C}) is the energy required to add one electron to the island. It is calculated as

$$E_{\mathrm{C}} = \frac{e^2}{2C_{\mathrm{Total}}} = \frac{\left(1.6 \times 10^{-19}\ \mathrm{C}\right)^2}{2\left(1 \times 10^{-18}\ \mathrm{F}\right)} = \frac{2.56 \times 10^{-38}}{2 \times 10^{-18}}\ \mathrm{C^2/F} = 1.28 \times 10^{-20}\ \mathrm{J} \tag{4.16}$$

since

$$\frac{\mathrm{C}^2}{\mathrm{F}} = \frac{\mathrm{C}}{\mathrm{F}} \times \mathrm{C} = \frac{\mathrm{C}}{\frac{\mathrm{C}}{\mathrm{V}}} \times \mathrm{C} = \mathrm{VC} = \frac{\mathrm{J}}{\mathrm{C}} \times \mathrm{C} = \mathrm{J} \tag{4.17}$$

To express this energy in electron volts (eV) for comparison, we utilize the conversion $1\ \mathrm{eV} = 1.6 \times 10^{-19}$ J. Then

$$E_{\mathrm{C}} = \frac{1.28 \times 10^{-20}}{1.6 \times 10^{-19}}\ \mathrm{eV} = 0.08\ \mathrm{eV} \tag{4.18}$$

The average thermal energy is determined as

$$\begin{aligned} E_{\mathrm{Thermal}} &= k_{\mathrm{B}}T = 1.38 \times 10^{-23}\ \mathrm{J/K} \times 77\ \mathrm{K} = 106.26 \times 10^{-23}\ \mathrm{J} = \frac{106.26 \times 10^{-23}\ \mathrm{eV}}{1.6 \times 10^{-19}} \\ &= 66.4125 \times 10^{-4}\ \mathrm{eV} \approx 0.0066\ \mathrm{eV} \end{aligned} \tag{4.19}$$

Let us compare the two energy values: $E_{\mathrm{C}} \approx 0.08$ eV, and $k_{\mathrm{B}}T \approx 0.0066$ eV. Since E_{C} is significantly larger than $k_{\mathrm{B}}T$ (approximately 12 times greater), the energy barrier created by the charge of a single electron is sufficiently high to prevent random thermal fluctuations from surmounting it. Therefore, Coulomb blockade is expected to occur effectively at 77 K for this SET device.

4.11 NEM SWITCH

NEM switches, especially the innovative graphene nanoelectromechanical (GNEM) switches, are at the forefront of next-generation nanoelectronics, poised to break through the power-consumption barriers of conventional CMOS technology (Szkopek and Martel 2021, Zheng et al. 2024). Graphene's remarkable qualities—its atomic-scale thinness, minimal adhesion energy, and feather-light mass density—make it the ultimate material for ultra-low-voltage NEM switching. A single graphene monolayer, just 0.335 nm thick, stands as the thinnest mechanical element ever engineered for nanostructures. While graphene boasts a formidable Young's modulus of 1.0 TPa, its elastic stiffness of 390 N/m is lower than any other mechanical plate or membrane, ensuring that the energy needed to actuate an NEM switch is kept to a minimum.

The main component of each GNEM switch is a suspended graphene nanoribbon (GNR) that springs into action, making contact with an actuation electrode when an electric field is applied (Figure 4.15). These devices often feature a sophisticated bilayer or multilayer structure, crafted from CVD-grown graphene and hBN (hexagonal boron nitride), to improve their performance. By leveraging graphene-to-hBN contacts, GNEM switches dramatically reduce stiction, thereby extending their operational lifespan. GNEM switches can be seamlessly integrated with CMOS

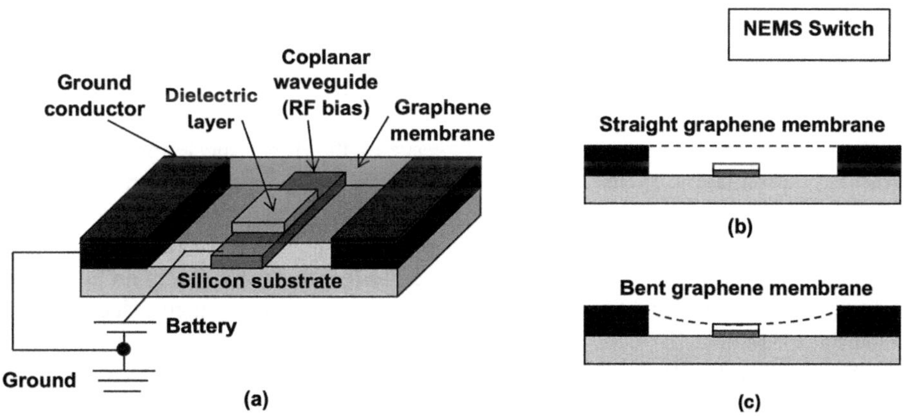

FIGURE 4.15 Graphene NEMS switch. (a) schematic cross-section, (b) OFF state of the switch, and (c) its ON state. (a) This switch features a pristine silicon base, a coplanar waveguide, a dielectric layer of hBN or HfO_2, and a graphene top electrode. The graphene film, serving as a slender suspended beam, hovers just a few nanometers above the dielectric, anchored firmly at both ends to the ground conductors. When a DC voltage is applied between the graphene beam and the central conductor, along with an RF signal, electrostatic forces come into play, pulling the beam into contact or releasing it. (b) In the OFF state, the graphene beam remains taut and elevated, breaking contact with the CPW and stopping current flow. (c) In the ON state, the graphene beam gracefully arches downward, drawn by the coplanar waveguide, and touches it, allowing current to pass through.

circuits, where NEMS deliver high-efficiency switching, and CMOS handles rapid operations. This synergy opens the door to a new era of non-volatile resistive switches, logic gates, and cross-bar memory devices.

4.12 DISCUSSION AND CONCLUSIONS

The semiconductor industry is advancing CMOS technology from the traditional FinFET structure to GAA designs to continue advancing Moore's Law towards complex 3D configurations. It is investigating fundamentally new device physics, such as 2D materials, spintronics, and neuromorphic systems, to address the substantial data requirements of AI and IoT. The FinFETs are three-dimensional, non-planar transistors, while VGAA (Vertical Gate-All-Around) FETs represent an evolution beyond FinFETs by enveloping the gate completely around a vertical channel, offering improved electrostatic control as chip dimensions progress to the sub-1 nm scales. Although FinFETs are still the primary technology for the 7 and 5 nm nodes, the emergence of GAAFETs within the 2 and 3 nm processes provides enhanced electrostatic control and improved power efficiency. Forksheet FETs and CFETs are promising devices. TFETs utilize quantum tunneling for switching, enabling a sharper subthreshold swing compared to conventional CMOS, making them well-suited for ultra-low-power applications. NCFETs incorporate ferroelectric materials in their gate stack to generate a "negative capacitance" effect, which serves as an internal voltage amplifier, facilitating faster and more efficient transistor switching. FeFETs utilize ferroelectric materials as a gate insulator to support non-volatile memory, enabling them to retain their state, either on or off, even when power is disconnected, which is advantageous for low-power memory applications. Materials such as molybdenum disulfide are being employed to develop ultra-scaled transistors.

Nanosheet FETs, a type of GAAFET, serve as the direct successors to FinFETs for mass production as the miniaturization of semiconductors progresses. Other technologies such as TFET, NCFET, FeFET, and SET mainly remain in the research phase, investigating alternative physics or memory integration for future specialized applications. Table 4.1 compares the main field-effect transistor (FET) architectures, which include innovative technologies currently in research stages, such as 2D-material FETs and SETs.

As we enter the "Beyond CMOS" era, scientists are pioneering new technologies that harness unconventional computational state variable tokens, aiming to overcome the limitations of conventional electronic charge. The transition from conventional electronic devices to spintronic logic devices indicates a shift from utilizing an electron's charge to harnessing its spin (intrinsic angular momentum) for information processing. This shift toward spintronics presents a significant advancement, allowing us to surpass the constraints of traditional charge-based transistors and utilize quantum mechanics to exceed the physical limitations of standard electronics, resulting in more powerful and energy-efficient computing. The following chapter explores spintronic logic devices.

TABLE 4.1
Comparison of FET Architectures and Other Switching Devices

Sl. No.	Technology	Architecture Type	Main Advantage	Present Status
1	FinFET	3D tri-gate with three-dimensional fin-shaped channel enclosed by a gate on three sides	High-drive current; established manufacturing	Dominant in 5nm/7nm; approaching scaling limits
2	GAAFET	3D: The channel is bordered by a gate on all four sides (nanosheet/ nanowire)	Complete gate control; 25%–35% power reduction compared to FinFET	In commercial production
3	Forksheet FET	Its standout feature is a dielectric wall separating the n-channel and p-channel devices	The dielectric wall neatly divides the NFET and PFET regions, eliminating the wide separation spacing required by photolithography in traditional nanosheets	Imec has already showcased this technology on 300mm wafers, proving its full electrical functionality. As research progresses, the focus has shifted from the original "inner wall" concept to an "outer wall" approach, addressing challenges such as mask alignment and preventing the dielectric wall from erosion during fabrication
4	CFET	CFETs stand out as the leading 3D transistor architecture poised to revolutionize CMOS nanoelectronics beyond the 2 nm node	Potentially shrinking the footprint of a CMOS inverter standard cell by about half	As these devices move through R&D and pilot testing, researchers are tackling formidable challenges, including mastering high-aspect-ratio etching, engineering a shared gate for stacked channels, and carefully controlling the thermal budget during sequential integration

(Continued)

TABLE 4.1 (Continued)

Sl. No.	Technology	Architecture Type	Main Advantage	Present Status
5	TFET	Tunneling based	In commercial production (Samsung 3 nm)	In research phase; potential for ultra-low-power IoT
6	NCFET	Negative capacitance	Reduced power usage through internal voltage gain	Prototypes; aims to improve current FinFET/GAA performance
7	FeFET	Ferroelectric gate	Combined non-volatile memory and logic functions	Niche applications; significant potential for AI/edge memory
8	2D FETs	Atomic layer (e.g., MoS_2)	Sub-1nm scaling; high mobility in very thin channels	Prototypes; likely to take over Si-based GAA FETs
9	SET	Single electron	Ultimate scaling; manipulation of individual electrons	Fundamental research; primarily focused on quantum computing
10	NEMS switch	Electrically actuated nano-relays	NEM switches outperform CMOS in demanding, leakage-sensitive, or extreme environments	Moving from lab prototypes to widespread adoption in foundries

REFERENCES

Beyer S., S. Dünkel, M. Trentzsch, J. Müller, A. Hellmich, D. Utess, et al. 2020 FeFET: A versatile CMOS compatible device with game-changing potential, *2020 IEEE International Memory Workshop (IMW)*, Dresden, Germany, 17–20 May, pp. 1–4, https://doi.org/10.1109/IMW48823.2020.9108150

Boukortt N. E. I. and D. Stroobandt 2026 Modeling and optimization of CFET structures for sub-3 nm CMOS scaling, *Micro and Nanostructures*, 210: 208506, https://doi.org/10.1016/j.micrna.2025.208506

Chen S., Y. An, S. Wang and H. Liu 2025 A review of tunnel field-effect transistors: Materials, structures, and applications, *Micromachines (Basel)*, 16(8): 881, https://doi.org/10.3390/mi16080881

Datta S., H. Liu and V. Narayanan 2014 Tunnel FET technology: A reliability perspective, *Microelectronics Reliability*, 54(5): 861–874, https://doi.org/10.1016/j.microrel.2014.02.002

Karimi K., A. Fardoost and M. Javanmard 2024 Comprehensive review of FinFET Technology: History, structure, challenges, innovations, and emerging sensing applications, *Micromachines (Basel)*, 15(10): 1187, https://doi.org/10.3390/mi15101187

Liao W., W. Qian, J. An, L. Liang, Z. Hu, J. Wang and L. Yu 2025 High-performance gate-all-around field effect transistors based on orderly arrays of catalytic Si nanowire channels, *Nano-Micro Letters*, 17: 154, https://doi.org/10.1007/s40820-025-01674-8

Liu L., H. Wang, Q. Wu, K. Wu, Y. Tian, H. Yang, C. M. Shen, L. Bao, Z. Qin and H.-J. Gao 2022 Ferroelectric-gated ReS_2 field-effect transistors for nonvolatile memory, *Nano Research*, 15: 5443–5449, https://doi.org/10.1007/s12274-022-4142-8

Maurya R. K. and B. Bhowmick 2022 Review of FinFET devices and perspective on circuit design challenges, *Silicon*, 14: 5783–5791, https://doi.org/10.1007/s12633-021-01366-z

Nasri, A., A. Boubaker and A. Kalboussi 2025 Principles design and applications of single electron transistors in memory devices logic gates neural networks and gas sensor, *Discover Electronics*, 2: 12. https://doi.org/10.1007/s44291-025-00053-2

Rahi S. B., S. Tayal and A. K. Upadhyay 2021 A review on emerging negative capacitance field effect transistor for low power electronics, *Microelectronics Journal*, 116: 105242, https://doi.org/10.1016/j.mejo.2021.105242

Rajput M., S. K. Mallick, S. Chatterjee, A. Shukla, S. Hwang, S. Sahoo, G. V. Pawan Kumar, and A. Rahman 2024 Defect-engineered monolayer MoS_2 with enhanced memristive and synaptic functionality for neuromorphic computing, *Communications Materials*, 5(1), https://doi.org/10.1038/s43246-024-00632-y

Robinson J. A. 2018 Perspective: 2D for beyond CMOS, *APL Materials*, 6, 058202, https://doi.org/10.1063/1.5022769

Ryckaert J., P. Schuddinck, P. Weckx, G. Bouche, B. Vincent, et al. 2018 The complementary FET (CFET) for CMOS scaling beyond N3, *2018 IEEE Symposium on VLSI Technology*, Honolulu, HI, 18–22 June, pp. 141–142, https://doi.org/10.1109/VLSIT.2018.8510618

Sahu S. K. and K. Mazumdar 2024 State-of-the-art analysis of single electron transistor, *2024 IEEE 3rd International Conference on Electrical Power and Energy Systems (ICEPES)*, Bhopal, India, 21–22 June, pp. 1–6, https://doi.org/10.1109/ICEPES60647.2024.10653610

Sarkar S. 2024 Understanding CFETs, a next generation transistor architecture. https://semiengineering.com/understanding-cfets-a-next-generation-transistor-architecture/

Srivastav S. and A. Acharya 2023 Challenges and future scope of gate-all-around (GAA) transistors: Physical insights of device-circuit interactions, In: Tayal S., B. Smaani, S. B. Rahi, S. Labiod and Z. Ramezani, *Device Circuit Co-design Issues in FETs*, CRC Press, Boca Raton, FL, 28.

Szkopek T. and E. Martel 2021 Suspended graphene electromechanical switches for energy efficient electronics, *Progress in Quantum Electronics*, 76: 100315, https://doi.org/10.1016/j.pquantelec.2020.100315

Tu L., X. Wang, J. Wang, X. Meng and J. Chu 2018 Ferroelectric negative capacitance field effect transistor, *Advanced Electronic Materials*, 4(11): 1800231, https://doi.org/10.1002/aelm.201800231

Verschueren L., G. Eneman, S. Yang, J. Bömmels, P. Matagne, K. B. Cahueñas, A. Sharma, D. Abdi, H. Mertens, G. Mirabelli and G. Hellings 2025 Extending the gate-all-around (GAA) era to the A10 Node: Outer wall forksheet enabling full channel strain and superior gate control, *2025 Symposium on VLSI Technology and Circuits (VLSI Technology and Circuits)*, Kyoto, Japan, 8–12 June, pp. 1–3, https://doi.org/10.23919/VLSITechnologyandCir65189.2025.11075021

Weckx P., J. Ryckaert, V. Putcha, A. De Keersgieter, J. Boemmels, P. Schuddinck, D. Jang, D. Yakimets, et al. 2017 Stacked nanosheet fork architecture for SRAM design and device co-optimization toward 3 nm, *2017 IEEE International Electron Devices Meeting (IEDM)*, San Francisco, CA, 2–6 December, pp. 20.5.1–20.5.4, https://doi.org/10.1109/IEDM.2017.8268430

Yu E., G. K. Kumar, U. Saxena and K. Roy 2024 Ferroelectric capacitors and field-effect transistors as in-memory computing elements for machine learning workloads, *Scientific Reports*, 14: 9426, https://doi.org/10.1038/s41598-024-59298-8

Zhang Q., Y. Zhang, Y. Luo and H. Yin 2024 New structure transistors for advanced technology node CMOS ICs, *National Science Review*, 11: nwae008.

Zheng L., J. Zhang, H. Wang and J. Liu 2024 RF NEMS switches based on graphene for low pull-in voltage and excellent RF performance, *Scientific Reports*, 14: 23346, https://doi.org/10.1038/s41598-024-74334-3

5 Spintronic Logic Devices

5.1 INTRODUCTION

Having explored charge-based logic devices in the foregoing chapter, we now turn our focus to the fascinating world of spin-based logic devices. Spintronic logic devices are an emerging area of nanoelectronics that leverage electrons' inherent spin, rather than solely their electrical charge, for information processing and storage. Information is represented by the spin orientation (e.g., spin-up for "1" and spin-down for "0") or by the magnetic alignment of different device layers. A current in which the majority of electrons possess a common spin direction is utilized to manipulate magnetic states. The "spin degree of freedom" enables the creation of hardware that is non-volatile, operates at high speeds, and is substantially more energy-efficient than conventional CMOS technology. Taking examples of an all-spin device, Figure 5.1 itemizes the defining features that set spintronic logic devices apart from the familiar charge-based devices.

Figure 5.2 draws a clear comparison between the spin-based and charge-based logic devices with reference to an all-spin device of the former category and a commonly used transistor belonging to the latter class.

5.2 MAGNETOELECTRIC SPIN-ORBIT (MESO) LOGIC DEVICES

The interaction between the spin and charge degrees of freedom is a highly intriguing field of study, both in basic science and practical applications, and it has the potential to transform the computing landscape (Husain et al. 2025). MESO devices leverage the magnetoelectric effect (ME) and spin-orbit coupling to perform logic and memory operations (Figure 5.3).

5.2.1 Magnetoelectric Effect

The ME is a phenomenon in which a magnetic field can influence the electrical characteristics of a material, or an electric field can influence its magnetic characteristics.

In the direct ME effect, a magnetic field (H) applied to the material produces an electric polarization (P) or voltage. In the converse ME effect, an applied electric field (E) induces magnetization (M) in the material. This interaction between electric and magnetic forces can be observed in certain single-phase materials called multiferroics, and it is typically stronger in artificial composites.

5.2.1.1 Multiferroic Materials

Multiferroic materials represent a category of "smart" substances that demonstrate at least two primary "ferroic" characteristics concurrently within the same phase (single-

DOI: 10.1201/9781003753469-5

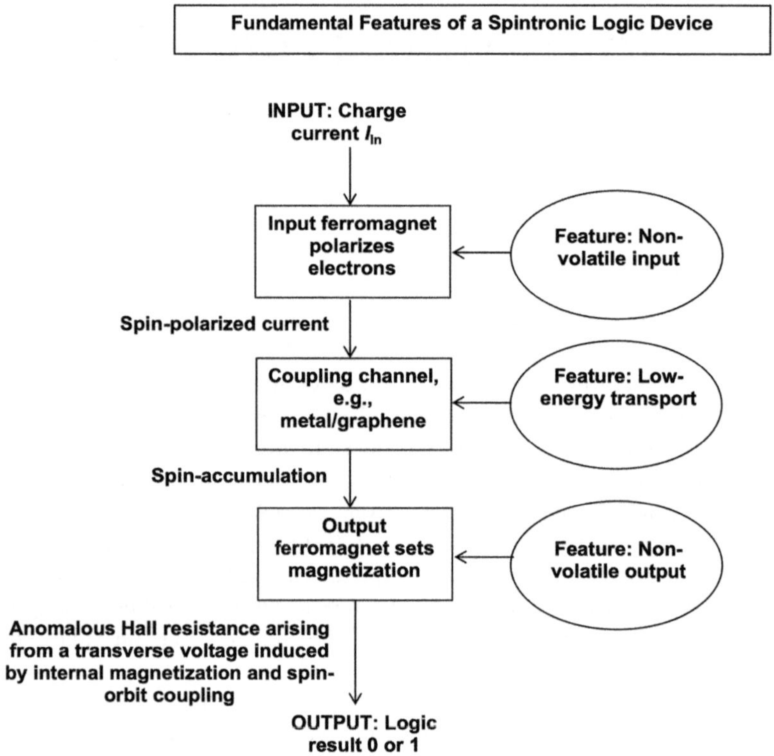

FIGURE 5.1 Diagram showcasing the salient features of spintronic logic devices, such as all-spin logic and spin-orbit torque devices, which harness electron spin rather than mere charge to deliver non-volatile, energy-efficient, and lightning-fast computation.

phase) or as a composite structure. The most extensively explored and technologically promising combination is the simultaneous presence of ferroelectricity (reversible electric polarization) and ferromagnetism/antiferromagnetism (reversible magnetization). Many of these multiferroic materials are complex oxide ceramics, such as bismuth ferrite ($BiFeO_3$), in which both bismuth and iron ions contribute to the ferroelectric and magnetic properties, respectively. Some natural examples include congalite and hnarite, but most practical applications rely on synthetic materials, such as thin films.

5.2.1.2 Spin-Orbit Coupling

Spin-orbit coupling (SOC) is a key relativistic interaction in quantum mechanics that describes how a particle's intrinsic spin angular momentum interacts with its orbital angular momentum. A physical way to grasp SOC is to consider an electron in an atom: it has an inherent magnetic moment due to its quantum spin (it behaves like a tiny magnet). As the electron moves in an orbit around the nucleus, from the electron's perspective, the positively charged nucleus seems to be revolving around it. This motion of the positive charge generates a local magnetic field. The coupling

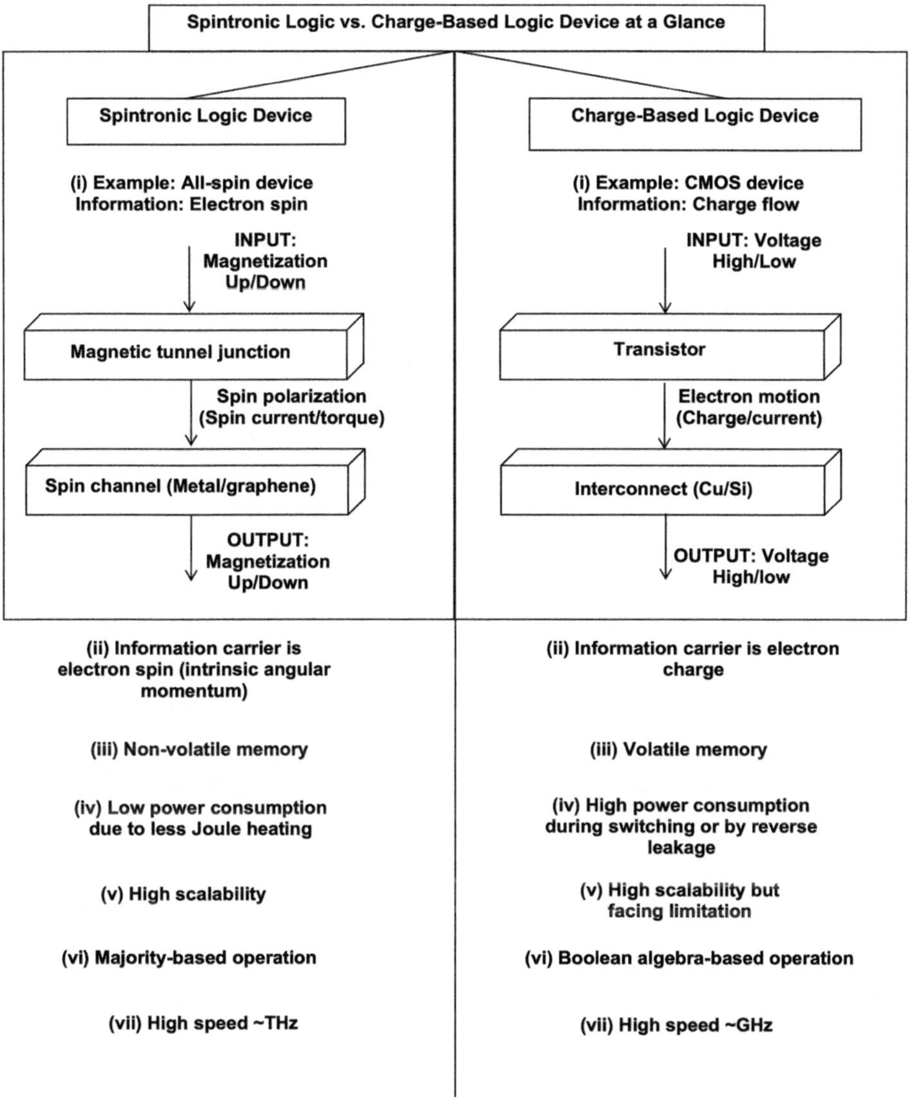

FIGURE 5.2 Comparative analysis of spin- and charge-based logic devices: Spin-based logic (spintronics) and charge-based logic (conventional CMOS) differ in how they represent, process, and store information. While spin-based logic utilizes the intrinsic angular momentum (spin) of electrons to define 0s and 1s, charge-based logic relies on the movement of electron charge. They are distinguished from the viewpoints of information carrier (electron spin/charge), memory retention after switching off power, energy consumption, scalability, operation (majority or Boolean), and speed.

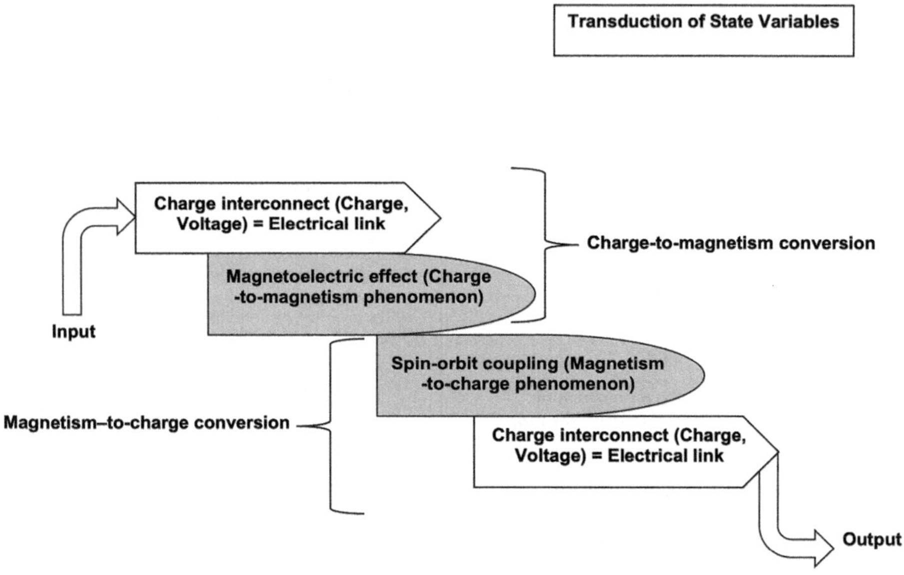

FIGURE 5.3 Two-step transduction of state variables in a MESO device. MESO devices transform charge into magnetism and back again. It all begins with a charge-based signal, like voltage or current, which energizes a magnetoelectric capacitor. This electrical signal flips the polarization of a multiferroic material such as $BiFeO_3$. The shift in polarization, through exchange coupling, nudges the magnetization of a neighboring ferromagnet, locking in a logic state as the nanomagnet's orientation. For output, the process reverses: a supply current flows through the ferromagnet into a spin-orbit material, like a topological insulator. Here, spin-to-charge conversion effects, such as the ISHE or the inverse Rashba-Edelstein effect, turn the spin-polarized current into a transverse charge current; the Rashba-Edelstein effect is a spintronic effect that unfolds at the surface, transforming spin accumulation into a flowing two-dimensional charge current. The direction of this current, positive or negative, signals the final logic output, ready to drive the next stage. Note that a charge interconnect, involving charge and voltage, is an electrical pathway made of a metal wire that transmits information between devices using electric current and voltage, rather than relying on spin current to move signals from one stage to another.

between the electron's intrinsic magnetic moment and the magnetic field generated by its orbital motion constitutes SOC. Depending on whether the inherent spin and orbital magnetic moments are aligned parallel or anti-parallel, the energy levels of the system are altered, which leads to the splitting of spectral lines, a phenomenon known as the fine structure in atomic spectra.

5.2.1.3 Spin Hall Effect, Inverse Spin Hall Effect, and the Interaction Between Them

The spin Hall effect (SHE) and its inverse, the ISHE, are interrelated phenomena that together transform charge and spin currents. SHE transforms a charge current into a perpendicular spin current, while ISHE turns a pure spin current into a transverse charge current or voltage. Collectively, they facilitate the manipulation and detection of spin currents in spintronic devices, effectively serving as converters between "spin-to-charge" and "charge-to-spin." In the SHE, a longitudinal charge current is introduced into a material exhibiting strong SOC. Electrons with differing spins are deflected in opposite directions, resulting in a transverse spin current and spin buildup at the material's edges. In the ISHE, a pure spin current is produced (for instance, via spin pumping or spin injection). As the spin current moves, the spin-orbit interaction deflects the electrons, leading to spin-dependent charge accumulation and generating a transverse charge current or voltage. The integration of SHE and ISHE enables a complete conversion cycle of reciprocal conversion. We begin with a charge current (SHE). This produces a spin current. We transfer this spin current to another layer, where ISHE converts it back into a charge current or voltage.

5.2.1.4 The MESO Logic Function

MESO devices operate using two primary elements that harness these effects to act as a logic gate, similar to a CMOS transistor. During input writing via the ME, an electric field (input voltage) is applied to a multiferroic material. The ME induces a change in the material's magnetization or alters its magnetic state. This establishes the device's logic state, e.g., "0" or "1," based on magnetic orientation. During output reading via SOC, current flows through an adjacent heavy metal layer with strong SOC to determine the logic state. The electron spins entering this layer are converted into a measurable charge current via the ISHE. The direction of the resulting charge-current output is determined by the magnetization direction (the stored logic state), supplying the output signal to the subsequent logic stage.

To understand the significance of the ME and SOC, we note that the ME mechanism is essential for input, facilitating the voltage control of magnetism. It enables highly energy-efficient switching of the magnetic state, requiring only a voltage (along with charge storage in a capacitor) rather than a sustained current, thereby representing a significant improvement over traditional spintronic devices. SOC plays a vital role in spin-charge interconversion (e.g., the SHE and its inverse). It serves as the "transducer" that transforms the information carried by electron spins (which hold the logic state) into a functional charge-based electrical signal, connecting the magnetic input component to the circuit's electrical output links. In deduction, the ME and SOC together create a low-voltage, energy-efficient computing environment by employing electric fields to write magnetic data and converting spins to read it, presenting a promising approach for electronics beyond the limitations of current CMOS technology.

5.2.1.5 MESO Device Structure

Figure 5.4 illustrates a MESO device that utilizes an ME material to convert an input voltage into magnetism and convert magnetism/spin into an output charge, based on the principles of electron spin-orbit (SO) physics (Liu et al. 2019).

The ME stack consists of a FM layer made from cobalt-iron (CoFe); a ferroelectric layer, typically a multiferroic material such as bismuth ferrite ($BiFeO_3$), which possesses both ferroelectric and antiferromagnetic (AFM) properties; and metal interconnects (Cu). An SO stack is composed of Ag/Cu and Bi/W layers. The layers of silver (Ag) and copper (Cu) serve to enhance the efficient conversion or transport of charge-to-spin currents within the device architecture. Bismuth exhibits strong SOC, which is critical for the functionality of spin-orbit logic devices.

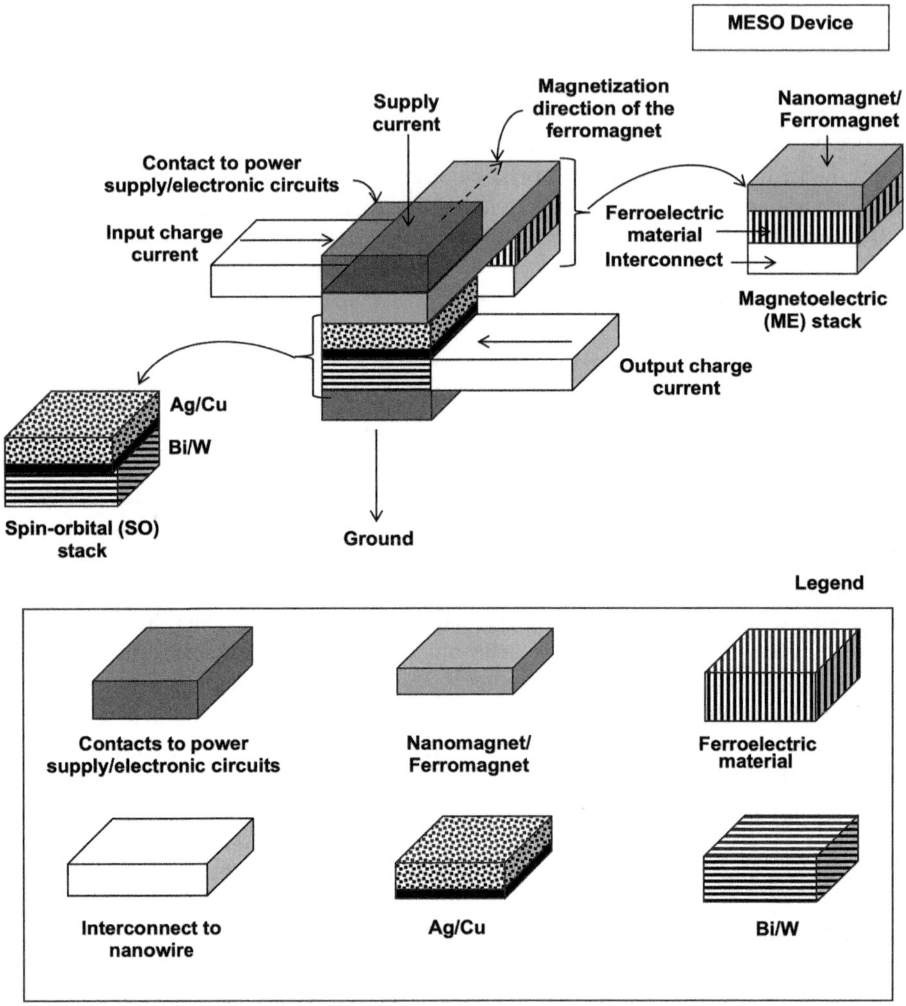

FIGURE 5.4 Structure of a MESO device. Top diagram shows the structural arrangement of the different layers constituting the device, while the bottom diagram explains the meanings of the different shading/patterns used in the top diagram for the various layers, e.g., contacts to power supplies (dark gray), nanomagnet/ferromagnet (light gray), ferroelectric material (vertical lines), interconnect to nanowire (white color filling), Ag/Cu (dotted/ black rectangle) and Bi/W films (horizontal lines). On the left-hand side of the top diagram, a spin-orbit (SO) stack is shown comprising Ag/Cu and Bi/W layers. On the right-hand side of the top diagram, a magnetoelectric (ME) stack is shown consisting of nanomagnet/ferromagnet, ferroelectric material, and interconnect layers. The middle diagram shows the composite structure of MESO device formed by combining the SO and ME stacks together. Input charge and output charge currents point in the inward direction. The direction of the supply current and magnetization direction of ferromagnet are indicated. The ground terminal is identified.

The tungsten layer serves as a heavy metal component used to generate or detect spin currents.

The FM material polarizes the injected charge current, resulting in a spin-polarized current. The transfer of the spin current to the SOC layer leads to the conversion of this spin information into a two-dimensional electron charge current, with its direction determined by the FM magnetization and the injected spin polarization, utilizing the inverse SOC (ISOC) effect at the interface of the SO stack to produce an output charge current. In this operation, the logical state of the incoming charge (current moving in the $+x$ direction) is reversed, resulting in the outgoing charge (current moving in the $-x$ direction). Energy for the gain is supplied from the power source (indicated by arrows). The dashed arrow indicates the magnetization direction of the ferromagnet. In contrast, full undashed arrows denote the electric currents at the input and output, as well as for the power supply and ground. The introduction of the power supply current enables energy gain, substantial signal gain, and the capacity to operate larger output devices (Manipatruni et al. 2019).

5.2.1.6 Advantages of MESO Logic Over CMOS

MESO circuits consume significantly less energy per switching event (approaching 1 aJ/bit), primarily due to voltage-driven ME switching. This results in extremely low power consumption. Moreover, every MESO device can execute logic functions while simultaneously serving as a non-volatile memory component, enabling ultra-low standby power consumption because the logic state persists without continuous power.

Outstanding scaling is attained because the energy required to switch a MESO device decreases cubically as device size is reduced. This scaling is better than what is possible in standard CMOS technology. High integration density is achieved because MESO devices offer greater logic density and can be arranged in a 3D structure. In addition, CMOS compatibility facilitates integration with current CMOS manufacturing processes and hybrid circuitry.

Example 5.1 A significant benefit of MESO logic is its capacity to execute intricate functions like a majority gate more effectively than CMOS technology. In standard CMOS, a five-input majority gate requires 62 transistors, whereas the same function can be implemented with just six MESO devices. Consider the following standard dimensions and performance parameters: Feature Size (F)—10 nm, CMOS Transistor Area—$4F^2$, and MESO Device Area—$4F^2$ (using a differential design for a fair comparison). The desired MESO Switching Energy is 1 aJ per switch at an operation voltage of 100 mV. Calculate the total area (in nm^2) for both configurations. Comment on the switching energies of the CMOS transistor vs. the meso device logic.

Solution: For CMOS Area: The area per transistor equals $4F^2$. It is

$$A_{\text{Per transistor}} = 4 \times (10\ \text{nm})^2 = 4 \times 100\ \text{nm}^2 = 400\ \text{nm}^2 \tag{5.1}$$

The overall CMOS Area is

$$A_{\text{Overall}} = 62\ \text{transistors} \times 400\frac{\text{nm}^2}{\text{transistor}} = 24{,}800\ \text{nm}^2 \tag{5.2}$$

For MESO Area: The area per device is

$$A_{\text{Per device}} = 4F^2 = 400 \text{ nm}^2 \tag{5.3}$$

Thus, the total MESO Area is calculated as

$$A_{\text{Total}} = 6 \text{ devices} \times 400 \text{ nm}^2/\text{device} = 2400 \text{ nm}^2 \tag{5.4}$$

The MESO approach provides 24,800/2400 = 10.33, i.e., approximately a tenfold reduction in area for this particular complex gate.

Now, let's evaluate the typical switching energies. Assuming that one operation entails a single switch event for each device/transistor:

Regarding MESO energy, the target energy for each switch is around 1 aJ, leading to a total energy for the five-input gate of roughly a few aJ (given that the devices are organized in stacks/pipelines). On the other hand, CMOS energy for complex gates tends to be considerably higher, frequently reaching the fJ range (femtojoules, 1000 aJ) or higher, depending on the technology node and voltage level. MESO devices can operate effectively at significantly lower voltages (approximately 100 mV) than advanced CMOS, which favorably reduces energy consumption $(E \propto V^2)$. Thus, MESO devices aim for switching energies in the attojoule (aJ) range, which are substantially lower than typical CMOS switching energies in the femtojoule (fJ) range, resulting in significantly enhanced energy efficiency for complex logic operations.

5.3 MAGNETIC TUNNEL JUNCTION LOGIC DEVICES

Magnetic tunnel junction (MTJ) logic devices employ voltage-controlled magnetic switching within MTJs for logic applications, offering low-power usage and non-volatility for both memory and logic functions. MTJs empower logic gates to shift their roles on the fly, e.g., seamlessly switching from AND to OR and performing multiple Boolean operations (NOT, AND, XOR). This polymorphic logic ability keeps attackers guessing, making it far tougher to uncover the circuit's real purpose, thereby shielding designs from reverse engineering (Kumar et al. 2025).

5.3.1 Structure of an MTJ Device

An MTJ logic device features a "sandwich structure" comprising two FM layers separated by a narrow insulating barrier, the tunnel layer (Joshi et al. 2020, Lv et al. 2024). Figure 5.5 shows the different layers constituting an MTJ device. One of these layers has a fixed magnetic direction. It is the pinned or reference layer (RL). The other layer can freely change its magnetic direction. This configuration creates two stable states: parallel (which indicates low resistance) and anti-parallel (indicating high resistance), corresponding to logic 0 and 1, respectively, depending on the orientation of the magnetic layers. Logic operations are achieved by altering the magnetic orientation of the FL, typically via mechanisms such as spin-transfer torque (STT) or spin-orbit torque (SOT), which switch the FL between states.

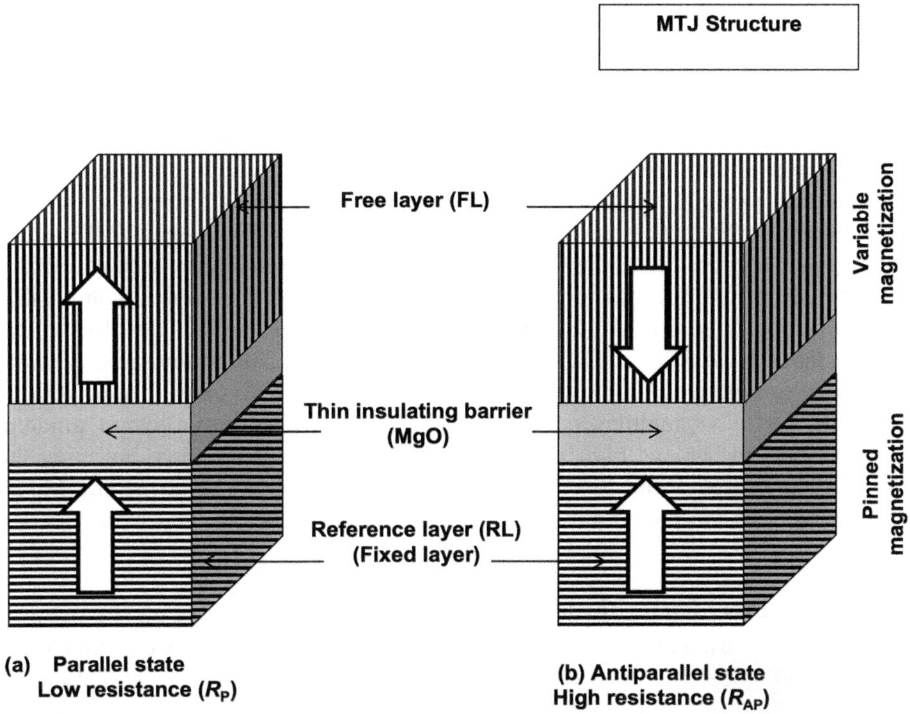

FIGURE 5.5 Triple layer MTJ device structure consisting of a free layer (FL), an insulating tunnel barrier (MgO) and a reference layer (RL); FL has a variable magnetization while that of RL is pinned: (a) low-resistance parallel state in which the magnetic orientation of free and reference layers are aligned in the same direction, and (b) high-resistance anti-parallel state when these orientations are aligned in opposite directions.

5.3.1.1 Pinned Layer

The pinned layer is usually enhanced in its magnetic strength through one of two prevalent methods. The first method is AFM exchange biasing. In this method, the RL is positioned next to an AFM layer (such as IrMn or PtMn). The intense exchange interaction at the interface between the FM RL and the AFM layer locks the RL's magnetization into a specific orientation.

The second method uses a synthetic antiferromagnet (SAF) structure. In this configuration, the RL is a composite design, typically composed of several FM layers separated by a thin non-magnetic spacer layer, commonly ruthenium (Ru). The interlayer coupling through the Ru spacer aligns the magnetizations of the various FM sub-layers in an anti-parallel configuration, creating a stable, flux-closed structure that is effectively pinned and resistant to external magnetic fields and thermal variations.

5.3.1.2 Free Layer

The FL of an MTJ generally consists of an FM material that allows easy modification (switching) of its magnetization direction by an external magnetic field or a spin-polarized current. One of the most widely used materials is the cobalt-iron-boron (CoFeB) alloy. This amorphous material is the leading choice for contemporary perpendicular MTJs (p-MTJs) due to their excellent spin polarization and robust perpendicular magnetic anisotropy, which is developed at the interface with an adjacent magnesium oxide (MgO) tunnel barrier upon annealing. Another material used is nickel-iron (NiFe, permalloy) alloy. Traditionally employed in in-plane MTJs, this material showcases low coercivity and minimal magnetocrystalline anisotropy. A synthetic free layer (SFL) or multi-interface free layer (MIFL) is a composite structure created to enhance performance characteristics such as thermal stability, switching speed, and energy efficiency. It typically comprises multiple thin FM layers divided by non-magnetic spacer layers such as W, Ta, Mo, or Ru. An instance is an SFL of FeCoB/W/FeCoB, which provides strong perpendicular magnetic anisotropy along with a low damping constant. Another method utilizes [Co/Pd] or [Co/Ni] multilayers, which may be incorporated into a hybrid structure with CoFeB to achieve high thermal stability in small dimensions. Selecting the appropriate material and layer configuration is essential for tailoring the MTJ's performance to specific applications, e.g., high-speed memory versus high-retention data storage.

5.3.1.3 Tunnel Barrier/Insulating Layer

The primary insulating materials used in MTJs are aluminum oxide (AlO_x) and magnesium oxide (MgO). Amorphous aluminum oxide (AlO_x) was the first barrier material widely adopted and continues to be used in specific applications. Although it offers decent tunneling magnetoresistance (TMR), its TMR ratio is typically lower than that of crystalline MgO barriers. Magnesium oxide (MgO) is currently the most commonly employed option, especially in devices such as MRAM, due to its capability to achieve a very high TMR ratio. This exceptional performance arises from coherent electron tunneling, enabled by MgO's crystalline structure when it is oriented in specific ways (e.g., the (001) orientation with CoFeB electrodes). Besides AlO_x and MgO, $SrTiO_3$ is used in select research applications, particularly with oxide-based magnetic electrodes such as $La_{0.7}Sr_{0.3}MnO_3$. Researchers are exploring atomically thin 2D Hexagonal boron nitride (hBN) layers for 2D-material-based MTJs. Zinc sulfide (ZnS) has been studied as a potential alternative barrier, offering a lower barrier height to achieve reduced resistance in high-density applications. Among high-entropy oxides, materials such as LiTiMgAlGaO are being investigated to achieve both high TMR ratios and low electrical resistance. The selection of the barrier material is crucial and is strongly influenced by the specific application requirements regarding TMR ratio, resistance, uniformity, and thermal stability.

5.3.1.4 Principle of Operation

The operating principle of an MTJ logic device relies on the tunnel magnetoresistance (TMR) effect, in which the device's electrical resistance varies with the magnetic alignment of its two FM layers (Figure 5.6). This variation in resistance is used to represent binary logic states "0" and "1." The fundamental operation encompasses two primary actions: writing (establishing the logic state) and reading (detecting the logic state).

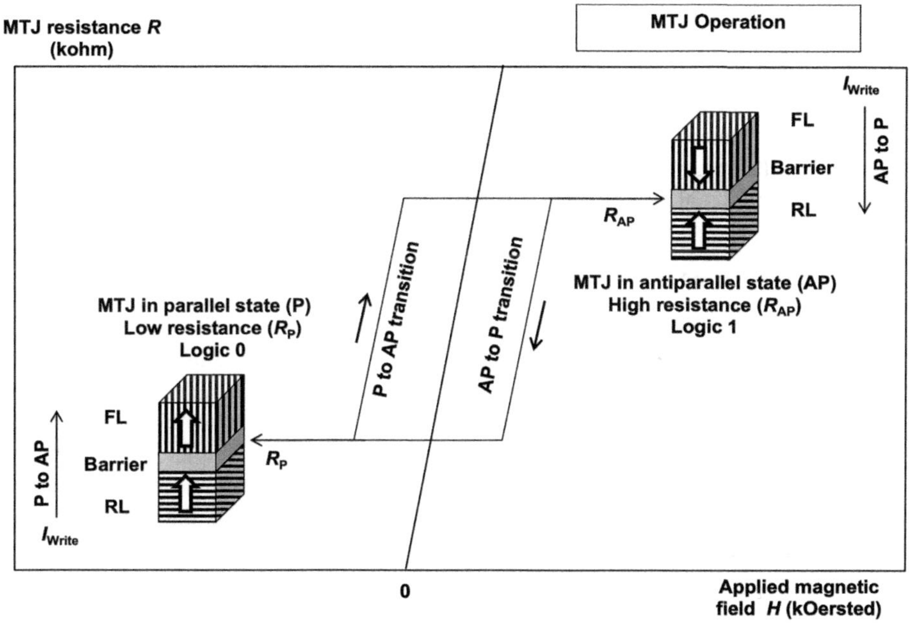

FIGURE 5.6 Operation of an MTJ logic device. To write data, spin-transfer torque (STT) applies a high-density spin-polarized current I_{Write} to set the free layer to either a parallel (P) or anti-parallel (AP) state. When I_{Write} flows from the RL to FL, the MTJ resistance changes from R_P to R_{AP}. Conversely, when I_{Write} flows from the FL to RL, the MTJ resistance changes from R_{AP} to R_P. During reading, a low-voltage current passes through the junction, and the resulting voltage or current is measured to identify the resistance state (R_P or R_{AP}).

During writing (establishing the logic state), the logic state is "written" by altering the magnetization direction of the FL in relation to the fixed layer. This is generally accomplished using current-based techniques such as STT or SOT, or through voltage-controlled magnetic anisotropy (VCMA). In the STT mechanism, a spin-polarized current flows through the MTJ stack itself. The angular momentum of the polarized electrons applies a torque on the FL's magnetization, which can switch its orientation if the current exceeds a critical level. The current direction determines the resulting magnetic alignment (parallel or anti-parallel). Regarding the SOT mechanism in a three-terminal device, an electric current passes through an adjacent heavy metal layer, producing a pure spin current via the SHE. This spin current then enters the FL, where it applies torque to switch its magnetization. SOT facilitates faster, more symmetrical switching while separating read/write pathways, thereby enhancing read stability.

The VCMA involves introducing a voltage across the junction, which modifies the energy barrier between magnetic states, enabling a low-power switch of the FL's orientation.

During reading (detecting the logic state), the logic state is "read" by assessing the electrical resistance of the MTJ. The parallel (P) state occurs when the magnetizations of the free and fixed layers are aligned; the MTJ shows low electrical resistance (R_P). This state typically signifies a logic "0." In an anti-parallel (AP) state, the magnetizations are oriented oppositely; the MTJ displays high electrical resistance (R_{AP}). This state usually represents a logic "1." The notable difference between R_P and R_{AP} (the TMR ratio) is detected by a peripheral circuit, such as a sense amplifier (SA) or a Wheatstone bridge, which converts the resistance variation into a standard digital voltage/current signal compatible with other logic circuits.

5.3.2 Types of MTJ Logic Implementation

Various methods are available, frequently utilizing hybrid MTJ/CMOS circuits. Implication logic (IMP) is a well-known "stateful" logic method that employs MTJs as both memory and logic components. Its operation relies on the material implication function "*p* IMP *q*," which translates to "NOT *p* OR *q*", allowing it to combine with other fundamental operations (like NOT) to create a complete logical basis. Hybrid CMOS/MTJ gates combine standard CMOS transistors and MTJs. In these logic gates, the MTJ components typically offer non-volatile data storage (one input), while the other input is sourced from the MOS logic tree, which is volatile. This is done by combining CMOS transistors to perform Boolean operations, e.g., AND, OR, NAND, NOR, and XOR. Logic functionality is achieved by designing circuits in which the combination of input signals and the MTJ's resistance state determines the final output. The majority logic technique involves configuring MTJs to implement the majority function (MAJ3 or MAJ5), a universal logic operation. In reconfigurable logic, a single MTJ device, often paired with additional circuits such as a multiplexer and an SA, can be dynamically configured to perform various Boolean operations by adjusting input voltages or magnetic fields, providing significant flexibility.

5.3.2.1 Advantages of MTJ Logic

Due to non-volatility, the logic state remains intact even when power is off, enabling "instant-on" systems and eliminating standby power consumption in the memory subsystem. Further, logic based on MTJs requires less energy compared to traditional CMOS, particularly by reducing leakage currents when powered down. It offers significant energy-efficiency benefits over conventional CMOS, particularly for Internet of Things (IoT) devices that require minimal power.

Additionally, MTJs can be stacked on top of CMOS logic layers, minimizing overall chip area and interconnect delays, enabling a high density.

5.3.2.2 MTJ/CMOS Logic-in Memory Structure

A hybrid MTJ/CMOS logic-in-memory (LIM) arrangement combines conventional CMOS elements with MTJ pairs to produce non-volatile logic circuits (Figure 5.7). The fundamental block diagram of a comprehensive hybrid MTJ/CMOS LIM design comprises three main sections: a logic network (LN), an MTJ writing block, and an SA. The LN/Pull-down Network (PDN) is a hybrid circuit that performs logical operations on incoming data. It merges CMOS logic trees (for volatile inputs) and MTJs (which retain non-volatile information). The current-controlled SA determines

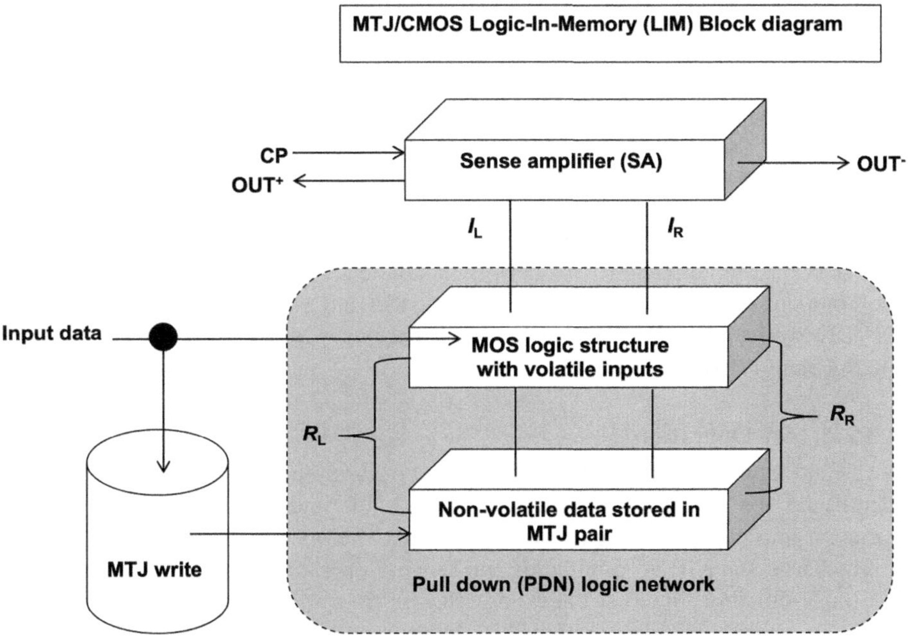

FIGURE 5.7 The generic LIM architecture offers a versatile blueprint for the development of hybrid CMOS/MTJ logic circuits. At its core, a MOS logic framework handles volatile inputs, while a paired MTJ performs non-volatile data storage. The dedicated MTJ writing circuit delivers the necessary current to reliably switch logic states.

the outcome of the PDN's logical operations. It generates the output and its complement based on variations in differential current flow. The MTJ switching/writing block component is responsible for switching the MTJs (from parallel to anti-parallel and vice versa) using STT when new data needs to be recorded or modified.

5.3.2.3 Primary Steps in Power Distribution Network Operation

The CMOS/MTJ PDN operates by interacting between fluctuating electrical inputs and persistent magnetic data, functioning in two stages: pre-charge and evaluation. The power distribution network (PDN) in this hybrid setup is typically configured to provide a path for current to flow toward ground (GND), pulling the output node to a logical low (0). By incorporating MTJs, this system enables the direct use of non-volatile memory in logic operations, reducing power consumption and minimizing data movement, as is typically seen in conventional computing architectures. The operating principle includes these primary steps:

During the pre-charge phase, the clock signal (CP) is generally low, and the output nodes (usually complementary, OUT+ and OUT−) are pulled up to a high voltage (V_{DD} or $V_{DD}/2$) by the pull-up network (PUN). Inputs to the PDN are applied in this phase. In the evaluation phase, the clock signal rises, activating the PDN to execute the logic operation based on the inputs from the volatile CMOS transistors and the non-volatile data stored in the MTJs.

The PDN comprises NMOS transistors and a pair of MTJs, which store a bit of information in either the high-resistance anti-parallel (R_{AP}) or the low-resistance parallel (R_P) state. Based on the inputs and the MTJ resistance state, one of the differential output paths (the left or right arm of the PDN) will discharge to ground more rapidly than the other. The variance in current flow (I_L and I_R) results in a voltage difference across the output nodes.

A SA, functioning as a current-controlled amplifier, identifies the difference in resistance/current between the two PDN arms. The SA converts this analog disparity into a stable digital output (0 or 1). The efficiency of the hybrid design lies in the compatibility between the MTJ's resistance range and the CMOS transistors' ON/OFF resistance, ensuring reliable performance without the need for complex intermediate conversion circuits.

5.3.2.4 SA Operation

A current-controlled SA in a CMOS/MTJ hybrid setup acts as a current comparator, identifying the resistance state of a pair of MTJs and translating this resistance variation into a standard digital voltage output. The operation generally starts from a pre-charge stage: Before initiating the sensing operation, the output nodes of the SA (and sometimes internal nodes) are charged to a specific voltage level, such as V_{DD} (the supply voltage). During this stage, input signals are applied to the pull-down logic network (PDN), which comprises CMOS transistors and a pair of MTJs. The MTJ pair retains a single bit of non-volatile data, with one MTJ in a low-resistance parallel (P) state and the other in a high-resistance anti-parallel (AP) state. In the succeeding evaluation (sensing) stage, the SA is activated by a control signal (commonly referred to as "SEN" or "CP") and starts the sensing procedure.

Due to the different resistances of the two MTJs in the pair, current flows through the left (I_L) and right (I_R) branches of the PDN at different rates, resulting in differential current flow. The SA, which serves as a current-mode or differential-voltage amplifier, enhances the slight difference between I_L and I_R (or the corresponding voltage difference). After voltage/current difference amplification, positive feedback within the amplifier quickly drives one output node to a logic "1" (V_{DD}) and the other to logic "0" (ground), based on which current is greater. This process converts the analog-domain resistance information from the MTJs into a stable, standard digital signal. The digital output is usable by subsequent CMOS logic stages. This functionality enables non-volatile data stored in MTJs to be read and effectively integrated with high-speed, low-power CMOS logic circuits, a fundamental principle of hybrid memory and computing architectures.

Example 5.2 Consider a straightforward hybrid inverter circuit that combines MTJ and CMOS, where an MTJ serves as a load resistor for an NMOS transistor. The MTJ can exist in either a low-resistance state (R_P) or a high-resistance state (R_{AP}). Given values include: Supply voltage $V_{DD} = 1$ V, low-resistance state $R_P = 5$ kΩ, high-resistance state $R_{AP} = 15$ kΩ, NMOS transistor resistance when activated $R_{ON} = 1$ kΩ, and NMOS transistor resistance when deactivated $R_{OFF} = \infty$ (theoretical). A logic "0" input deactivates the NMOS transistor, while a logic "1" input activates it. For this operation, the MTJ remains in a fixed state (e.g., R_{AP} to act as an inverter). Determine the inverter's output voltage (V_{OUT}) for both logic "0" and "1" inputs.

Solution: The circuit functions as a simple voltage divider when the NMOS is activated. When the NMOS is deactivated, the MTJ pulls the output voltage high.
Case 1: Input = Logic "0." The NMOS transistor is deactivated, meaning its resistance (R_{OFF}) is infinite (or significantly high). No current flows through the circuit. The output node is raised to the supply voltage via the load of the MTJ. Thus,

$$V_{OUT} = V_{DD} = 1 \text{ V } (\text{Logic}'1') \tag{5.5}$$

Case 2: Input = Logic "1." The NMOS transistor is activated, which gives it a resistance of $R_{ON} = 1$ kΩ. The circuit then functions as a voltage divider between the MTJ ($R_{AP} = 15$ kΩ) and the NMOS (R_{ON}). The output voltage can be calculated using the voltage divider principle:

$$V_{OUT} = V_{DD} \times \frac{R_{ON}}{R_{AP} + R_{ON}} \tag{5.6}$$

This leads to

$$V_{OUT} = 1 \text{ V} \times \frac{1 \text{ k}\Omega}{15 + 1 \text{ k}\Omega} = \frac{1}{16} \approx 0.0625 \text{ V}(\text{Logic}'0') \tag{5.7}$$

This scenario illustrates how the MTJ's resistance characteristics (specifically R_{AP}) enable it to serve as an effective load for CMOS logic, facilitating non-volatile logic-in-memory architectures.

5.4 SPIN HALL EFFECT (SHE) LOGIC

The SHE logic is a category of spintronic devices that utilize the SHE to electrically generate and manage electron spins for computing and data storage, providing a prospective substitute for traditional CMOS technology in low-power, non-volatile LIM systems (Wan et al. 2017, Zhang et al. 2017).

5.4.1 Fundamental Concepts and Operational Mechanisms

The SHE is a transport phenomenon in which an electric charge current traveling through a material with significant SOC produces a transverse "pure" spin current (a movement of spin angular momentum without any net charge movement). Spins of opposite polarization accumulate on different lateral surfaces of the material (Sinova et al. 2015). The generated spin current can be injected into a neighboring FM layer, where the transfer of spin angular momentum exerts a torque, the SOT, on the local magnetization. This torque enables efficient alteration of the magnetization direction in the FM layer, serving as the memory element (logic states 0 or 1).

When an electric charge current flows through a material with substantial SOC, the SHE directs spin-up electrons to one side and spin-down electrons to the opposite side, converting a charge current into a spin current (Figure 5.8). The spin accumulation process generates a spin current, leading to the accumulation of spins with opposite polarization at the material's perpendicular edges. Spin manipulation takes

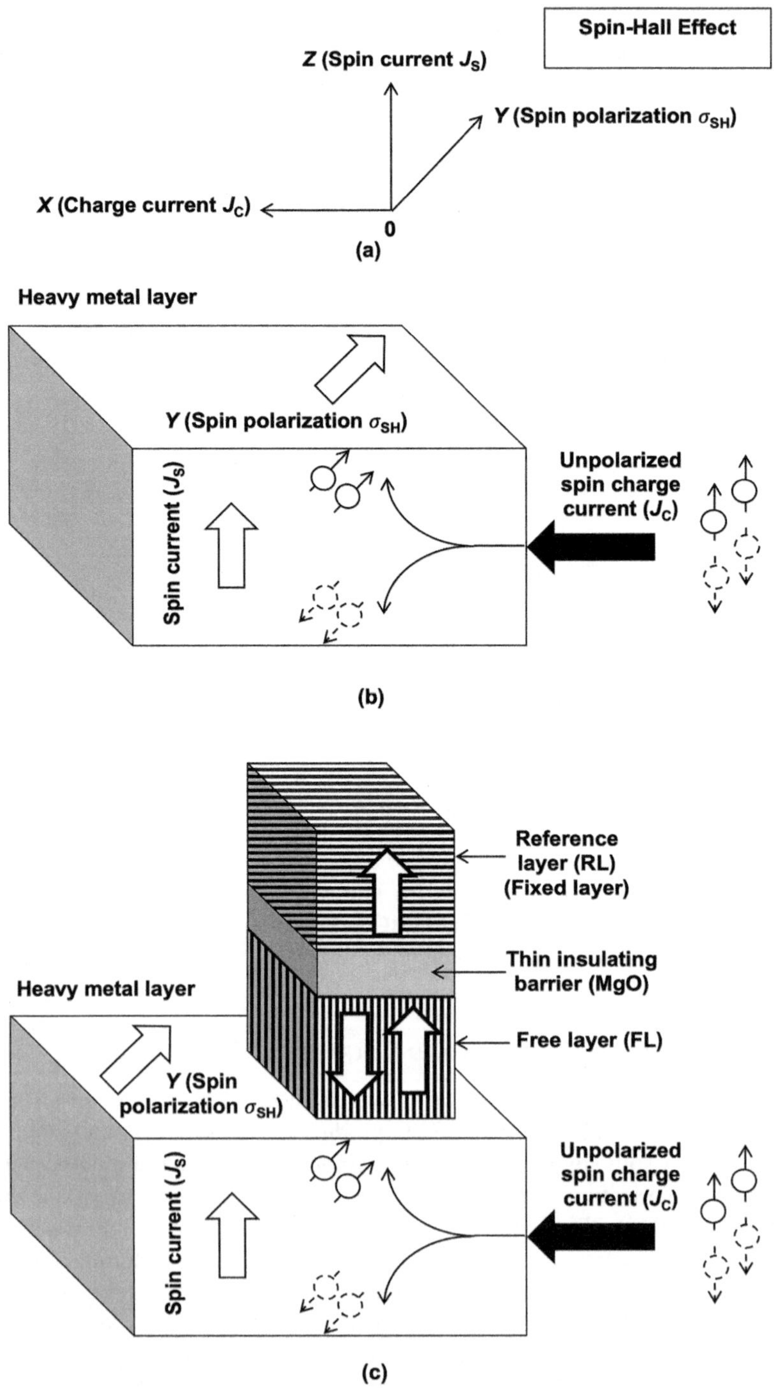

FIGURE 5.8 SHE and SHE-assisted switching of MTJ: (a) The charge-current-induced spin Hall effect (SHE). It occurs when a charge current J_C flowing through the heavy metal bar

(Continued)

FIGURE 5.8 (Continued) along the x direction causes up spins polarized along the y-axis and down spins polarized along the $-y$-axis to accumulate toward the $+z$- and $-z$-axes, respectively, generating a spin current J_s in the y-direction with polarization parallel to the z-axis. The charge current J_C and the electron spin polarization direction ($\pm y$-axis) are always perpendicular to each other. (b) SHE/SOT switching mechanism. Switching an MTJ with the spin Hall effect (SHE) uses a three-terminal device that separates read and write current paths. The free layer (FL) is in contact with a heavy metal layer. In this approach, a charge current passes through a heavy metal (e.g., Ta, Pt, W) layer, generating a pure spin current via spin-orbit torque (SOT). The MTJ stack is placed on a heavy metal strip, known as the SOT channel. When a write current (J_c) is applied along the channel, strong spin-orbit coupling in the heavy metal deflects electrons with opposite spins in different directions. This produces a pure spin current (J_s) that flows vertically into the magnetic free layer. The resulting spin injection applies a spin-orbit torque to the free layer's magnetization, enabling switching. By keeping the high-density write current away from the delicate MgO tunnel barrier, electrical stress drops, and the device enjoys a longer lifespan. This separation also lets engineers fine-tune the read and write channels for both speed and reliability.

place. The spin current applies an SOT on an adjacent magnetic layer, flipping its magnetization. The state of this magnetic layer can represent a logical "1" or "0."

5.4.2 Benefits and Uses

Unlike traditional volatile logic, which requires a continuous power supply to retain its state, SHE logic devices that use MTJs are non-volatile, meaning they retain their data (magnetization state) when power is cut off (Barla et al. 2021). Further, numerous SHE-driven devices, such as three-terminal MTJs, have distinct write (charge current in the heavy metal layer) and read (voltage across the MTJ) pathways. The physical separation due to decoupled read/write paths enhances reliability and allows for the direct serial cascading of logic gates, addressing a significant challenge faced by earlier spintronic designs grounded solely in STT. Besides non-volatility and path decoupling, devices based on SHE also emphasize energy efficiency, with some designs achieving considerable reductions in power consumption compared to STT-based logic. High speed is another advantage. SHE devices have shown rapid switching times, with current pulses as brief as 1 nanosecond capable of reversing magnetization. A key benefit is the CMOS compatibility, the ability to integrate with current CMOS technology, which could alleviate the "memory wall" challenge in contemporary computing by merging memory and logic functionalities on a single chip. Researchers have successfully demonstrated the realization of various Boolean logic gates (AND, OR, NOT, NAND, NOR) within a single SHE-based device by altering input current combinations, enabling programmable "logic-in-memory." SHE is also being investigated for nontraditional computing models, such as Ising model computation, to tackle complex optimization problems. Computation related to the Ising model generally entails simulating a grid of spins ($s_i = \pm 1$) to identify the configuration that lowers the overall energy (the Hamiltonian) or to calculate statistical averages such as magnetization and susceptibility.

5.4.3 Materials and Challenges

Heavy metals known for strong SOC, such as platinum (Pt), tantalum (Ta), and tungsten (W), are frequently used as spin Hall materials. The essential performance metric is the "spin Hall angle," which quantifies the efficiency of charge-to-spin current conversion.

Challenges include accurately measuring material-specific parameters (spin Hall angle, spin diffusion length) and distinguishing between intrinsic (material band structure) and extrinsic (impurity scattering) mechanisms of the effect. Theoretical models are continually improved to provide a comprehensive understanding and precise predictions of device performance.

Example 5.3 A device exhibiting the Spin Hall Effect (SHE) employs a non-magnetic material with a spin Hall angle θ_{SH} of 0.1 and a longitudinal conductivity σ_{xx} of $1 \times 10^5\ (\Omega\ \mathrm{cm})^{-1}$. Calculate the spin Hall conductivity σ_{SH} of this material. For a charge-current density of 5×10^4 A/m^2, what is the transverse spin current density?

Solution: We need to understand the fundamental connections between charge current, spin current, spin Hall angle, and conductivities. To determine the spin Hall conductivity σ_{SH}, we can use the spin Hall angle, which is defined as the ratio between spin Hall conductivity and longitudinal conductivity:

$$\theta_{SH} = \frac{\sigma_{SH}}{\sigma_{xx}} \tag{5.8}$$

By rearranging this equation, we can isolate σ_{SH}:

$$\sigma_{SH} = \theta_{SH} \times \sigma_{xx} \tag{5.9}$$

Given that $\theta_{SH} = 0.1$ (dimensionless) and $\sigma_{xx} = 1 \times 10^5\ (\Omega\ \mathrm{cm})^{-1}$, we find:

$$\sigma_{SH} = 0.1 \times 1 \times 10^5\ (\Omega\ \mathrm{cm})^{-1} = 1 \times 10^4\ (\Omega\ \mathrm{cm})^{-1} \tag{5.10}$$

Thus, the calculated spin Hall conductivity is $1 \times 10^4\ (\Omega\ \mathrm{cm})^{-1}$. In the context of spintronics, it is frequently represented as $\frac{\hbar}{2e}\sigma_{SH}$. Next, we calculate the transverse spin current density (J_s). The spin Hall angle also provides insights into the efficiency of converting charge-current density (J_c) into transverse spin current density (J_s):

$$\theta_{SH} = \frac{J_s}{J_c} \tag{5.11}$$

Rearranging this equation enables us to solve for J_s:

$$J_s = \theta_{SH} J_c \tag{5.12}$$

Using $\theta_{SH} = 0.1$ and $J_c = 5 \times 10^4\ \mathrm{A/cm^2}$, we calculate:

$$J_s = 0.1 \times 5 \times 10^4 \mathrm{A/cm^2} = 5 \times 10^3 \mathrm{A/cm^2} \tag{5.13}$$

Therefore, the resulting transverse spin current density is 5×10^3 A/cm^2. It is important to note that J_s is expressed in units of Amperes per area, indicating the movement of spin angular momentum (which is often depicted with a factor of $\frac{\hbar}{2e}$) in the case of a pure spin current.

5.5 NANOMAGNETIC LOGIC

The Nanomagnetic Logic (NML) method harnesses nanomagnets to represent and process data, showing potential for dense integration (Gypens et al. 2021, Misra and Bhoi 2023).

5.5.1 Elementary Concept

The fundamental concept of NML is to use the stable magnetic orientation of bistable nanoscale magnets to convey binary information ("0" and "1") and to process it through magnetic dipole-dipole interactions among adjacent magnets, thereby simulating logic gates (Csaba and Becherer 2020). Data is encoded by the magnetization direction (e.g., up or down), and logic operations are executed by altering these states using a global magnetic clocking field in conjunction with the magnets' magnetic fields. Information travels through the circuit via the magnetostatic (dipole) fields generated by neighboring components, eliminating the need for an electric current to pass through the logic elements.

5.5.2 Essential Principles

Several essential principles guide this field-coupled computing framework. Binary representation is used. Each single-domain nanomagnet possesses two stable magnetization orientations (e.g., left or right, or up or down), which correspond to the binary logic states "0" and "1." Field-coupled computing is done. The magnets are configured in specific geometric arrangements (such as chains or arrays) so that the magnetic field from one magnet affects the magnetization direction of its neighbors. This relationship enables information transfer without electric current flowing between logic elements, potentially reducing power consumption (Figure 5.9).

Various logic functions (such as majority gates that can be adapted to form AND, OR, NAND, NOR gates) are realized through the physical arrangement and interactions of the nanomagnets. The output magnet tends to settle into a low-energy state defined by the layout of the inputs. An external clocking signal is crucial for synchronizing the information flow and guaranteeing deterministic, directional signal transmission throughout the circuit. This clocking mechanism typically involves applying an external field (whether magnetic, electric, through the magneto-elastic effect, or STT) that briefly pushes the magnets into an unstable intermediate "reset" state, lowering the energy barrier. Upon removing the clock field, the magnets revert to their appropriate stable states, determined by their neighbors.

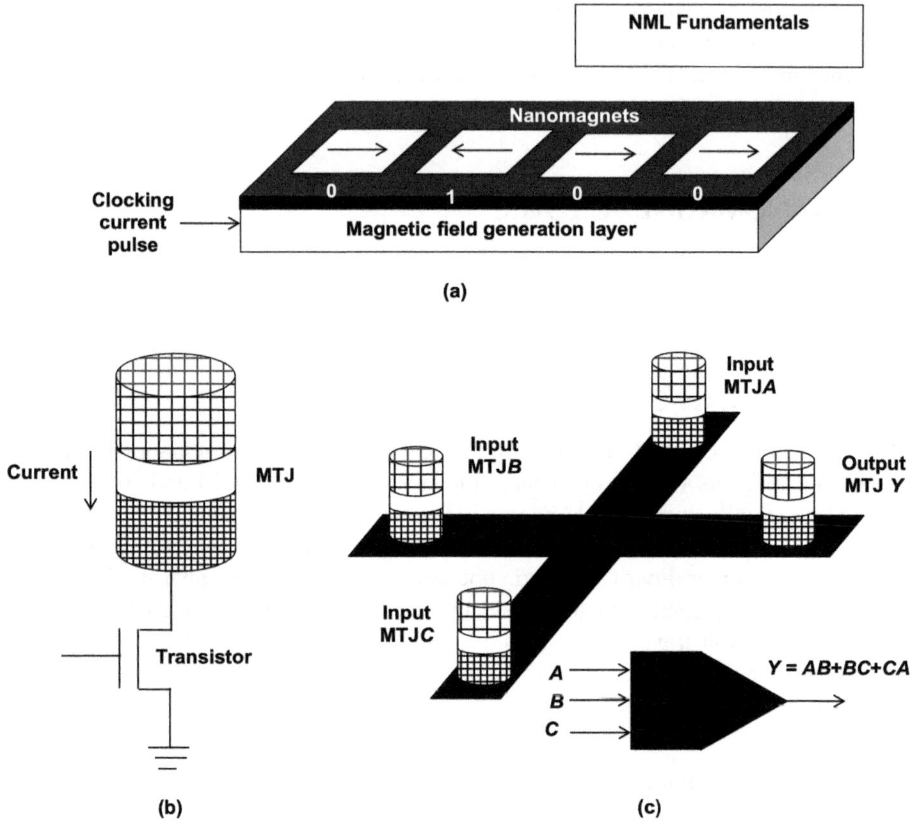

FIGURE 5.9 Nanomagnet logic basics: (a) The NML logic is initiated with single-domain nanomagnets representing the familiar logic values of zero and one. Traditionally, a current running through a wire beneath the magnets creates a magnetic field that acts as the clock signal. Diagram shows miniscule nanomagnet elements arranged along a clock line, where dipolar coupling transmits the signal. (b) Clocking by spin-transfer torque (STT): In a more advanced approach, STT currents drive the clocking, with magnetic tunnel junctions serving as the building blocks and the current itself setting the pace. (c) The three-input majority gate is an arrangement shaped like a cross of five dots. It is pictured as a central dot encircled by four others, each standing for an input (A, B, C) or the output. The output emerges from the magnetic interplay, reflecting the majority magnetization among the three inputs. Here, a three-input majority gate is built from four MTJs using spin-transfer torque technology. Here, three gate inputs program three of the STT-MTJs, while the fourth MTJ reveals the output via the direction of its free layer's magnetization.

5.5.3 Advantages of NML Logic

NML offers potential advantages over conventional CMOS technology, such as non-volatility (data retention without power). A significant benefit of NML is its non-volatile nature, meaning the stored information persists even when power is cut off. This characteristic allows for the combination of logic and memory functionality within the same device architecture, potentially alleviating the von Neumann bottleneck (the delay in transferring data between the processor and memory). Furthermore, the NML process can be performed without passing electric current, making it a suitable choice for ultra-low-power applications. Hence, NML provides reduced power consumption. In addition, NML can integrate logic and memory functions into a single device. NML also holds promise for higher integration density, especially with advancements in 3D structures. Apart from a high density, NML is more radiation-resistant, making it suitable for extreme environments, including space.

5.5.4 Challenges and Future Areas of Investigation

Reliability of NML is an issue. Mistakes may arise, and research is underway to develop strategies to minimize these errors and ensure reliable operation across large circuits. The speed of NML needs to be improved. Although initial speeds have been achieved, further improvements are required to accelerate complex calculations. The scalability of NML is difficult. Constructing and managing extensive, intricate NML circuits presents a considerable engineering challenge. 3D integration of NML is being investigated. Researchers are exploring 3D designs that combine in-plane and out-of-plane nanomagnets to boost density and enable more versatile configurations.

Example 5.4 A particular kind of permalloy nanomagnet is utilized in research. Micromagnetic simulations indicate that the average time it takes for one of these nanomagnets to switch from a "reset" position to a stable binary state (indicating "0" or "1") is roughly 120 picoseconds (ps). In a basic NML wire (a series of magnets) within a clock zone, 15 nanomagnets must switch one after another to transmit a signal, resulting in a total propagation delay through a single clock zone. Calculate this total propagation delay and express it in nanoseconds (ns).

Solution: Setting the known values—Average switching time for each magnet (t_{Switch}): 120 ps and Number of magnets in the sequence (N): 15.
Determining the total delay: The total propagation delay $\left(t_{\text{Delay}}\right)$ is equal to the number of magnets N multiplied by each magnet's switching time t_{Switch}, since they change states one after the other.

$$t_{\text{Delay}} = N t_{\text{Switch}} = 15 \times 120 \text{ ps} = 1800 \text{ ps} \tag{5.14}$$

Converting the total delay into nanoseconds (ns): Given that 1 ns equals 1000 ps,

$$t_{\text{Delay}} = \frac{1800}{1000} \text{ ns} = 1.8 \text{ ns} \tag{5.15}$$

The total propagation delay through one clock zone amounts to 1.8 ns.
Devices based on the Spin Hall Effect (SHE) make use of the accumulation of spin-polarized electrons to execute logic operations, presenting another path to spin logic with high energy efficiency.

5.6 TUNNELING SPIN-FET

A tunneling spin-FET (STFET) represents an advanced hybrid device beyond conventional CMOS technology, merging the low-power operation of a TFET with the spin-sensitive transport characteristics of a spin-FET (Jiang et al. 2019, Niknezhad and Touski 2025). It aims to surpass the "60 mV/decade" subthreshold swing limit of traditional MOSFETs while incorporating non-volatile or reconfigurable spin logic.

5.6.1 Core Structural Components

The STFET is typically composed of multiple integrated layers and regions. FM source/drain (contacts) are constructed from FM materials such as iron, cobalt, or Heusler alloys, unlike standard FETs. One contact serves as a spin polarizer (injecting spin-polarized electrons), whereas the other functions as a spin filter (detecting alignment). A tunneling barrier/junction is a vital interface (often implemented as a thin insulating layer, such as HfO_2 or hBN) that exists between the contacts and the channel. Some designs feature a BTBT junction where charge carriers tunnel from the valence band of the source to the conduction band of the channel. A nanoscale channel is commonly formed from 2D materials (such as graphene, antimonene, or transition-metal dichalcogenides) or high-mobility III–V semiconductors (such as InAs). These materials exhibit long spin-relaxation lengths and efficient tunneling. The gate electrode is a metallic gate electrode positioned above the channel and separated from the channel by a high-*k* dielectric (gate oxide). The gate voltage V_G regulates the tunneling probability by adjusting the energy bands to facilitate or BTBT, and spin precession by employing the Rashba effect, where the electric field generated by the gate behaves as an effective magnetic field, causing the spins of electrons to precess as they travel through the channel.

The Datta–Das spin field-effect transistor (spin-FET) is built with FM source and drain contacts connected by a sleek, one-dimensional channel (Datta and Das 1990, Datta 2018, Ciorga 2025). As shown in Figure 5.10, it borrows much of its design from classic FETs. At its core is a two-dimensional semiconductor channel, formed in a heterostructure (Shen et al. 2005) cradled between the source (S) and drain (D) electrodes, all of which are layered together and capacitively coupled to a metallic gate by a thin insulating film. What truly distinguishes the spin-FET is its use of FM contacts at both ends, injecting a current of electrons whose spins are all aligned. Rather than flipping ON and OFF by shuttling charge, the spin-FET operates by orchestrating electron spins.

When spins are injected, aligned with the magnetization of the contacts, they twirl through the YZ plane, guided by an effective spin-orbit field pointing along X. The gate voltage tunes the Rashba SOC and thus the precession angle. In the ON state, spins reach the drain perfectly aligned with its magnetization, while in the OFF state, their precession flips them to arrive anti-parallel.

The sFET relies on spin-polarized tunneling at the crucial boundaries where FM electrodes meet the semiconductor channel. Specifically, the key locations for tunneling are source-channel interface, where tunneling serves as a spin-polarized

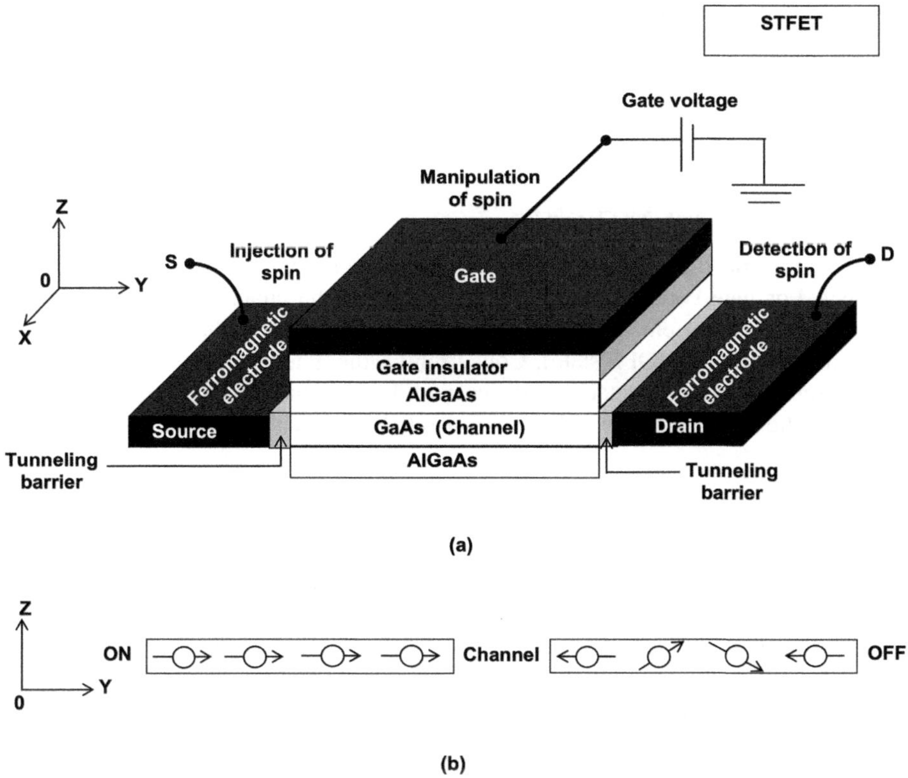

FIGURE 5.10 Schematic diagram of tunneling spin-FET and its ON and OFF states: (a) The tunneling sFET consists of a ferromagnetic source and a ferromagnetic drain at the two ends. The device channel is a quantum well of AlGaAs/GaAs/AlGaAs heterostructure or some other suitably chosen heterostructure. Gate insulator and gate electrode sit above this heterostructure. Tunneling barriers are introduced between source contact and channel, and drain contact and channel. The rectangular coordinate axes X, Y, and Z are shown. (b) During the ON state, the electron spin orientations in the channel are aligned in the same direction, whereas in the OFF state they point in random directions. In both cases, these orientations are directed in the YZ plane in which the electrons precess.

injector, transferring spin-polarized electrons from the FM metal into the semiconducting two-dimensional channel, and drain-channel interface, where tunneling serves as a detector, collecting spin-polarized electrons. Tunneling barriers (e.g., Fe/GaAs or Fe/Ge) or using efficiency-enhancing step-doping profiles are essential for overcoming the "conductivity mismatch" that limits efficient spin injection from a metal into a semiconductor. At the source, a tunneling barrier bridges the gap between the FM contact and the semiconductor, ensuring spin-polarized electrons are injected efficiently. Without this barrier, the so-called conductivity mismatch problem would scramble the spins before they even enter the channel. Similarly, at the drain, a tunneling barrier serves as a spin-selective gatekeeper, letting through only those electrons whose spins match the drain's magnetization, i.e., it acts as a spin-selective filter, allowing only electrons with spin orientation aligned to the drain's magnetization to exit the channel and contribute to the output current. To create these precise filters, materials such as alumina or magnesium oxide are sandwiched between the FM and semiconductor layers, forming robust tunnel barriers that tackle injection inefficiencies.

5.6.2 Operational Mechanism

During spin injection, spin-polarized carriers tunnel from the FM source into the channel. For gate control, the gate voltage adjusts the channel's potential. If the energy bands align, electrons tunnel through (ON state); if they are misaligned, tunneling is inhibited (OFF state). Concurrently, the gate also influences the spins' orientation. During spin detection, the current flowing through the drain depends on the magnetic alignment (parallel vs. anti-parallel) between the source and drain.

5.6.3 Advantages in Beyond-CMOS Nanoelectronics

The STFET can achieve a subthreshold swing (SS) below 60 mV/dec, enabling significantly lower supply voltages ($V_{DD} < 0.5$ V) and ultra-low power than silicon MOSFETs. The magnetic configuration of the contacts can retain information without power. Non-volatility enables "normally-off, instantly-on" computing. Reconfigurability is achieved by logic functions that can be modified by reversing the magnetization direction of the source or drain.

Example 5.5 A tunneling spin-FET, which utilizes a magnetic tunnel junction as its primary element, exhibits a conductance of G_P when the magnetic configurations of the source and drain ferromagnetic contacts are aligned in parallel (P), and a conductance of G_{AP} when their alignment is anti-parallel (AP). Provided values: Parallel conductance, $G_P = 120$ μS (microsiemens), and anti-parallel conductance, $G_{AP} = 40$ μS. Calculate the tunneling magnetoresistance (TMR) ratio, typically expressed as a percentage.
Solution: To determine the TMR ratio, we use the following formula:

$$\text{TMR} = \frac{G_P - G_{AP}}{G_{AP}} \tag{5.16}$$

Let's substitute the given values into the equation:

$$\text{TMR} = \frac{120 - 40\ \mu\text{S}}{40\ \mu\text{S}} = \frac{80\ \mu\text{S}}{40\ \mu\text{S}} = 2 \tag{5.17}$$

Finally, we convert the ratio into a percentage: TMR $= 2 \times 100\% = 200\%$. Therefore, the tunneling magnetoresistance (TMR) ratio is 200%.

5.7 ALL-SEMICONDUCTOR/ALL-ELECTRIC SPIN-FET

The all-semiconductor/all-electric spin-FET signifies a significant evolution from conventional Datta-Das configurations by substituting FM contacts with entirely electrical components (Chuang et al. 2015, Chen et al. 2016).

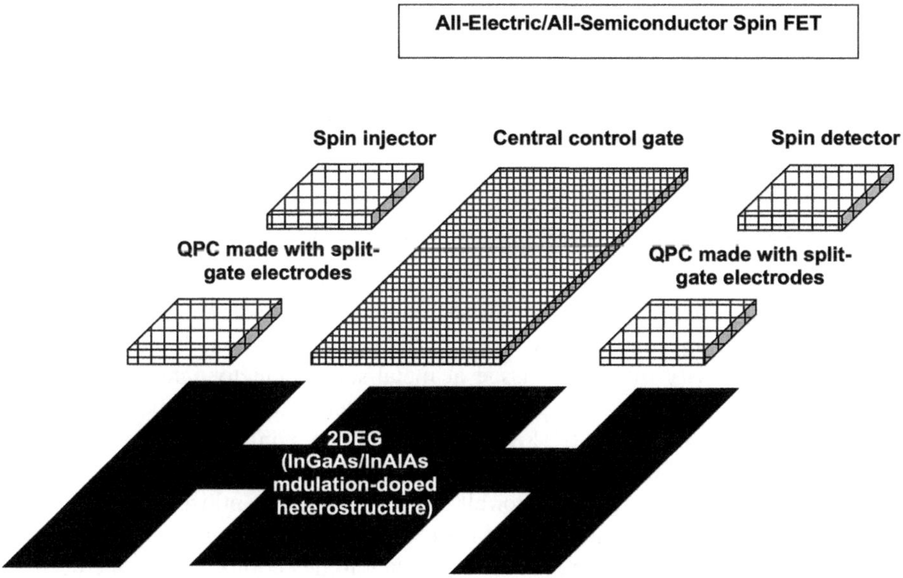

FIGURE 5.11 The all-electric, all-semiconductor spin-FET. It is crafted on a sophisticated $In_{0.75}Ga_{0.25}As/In_{0.75}Al_{0.25}As$ heterostructure, which creates a sleek two-dimensional electron gas (2DEG) channel for electrons to transport ballistically beneath a metallic middle gate and between two precisely engineered gate-defined quantum point contacts (QPCs). These QPCs are slender and short one-dimensional constrictions. These pathways are sculpted by applying voltages to split-gate patterns on the semiconductor's surface. When the split gates of the left or right QPC are asymmetrically biased, they transform into spin injectors or detectors, conjuring a lateral inversion asymmetry that results in a spin-orbit (SO) effective magnetic field.

5.7.1 Core Structure and Components

Typically, the device employs a one-dimensional (1D) semiconductor channel integrated within a high-mobility heterostructure, such as InAs or InGaAs (Figure 5.11).

The spin injector and detector are two quantum point contacts (QPCs) constructed with split-gate electrodes that constrict electron flow into a 1D subband, rather than using FM metals. By applying asymmetric voltages to these gates, a lateral electric field is established, which, through spin-orbit (SO) coupling, generates an effective magnetic field that polarizes the electron spins. The central semiconductor channel is a gated semiconductor layer, either a 2D Electron Gas or 1D wire, that links the injector and detector. This channel is engineered to facilitate coherent spin transport over necessary distances. A control gate is a central metallic gate, often located at the top or in the middle, that sits above the channel. By applying voltage to this gate, the Rashba SOC is modulated, controlling the precession (rotation) of the electron spins as they migrate from the injector to the detector.

5.7.2 Operational Principles

As electrons traverse the first QPC, they become spin-polarized due to the lateral SO field, leading to all-electric spin injection. Spin manipulation is achieved by adjusting the voltage on the central channel, which alters the internal electric field and causes the spins to precess by a defined angle. For all-electric spin detection, the second QPC acts as a selective filter. If an electron's spin aligns parallel to the QPC's polarization, it will pass through; if it is anti-parallel, it will be obstructed, resulting in either a high or low current (ON/OFF state).

5.7.3 Advantages for Beyond-CMOS Integration

CMOS compatibility is attained by eliminating FM materials, enabling the device to be more readily integrated into established semiconductor manufacturing processes. All-electric QPC injectors achieve close to 100% spin-injection efficiency, addressing the "conductivity mismatch" issue at metal-semiconductor interfaces. The all-semiconductor/all-electric spin-FET devices are being investigated as part of the IRDS to facilitate high-density, low-power logic and non-volatile computing that transcends the limitations of traditional scaling (Obeng et al. 2025).

Summarily, an all-semiconductor/all-electric spin-FET utilizes QPCs as spin injectors and detectors in a semiconductor channel, employing SOC for electrical control over electron spins (injection, manipulation, detection) without the need for magnetic materials, providing CMOS compatibility and the potential for non-volatile, energy-efficient spintronics by creating spin-polarized 1D subbands through gate voltages to manage spin states, achieving significant on/off ratios.

Example 5.6 The effectiveness of a spin-FET as a switch in logic applications is typically measured by its conductance or the ratio of current in the on/off states, which is greatly influenced by the efficiencies of spin injection and detection η_S and η_D. The conductance ratio for the "ON" state (when magnetizations are parallel, G_{On}) compared to the "OFF" state G_{Off} (when magnetizations are anti-parallel) is given by the formula

$$\frac{G_{On}}{G_{Off}} = \frac{1 + \eta_S \eta_D}{1 - \eta_S \eta_D} \tag{5.18}$$

A novel all-electric spin-FET incorporates materials that have a spin-injection efficiency η_S of 80% (or 0.8) and a spin detection efficiency η_D of 75% (or 0.75). Calculate the anticipated conductance on/off ratio G_{On}/G_{Off} for this device. Comment on its low value, and suggest suitable remedies for improvement.

Solution: Converting the efficiencies into decimal format:

$$\eta_S = 80\% = 0.8; \eta_D = 75\% = 0.75 \tag{5.19}$$

Inserting the values into the equation and carrying out the multiplications:

$$\frac{G_{On}}{G_{Off}} = \frac{1 + \eta_S \eta_D}{1 - \eta_S \eta_D} = \frac{1 + 0.8 \times 0.75}{1 - 0.8 \times 0.75} = \frac{1 + 0.6}{1 - 0.6} = \frac{1.6}{0.4} = 4 \tag{5.20}$$

The anticipated conductance ON/OFF ratio for this spin-FET is 4 (or 4:1). This indicates that the "ON" current is four times greater than the "OFF" (leakage) current.

While traditional charge-based FETs can attain exceptionally high on/off ratios (for instance, 10^6–10^{11} in cutting-edge 2D materials), all-electric spin-FETs often find it difficult to achieve similar levels, particularly at ambient temperature. The limited on/off ratio in all-electric spin field-effect transistors (spin-FETs) is mainly due to difficulties with effective spin injection and detection, as well as with preserving spin coherence within the semiconductor channel at room temperature.

Utilizing quasi-one-dimensional (1D) channels, such as quantum wires or nanowires, is expected to enhance performance, as current modulation from spin precession is less prone to being diminished than in two-dimensional transport. Researchers are exploring materials with significant SOC or low spin relaxation, including graphene (which has low SOC), transition-metal dichalcogenides (TMDs), and certain III–V semiconductors such as InP. Employing MTJs as contacts can offer high spin polarization, though they may also pose challenges, such as elevated impedance. Sophisticated gate designs, such as dual-gate or triple-metal gate configurations, provide improved electrostatic control over the channel, helping suppress OFF-state leakage and adjust the threshold voltage to optimize the ON/OFF ratio.

5.8 FROM SPIN-FET TO SKYRMION LOGIC

The spin-FET marked a bold leap beyond classic electronics. Imagine electrons as tiny dancers, each with a unique spin, gliding down a channel. A voltage acts as the choreographer, twisting their spins mid-performance. If their final pose matches the collector's, the gate is ON; if not, it stays OFF. The catch? Keeping these spins in synchronization is tricky. Heat and imperfections easily throw them off rhythm, making spin-FETs delicate. They also struggle to inject spins efficiently and keep them coordinated over long distances.

Magnetic skyrmions are nanoscale quasiparticles with topological protection, making them ideal for information transmission due to their exceptional stability and mobility (Shu et al. 2026). Measuring as small as 1 nm, skyrmions can be created, moved, and annihilated at very low current densities. Unlike devices that rely on a single electron's spin, skyrmion logic employs stable, particle-like magnetic vortices as information carriers. Their topological protection ensures they remain intact and highly resistant to defects or thermal fluctuations, offering greater stability than ordinary domain walls.

Imagine guiding miniature whirlpools—skyrmions—along microscopic racetracks to perform logic. For AND and OR gates, scientists harness the repulsion between skyrmions, nudging them onto diverging paths to solve problems. In NOT gates, a skyrmion either blocks or lets a signal through, flipping the signal from high to low or the other way around. Shifting from spin-FETs to skyrmions is like trading a spinning compass for a swirling magnetic whirlpool.

Skyrmions spring to life under the influence of SOT from a heavy metal base like Pt, Co, or Ta. With VCMA, their motion and pinning can be locally steered with minimal power. Unlike domain walls, skyrmions glide along at much lower current densities, and logic powered by VCMA barely sips energy. These nanoscale marvels can be crafted smaller than 10–50 nm, paving the way for ultra-dense racetrack memory and logic hybrids.

5.9 DISCUSSION AND CONCLUSIONS

Critical elements and structures in the domain of spintronics involve utilizing the spin of an electron (its magnetic moment) for information processing and storage, rather than relying solely on its electric charge. MESO (Magnetoelectric spin-orbit) devices function by applying an electric field to trigger a magnetization change (the ME), which is then converted to an electric charge via SOC. MTJ is a commonly used spintronic component that features two FM layers separated by a skinny insulating layer. Electrons pass through this barrier, and the device's electrical resistance varies depending on the alignment of the magnetizations in the two FM layers (either parallel or anti-parallel), a behavior known as TMR. The SHE is a relativistic SOC effect used in spintronic devices to generate or detect spin currents and to apply SOTs to nearby magnetic layers, facilitating efficient magnetization switching. NML represents a computational technology that relies on the interactions of nanoscale magnets to process information. Information is defined by the orientation of magnetization in these single-domain magnets. The tunneling spin-FET (tunneling spin field-effect transistor) is a type of spintronic field-effect transistor that leverages quantum-mechanical tunneling of spin-polarized electrons. The all-semiconductor/all-electric spin-FET is a design for a spin field-effect transistor (spin-FET) in which all elements, including the source, drain, and channel, are constructed from semiconductor materials, and electron spin manipulation is achieved solely through electrical methods.

Table 5.1 summarizes the operational principles, performance metrics, and potential applications of spintronic devices, including MESO, MTJ, SHE, NML, tunneling spin-FET, and all-electric spin-FET. The advancement of these technologies aims to address the shortcomings of traditional charge-based devices, emphasizing enhanced energy efficiency, non-volatile operation, and scalability.

Skyrmions have captured the spotlight for their ability to function at room temperature and power innovative hybrid systems that could one day replace electronic logic with ultra-efficient spin-based processing. Researchers have already built reversible logic gates using skyrmion interactions, hinting at the dawn of entirely new computing paradigms. The race between spin-FETs and skyrmionic logic signals a bold move toward post-CMOS, non-volatile, and energy-saving technologies. Spin-FETs, which steer spin-polarized currents through semiconductors, challenge the dominance of MOSFETs, while skyrmionic logic harnesses the unique properties of magnetic solitons to promise unmatched scaling and density. Yet, spin-FETs face hurdles in preserving spin polarization over distance and at room temperature, while skyrmionic devices must overcome the skyrmion Hall Effect, where skyrmions veer off course and risk being lost at device edges. To tackle this, scientists are experimenting with SAFs and clever notch designs. While spin-FETs are strong contenders for the next generation of transistors, their reliance on charge transport may still run into familiar thermal barriers. Skyrmions, on the other hand, can shrink to just a few nanometers and remain stable, opening the door to unprecedented integration density.

TABLE 5.1
Comparison of MESO, MTJ, SHE, NML, Tunneling Spin-FET, and All-Electric Spin-FET

Sl. No.	Device Type	Operating Principle	Salient Features	Primary Uses
1	MESO (magnetoelectric spin-orbit)	Utilizes an electric field (voltage) to manage magnetism, potentially through magnetoelectric coupling in multiferroic materials.	Offers extremely low-power switching (driven by voltage) and high energy efficiency (capacitive).	Applications for ultra-low power memory and logic.
2	MTJ (magnetic tunnel junction)	Electron tunneling through a thin insulating barrier is contingent on the relative magnetization of two ferromagnetic layers.	Non-volatile, high-density, scalable memory element with low-power consumption.	MRAM (magnetic random access memory) serves as a universal data storage device.
3	SHE (spin hall effect) device	A charge current in a heavy metal layer produces a pure spin current through spin-orbit coupling, which can then switch the magnetization of an adjacent MTJ layer.	A three-terminal architecture separates reading and writing paths, enhancing reliability and minimizing write energy in comparison to STT–MTJ.	High-endurance, low-power MRAM.
4	NML (nanomagnetic logic)	Employs the magnetic states of interacting nanomagnets for information processing via domain wall motion.	Non-volatile logic with theoretical inherent cascading potential and zero standby power consumption.	Circuits and architectures for non-volatile logic beyond CMOS technology.
5	Tunneling spin-FET	Resembles a standard FET but employs ferromagnetic source/drain electrodes and a gate voltage to control the spin-polarized current in a semiconductor channel through a tunneling barrier.	Integrates the benefits of spintronics with semiconductor FET technology, providing enhanced spin injection/filtering compared to typical spin-FETs when half-metallic electrodes are utilized.	Potential for high-efficiency spin injection/ detection and integration with existing electronics.

(Continued)

TABLE 5.1 (Continued)

Sl. No.	Device Type	Operating Principle	Salient Features	Primary Uses
6	All-semiconductor/ all-electric spin-FET	All components (injectors, channel, detectors) are based on semiconductors and utilize electric fields and spin-orbit coupling for spin control.	Fully compatible with large-scale semiconductor integration, it employs purely electrical means to manipulate spin, with high potential for general-purpose logic.	Future energy-efficient microprocessors and integrated logic circuits.

Interestingly, spintronic logic and non-volatile memory are natural allies. Devices like MTJs store data in magnetic spin states rather than electrical charges, allowing them to retain information even when the power is off while still handling lightning-fast logic operations. This powerful combination fuses storage and processing, reducing the need for energy-hungry data transfers. With this in mind, the next chapter will dive into the world of non-volatile memories.

REFERENCES

Barla P., V. K. Joshi and S. Bhat 2021 Design and analysis of SHE-assisted STT MTJ/CMOS logic gates. *Journal of Computational Electronics*, 20: 1964–1976, https://doi.org/10.1007/s10825-021-01759-8

Chen T.-M., S.-C. Ho and P. Chuang 2016 All-electric spin field effect transistor US20150364583A1, 2016-06-21. Application granted.

Chuang P., S. C. Ho, L. W. Smith, F. Sfigakis, M. Pepper, C. H. Chen, J. C. Fan, J. P. Griffiths, I. Farrer, H. E. Beere, G. A. Jones, D. A. Ritchie and T. M. Chen 2015. All-electric all-semiconductor spin field-effect transistors. *Nature Nanotechnology*, 10(1): 35–39, https://doi.org/10.1038/nnano.2014.296

Ciorga M. 2025 Perspective on the spin field-effect transistor, *Journal of Physics D: Applied Physics*, 58(1): 012001, https://doi.org/10.1088/1361-6463/ad8003

Csaba G. and M. Becherer 2020 Nanomagnet logic: Computing by magnetic ordering, *IEEE Nanotechnology Magazine*, 14(1): 6–13, https://doi.org/10.1109/MNANO.2019.2952232

Datta S. 2018 How we proposed the spin transistor, *Nature Electronics*, 1: 604–606, https://doi.org/10.1038/s41928-018-0163-4

Datta S. and B. Das 1990 Electronic analog of the electro-optic modulator, *Applied Physics Letter*, 56: 665–667, https://doi.org/10.1063/1.102730

Gypens P., B. V. Waeyenberge, M. D. Ventra, J. Leliaert and D. Pinna 2021 Nanomagnetic self-organizing logic gates, *Physical Review Applied*, 16: 024055, https://doi.org/10.1103/PhysRevApplied.16.024055

Husain S., Z. Yao, R. Ramesh 2025 Enabling magnetoelectric spin-orbit logic and memory, *Newton*, 1(1): 100026, https://doi.org/10.1016/j.newton.2025.100026

Jiang S., L. Li, Z. Wang, J. Shan and K. F. Mak 2019 Spin tunnel field-effect transistors based on two-dimensional van der Waals heterostructures, *Nature Electronics*, 2: 159–163, https://doi.org/10.1038/s41928-019-0232-3

Joshi V. K., P. Barla, S. Bhat and B. K. Kaushik 2020 From MTJ device to hybrid CMOS/MTJ circuits: A review, *IEEE Access*, 8: 194105–194146, https://doi.org/10.1109/ACCESS.2020.3033023

Kumar R., Y. Sharma and A. K. Goyal 2025 A hybrid CMOS-MTJ polymorphic logic for secure and versatile IC design, *Magnetochemistry*, 11(12): 108. https://doi.org/10.3390/magnetochemistry11120108

Liu H., S. Manipatruni, D. H. Morris, K. Vaidyanathan, D. E. Nikonov, T. Karnik and I. A. Young 2019 Synchronous circuit design with beyond-cmos magnetoelectric spin–orbit devices toward 100-mV logic, *IEEE Journal on Exploratory Solid-State Computational Devices and Circuits*, 5(1): 1–9, https://doi.org/10.1109/JXCDC.2019.2897598

Lv Y., B. R. Zink, R. P. Bloom, H. Cilasun, P. Khanal, S. Resch, Z. Chowdhury, A. Habiboglu, W. Wang, S. S. Sapatnekar, U. Karpuzcu and J.-P. Wang 2024. Experimental demonstration of magnetic tunnel junction-based computational random-access memory, *npj Unconventional Computing*, 1: 3, https://doi.org/10.1038/s44335-024-00003-3

Manipatruni S., D. E. Nikonov, C. C. Lin, T. A. Gosavi, H. Liu, B. Prasad, Y. L. Huang, E. Bonturim, R. Ramesh and I. A. Young 2019 Scalable energy-efficient magnetoelectric spin-orbit logic, *Nature*, 565(7737): 35–42. https://doi.org/10.1038/s41586-018-0770-2

Misra N. K. and B. K. Bhoi 2023 Efficient architecture for arithmetic designs using perpendicular NanoMagnetic Logic, *Nano Communication Networks*, 36: 100454, https://doi.org/10.1016/j.nancom.2023.100454

Niknezhad H. N. and S. B. Touski 2025 Quantum simulation of spin tunneling field-effect transistor based on antimonene, *Physical Review*, B111, 245430, https://doi.org/10.1103/6xkp-h3wl

Obeng Y., E. Mansfield, A. Davydov, B. Barnes and A. Vladar 2025 International roadmap for devices and systems TM 2024 metrology, International Roadmap for Devices and Systems (IRDS™) 2024 Edition, [online], https://tsapps.nist.gov/publication/get_pdf.cfm?pub_id=959302, https://irds.ieee.org/editions/2024

Shen M., S. Saikin and M.-C. Cheng 2005 Spin injection in spin FETs using a step-doping profile, *IEEE Transactions on Nanotechnology*, 4(1): 40–44, https://doi.org/10.1109/TNANO.2004.840150

Shu Y., Q. Li, W. Zhang, Y. Peng, P. Lai and G. Zhao 2026 Logic gates based on skyrmions, *Nanomaterials (Basel)*, 16(2): 135, https://doi.org/10.3390/nano16020135

Sinova J., S. O. Valenzuela, J. Wunderlich, C. H. Back and T. Jungwirth 2015 Spin hall effects, *Reviews of Modern Physics*, 87: 1213, https://doi.org/10.1103/RevModPhys.87.1213

Wan C., X. Zhang, Z. Yuan, C. Fang, W. Kong, Q. Zhang, H. Wu, U. Khan and X. Han 2017 Programmable spin logic based on spin hall effect in a single device, *Advanced Electronic Materials*, 3(3): 1600282, https://doi.org/10.1002/aelm.201600282

Zhang H., W. Kang, L. Wang, K. L. Wang and W. Zhao, 2017 Stateful reconfigurable logic via a single-voltage-gated spin hall-effect driven magnetic tunnel junction in a spintronic memory, *IEEE Transactions on Electron Devices*, 64(10): 4295–4301, https://doi.org/10.1109/TED.2017.2726544

6 Non-Volatile Memories

6.1 INTRODUCTION

As we delved deeper into spintronic logic in the last chapter, we observed that non-volatile memory (NVM) emerges as its indispensable ally. NVM not only decreases power leakage to zero but also breaks through the "memory wall" by letting computation happen right where the data lives. With data safely stored even when the power is off, entire systems can power down during idle moments and spring back to life instantly, sidestepping the slow, energy-hungry process of reloading from a hard drive. These advanced memories bring fast boot-ups, drain less power, pack in more data, and boost security for IoT and AI. In the race to move beyond CMOS, NVM is key to overcoming the high standby power, volatility, and scaling roadblocks that plague traditional memories like SRAM and DRAM, which constantly waste energy just to keep data alive.

Resistive random-access memory (ReRAM), magnetic random-access memory (MRAM), phase-change memory (PCM), and ferroelectric random-access memory (FeRAM) are NVM technologies that present significant benefits compared to conventional flash memory, including greater density, faster speeds, and reduced power usage (Hellenbrand et al. 2024, Katehi et al. 2025, Sharma et al. 2026).

6.2 RESISTIVE RANDOM-ACCESS MEMORY

6.2.1 Motivation for ReRAM

Resistive Random-Access Memory (ReRAM) is crucial in advancing "beyond-CMOS" nanoelectronics, acting as a high-capacity NVM solution and a significant driver for innovative IMC architectures, especially in AI and neuromorphic systems (Yu 2016, Zahoor et al. 2020, Ielmini and Pedretti 2025). ReRAM saves data by modifying the electrical resistance of a substance, typically by forming or dissolving conductive filaments within a thin oxide layer.

6.2.2 ReRAM Cell

At its core, a ReRAM cell features a metal-insulator-metal structure (MIM), with each layer playing a distinct role (Figure 6.1). The platinum electrode serves as a passive, unreactive exterior covering that does not easily oxidize or provide ions, providing an inert surface for metal ion reduction and electrical contact; while the HfO_2 layer, which is a high-*k* dielectric material, serves as the solid electrolyte, providing a dynamic playground for the formation of conductive filaments. The copper electrode, eager to participate in cell function, serves as the "active" or oxidizable electrode. When a

DOI: 10.1201/9781003753469-6

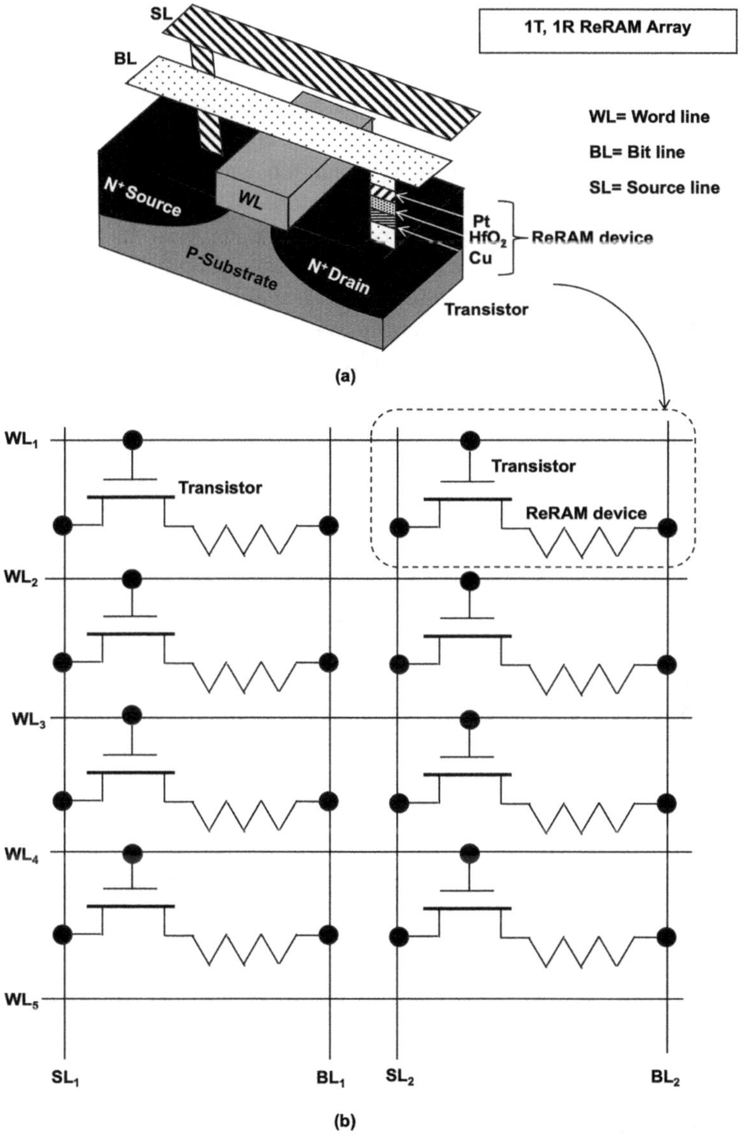

FIGURE 6.1 1T1R ReRAM array: (a) ReRAM cell comprising a Pt electrode, an HfO_2 layer, and a Cu metal electrode connected to a transistor with the Cu electrode joined to the N^+ drain of the transistor and the Pt electrode to the bit line (BL). The source line (SL) is connected to the N^+ source while the word line (WL) passes over the gate of the transistor. (b) The ReRAM array in which the ReRAM cells are organized in a rectangular grid with gates of the transistors connected to word lines WL_1, WL_2, WL_3, WL_4, and WL_5 running horizontally. The source terminals of the transistors are connected to the source lines SL_1 and SL_2, placed vertically. Their drain terminals are attached to one end of the ReRAM cells. The opposite ends of the ReRAM cells are joined to the bit lines BL_1 and BL_2 running in a vertical direction.

voltage is applied, copper atoms spring into action, oxidizing and journeying into the HfO_2, where they weave a slender metallic bridge to the platinum side. This tiny filament, just a few nanometers wide, is the key to the cell's memory.

This cell operates as a conductive bridge RAM (CBRAM). Its switching mechanism relies on the formation and dissolution of a copper conductive filament within the HfO_2 layer. During the SET state, oxidation and migration processes occur. A positive bias on the Cu electrode causes Cu atoms to oxidize into Cu^{2+} ions. These ions migrate through the HfO_2 and are reduced at the Pt electrode, forming a metallic bridge (filament). This results in a low-resistance state (LRS). Thus, in the SET state, copper ions migrate and assemble the bridge, dropping resistance and storing a "1."

During the RESET state, metal dissolution takes place. A reverse bias (negative on Cu, positive on Pt) causes the metallic filament to dissolve or rupture due to electrochemical oxidation and Joule heating. This returns the cell to a high-resistance state (HRS). Therefore, to RESET, a reversed voltage dissolves the bridge, breaking the connection and restoring high resistance, ready for the next cycle.

6.2.3 ReRAM Array Structure

The backbone of ReRAM memory is the one-transistor and one-resistor (1T1R) array, in which each memory cell is paired in series with its own selection transistor. This transistor acts as a gatekeeper, isolating the cells to ensure that only the designated cell responds while others remain silent. The word line (WL), linked to the transistor's gate, selects entire rows with a simple voltage tweak, precisely controlling the current that reaches each cell. The top of the ReRAM cell connects to the bit line (BL), while its bottom links to the contact through the transistor's drain. The source line (SL) connects to the transistor's source, completing the circuit and enabling fast, reliable memory operations.

6.2.4 The Rationale Behind the Requirement of a Bit Selector

Every bit cell in a memory chip relies on a selector, which serves as a switch isolating each cell to ensure precise reading and writing while shielding it from neighboring interference (Handy 2018). The selector channels the cell's data onto a bus, linking it to the chip's pins for seamless data transfer. In Figure 6.2(a), a simplified top-down view of a resistive RAM (ReRAM) array reveals each bit cell as a square where WLs and BLs cross. In this illustration, selectors are left out. The WL chooses which row to access, while the BL either retrieves the bit or lets current flow to program it.

A bit exists in one of two states: high resistance or low resistance. In the diagrams, a black square represents high resistance, blocking current, while a white square marks low resistance, letting current pass through.

If selectors are missing, activating a WL when just one bit is in an LRS lets current travel from the WL to the BL for that cell, as shown in Figure 6.2(b). The other BLs do not receive current because their cells are locked in high resistance. If we activate a different WL, no current moves at all, since those cells still resist.

But issues arise when several bits slip into an LRS, as Figure 6.2(c) illustrates. Output current can sneak up the BL to another low-resistance cell, following the striped arrow, and then double back onto a different WL. This can trick the system into thinking a bit is in an LRS when it is not. The selector steps in to block this confusion.

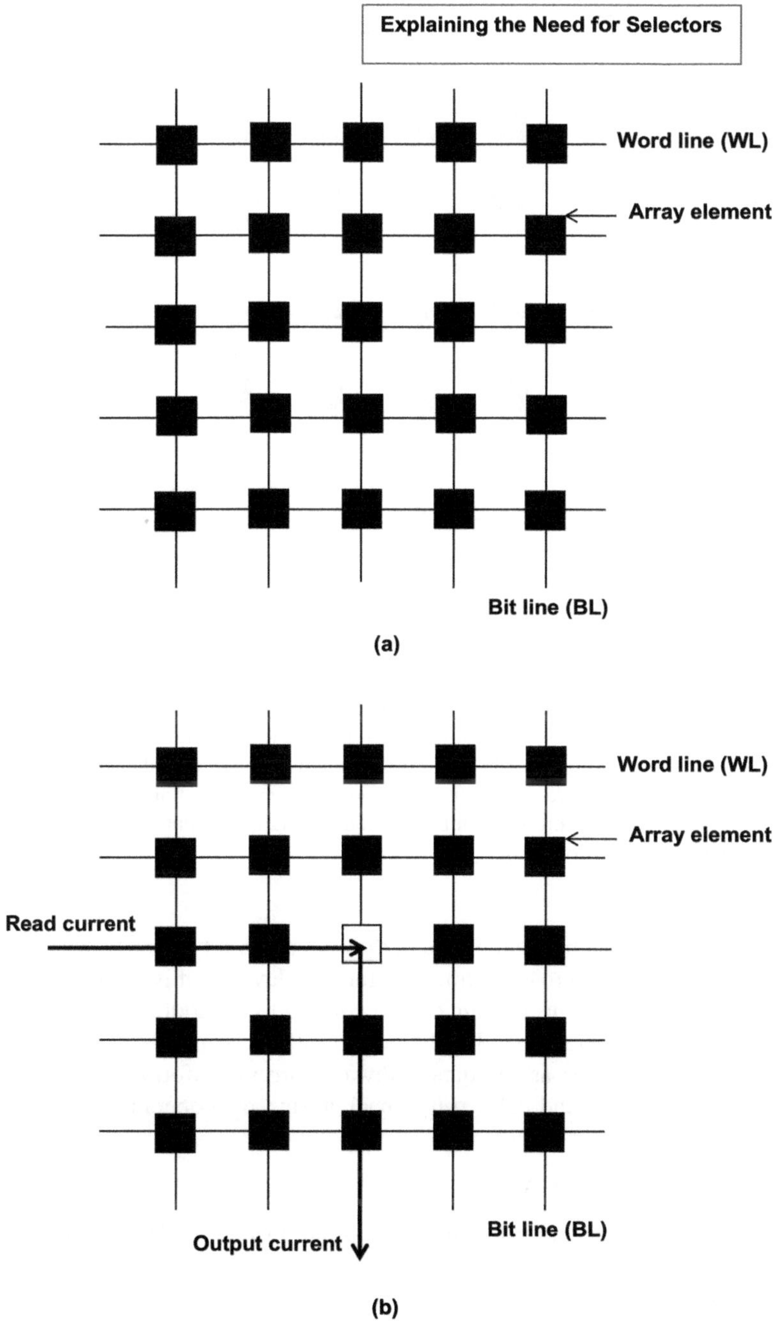

FIGURE 6.2 Illustrating the need for selectors in building a memory chip array: (a) A simplified overhead view of a resistive RAM (ReRAM) array, where each bit cell appears as a dot at the intersection of a word line and a bit line. Selectors are omitted in this illustration.

(Continued)

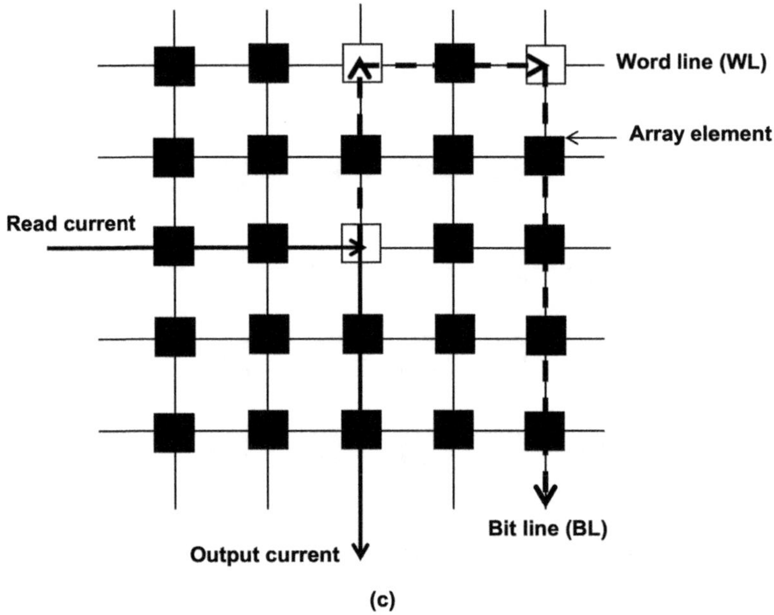

FIGURE 6.2 (Continued) The word line selects the row to be read or written, while the bit line either reads the bit or allows current to program it. (b) Without selectors if a single bit is in a low-resistance state and its word line is activated, current flows from the word line to the bit line for that cell. Other bit lines do not receive current because their cells are in a high-resistance state. (c) Trouble starts if multiple bits are in a low-resistance state. Output current can flow up the bit line to another low-resistance cell, as indicated by the striped arrow. This allows current to flow backward onto another word line, and any bit along that word line can misguide the stray current onto its own bit line, falsely signaling that another bit is in a low-resistance state when it truly is not.

Transistor selectors are used, although they occupy a lot of real estate, since both WLs and SLs must crisscross the array. Figure 6.3 lays out this setup. The WL acts like a conductor, energizing specific cells according to the address input during both reading and writing operations. The SL connects all the sources to a common ground potential. The BL serves as the data pathway, carrying information into memory during write operations and delivering it back during read operations.

6.2.5 Working of the Array

Let us now return to Figure 6.1 to understand the operation of ReRAM array.

(i) Setting (Writing 1): To write a "1," a positive voltage (e.g., 1.5 V) energizes the BL, whereupon the WL springs to life, and the SL is grounded. This electric choreography encourages a conductive filament (oxygen vacancies) to form within the dielectric, transforming the cell swiftly from a stubborn HRS to an LRS.

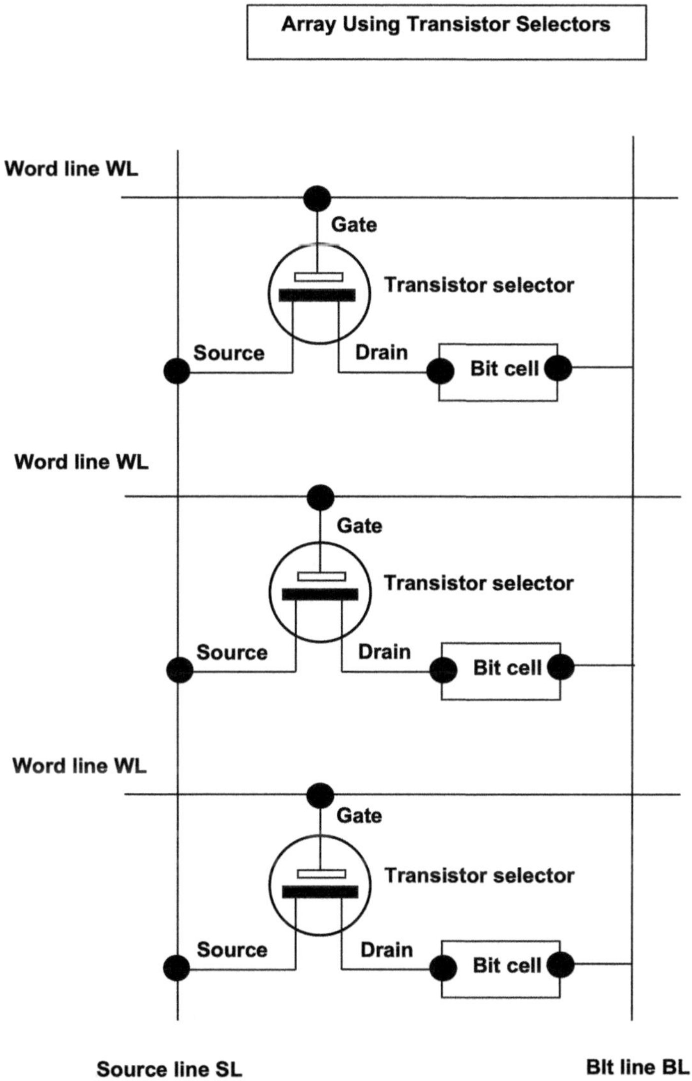

FIGURE 6.3 Array arrangement using transistor selectors with word lines and source lines crossing each other in the array. Source, gate, and drain terminals of transistor selectors are indicated. The bit cells are also shown.

(ii) Resetting (Writing 0): To erase or write a "0," a negative voltage or a pulse of opposite polarity is sent through the BL, or a high voltage is applied to the SL. This action snaps the conductive filament, returning the cell to its original HRS.

(iii) Reading: To read the cell, a gentle voltage is applied to the BL, and the transistor is switched on. The resulting current reveals the cell's secret: a strong current signals a "1," while a faint current means a "0."

(iv) Isolation: For cells not in use, the transistor stands guard in the off position to block unwanted "sneak path" currents and ensure that only the intended cell is read or written with precision.

6.2.6 ReRAM's Contribution to Beyond-CMOS Nanoelectronics

ReRAM provides a variety of advantages to the realm of nanoelectronics. It serves as a replacement for NVM, standing out as a viable alternative to conventional memory innovations such as DRAM and Flash, owing to its exceptional features like ultimate scalability (below 10 nm), impressive endurance, rapid operation speed, and minimal power consumption. Its non-volatile characteristic allows it to preserve data without the need for constant power, leading to increased energy efficiency.

ReRAM facilitates IMC. Traditional computing frameworks encounter the "von Neumann bottleneck," in which data exchange between distinct memory and processing units restricts speed and leads to energy wastage. ReRAM enables simultaneous data storage and computation in the same physical location (in-memory computing), markedly alleviating this bottleneck.

ReRAM serves as an essential component of neuromorphic computing hardware. Particularly for ReRAM devices utilizing metal oxides such as HfO_2 and TiO_2, they can replicate the synaptic weight alterations found in biological brains by providing varied, adjustable resistance levels. This analog switching capacity renders them suitable for creating energy-efficient hardware utilized in neural networks and AI processors.

ReRAM is the subject of intensive research and development for application in logical operations. ReRAM devices extend their capabilities beyond mere memory functions by executing logic operations and arithmetic directly within the memory arrays, showcasing their potential in the design of non-von Neumann computing systems.

ReRAM provides compatibility benefits and can be integrated with CMOS. A key benefit of ReRAM is its alignment with current CMOS manufacturing techniques. It can seamlessly integrate into the BEOL of standard CMOS chips, allowing for cost-effective amalgamation of high-density memory units with existing CMOS circuitry.

The functional diversification offered by ReRAM holds great potential. ReRAM supports "More-than-Moore" (MtM) strategies by introducing new functionalities to chips, transcending simple dimensional scaling. This heterogeneous integration permits the development of specialized, high-performance, and energy-efficient systems tailored for applications in IoT, edge AI, and autonomous vehicles.

Ultimately, ReRAM serves as a foundational element for the next generation of nanoelectronics, providing not just enhanced memory solutions but also a route toward entirely new, more efficient computing paradigms that tackle the basic scaling limits of conventional CMOS technology.

Example 6.1 A ReRAM device designed for neuromorphic computing exhibits a high-resistance state (HRS) with ($R_{\text{Off}} = 800\ \text{k}\Omega$) and a low-resistance state (LRS) of ($R_{\text{On}} = 50\ \text{k}\Omega$). To identify its logic state, a read voltage (V_{Read}) of 0.15 V is applied across the device. Calculate the current flowing through the device while it is in the LRS

(I_{On}), the current flowing through the device when it is in the HRS (I_{Off}), and the ON/OFF ratio of the device's resistance, often referred to as the switching window.

Solution: To find the current, we can apply Ohm's Law, which states

$$I = \frac{V}{R} \tag{6.1}$$

Current flowing in LRS (I_{On}) is

$$I_{On} = \frac{V_{Read}}{R_{On}} = \frac{0.15\text{ V}}{50\text{ k}\Omega} = \frac{0.15\text{ V}}{50 \times 1000\ \Omega} = 0.000003\text{ A} = 0.000003 \times 10^6\ \mu\text{A} = 3\ \mu\text{A} \tag{6.2}$$

Current flowing in HRS (I_{Off}) is

$$I_{Off} = \frac{V_{Read}}{R_{Off}} = \frac{0.15\text{ V}}{800\text{ k}\Omega} = \frac{0.15\text{ V}}{800 \times 1000\ \Omega} = 1.875 \times 10^{-7}\text{ A} = 1.875 \times 10^{-7} \times 10^6\ \mu\text{A}$$
$$= 0.1875\ \mu\text{A} = 0.1875 \times 1000\text{ nA} = 187.5\text{ nA} \tag{6.3}$$

The resistance ratio for ON/OFF states (switching window) is

$$\text{Ratio} = \frac{R_{Off}}{R_{On}} = \frac{800\text{ k}\Omega}{50\text{ k}\Omega} = 16 \tag{6.4}$$

The device features a resistance ratio of 16 between its high and low states, enabling distinct differentiation when a read voltage of 0.15 V is utilized.

6.3 MAGNETIC RESISTIVE RANDOM-ACCESS MEMORY

6.3.1 Significance of MRAM

Magnetic resistive random-access memory (MRAM) is the leading candidate in beyond-CMOS nanoelectronics (Zhu 2008, Hu et al. 2011, Nguyen et al. 2024). It mainly serves as a flexible NVM that can be combined with traditional CMOS logic to develop high-performance, energy-efficient systems. Similar to SRAM and DRAM, it addresses the scaling and power challenges encountered by standard charge-based memories (like SRAM, DRAM, and Flash) through the use of spintronics and finds applications in both memory and emerging logic sectors.

6.3.2 MRAM Cell

Imagine an MRAM cell as a high-tech sandwich, layered with precision (Figure 6.4). The free layer, made of a ferromagnetic material like CoFeB, can flip its magnetic orientation to store a 0 or 1. Next comes the tunnel barrier, an ultra-thin insulator, often an MgO film, that lets electrons flow by quantum mechanical tunneling between layers. The pinned layer, another ferromagnetic sheet, maintains its

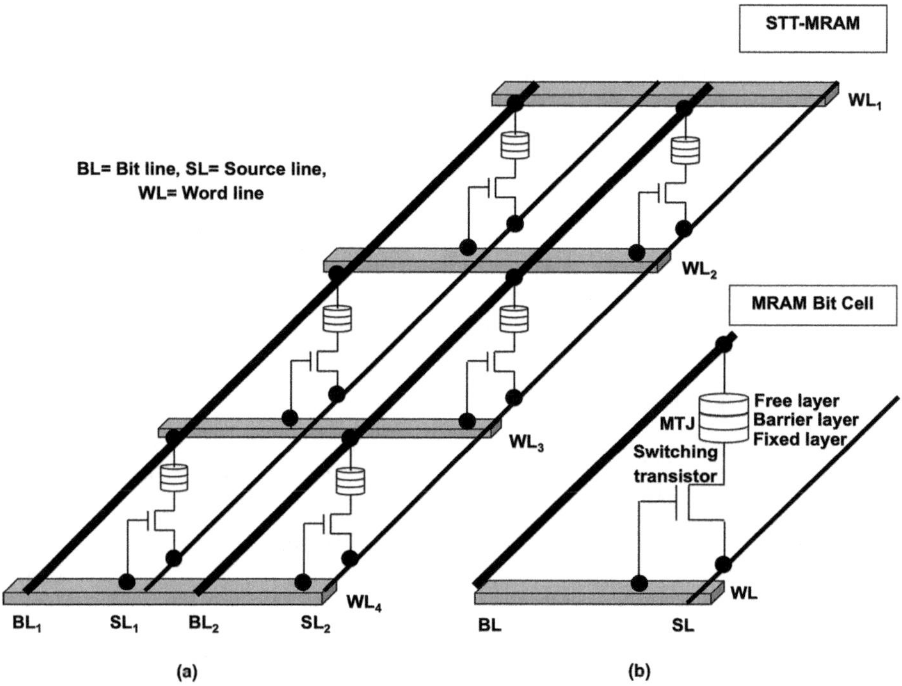

FIGURE 6.4 STT-MRAM. (a) Array containing MRAM bit cells whose construction is shown in (b). Each bit cell in (b) consists of an MTJ with free layer/barrier layer/fixed layer structure and a switching transistor whose source terminal is connected to the source line (SL), gate terminal to word line (WL), and drain terminal to the fixed layer of MTJ. The opposite free layer of MTJ is tied to the bit line (BL). In the array drawn in (a), the word lines WL_1, WL_2, WL_3, and WL_4 are laid out horizontally, whereas the bit lines BL_1 and BL_2 and source lines SL_1 and SL_2 are vertically placed running over the word lines.

magnetic direction to act as a reference for the free layer's state. The switching event occurs during alignment of magnetic fields: in the parallel state (low resistance of MRAM cell), when the magnetic fields of the free and pinned layers point in the same direction, current flows easily, signaling a binary 0. In the anti-parallel state (high resistance of MRAM cell), when these magnetic fields oppose each other, resistance spikes, setting the output to binary 1.

6.3.3 Array Architecture

To integrate these junctions for building a memory array, the MRAM cells are paired with transistors that act like gatekeepers, controlling which cell is accessed. The access transistors are CMOS transistors. Each CMOS transistor selects a specific cell and channels current through the magnetic tunnel junction (MTJ) for reading or writing data. In STT (spin-transfer torque) MRAM, a spin-polarized current acts as a precise switch, flipping the free layer directly. This approach boosts both scalability and energy efficiency.

6.3.4 Operational Mechanism of the Array

(i) Writing: Writing data in the array is like sending a stream of electrons on a mission. As the write current flows through the MTJ stack, electrons pass through the fixed (pinned) layer and emerge spin-polarized, all aligned in one direction. These polarized electrons tunnel through the insulating barrier and reach the free layer, where they impart a stimulus of angular momentum (torque) to the free layer's magnetization, nudging its magnetization. If the current is strong enough ($>10^6$ A/cm^2), this nudge flips the magnetization direction of the free layer. To write a "0" (parallel-low resistance), current flows from the free layer down to the reference layer, making the magnetization of the free layer parallel (P) to the fixed layer to align the layers. To write a "1" (anti-parallel—high resistance), the current reverses direction, flowing from the reference layer up to the free layer, flipping the free layer to an AP state, which is in opposition with the reference layer.

(ii) Reading: Reading from the array is a delicate process of measuring the tunneling magnetoresistance (TMR) of the MRAM cell. A gentle voltage is applied via the WL, coaxing a current through the cell to measure its resistance for distinguishing between high- and low-resistance states without disturbing the cell's magnetic state. If the layers are aligned, current flows easily, signaling a low resistance. If they are opposed, resistance rises. In other words, parallel alignment equals low resistance and anti-parallel alignment equals high resistance. A sense amplifier (SA) then compares this signal to a reference, decoding the final stored bit value.

In tightly packed, high-density arrays, read currents can approach write levels, risking accidental bit flips and leading to read-disturb errors. To guard against read failures, engineers modify the surrounding circuitry. By using smart read strategies in the peripheral circuitry that alternate current levels (e.g., high current vs. low current), they are able to keep data safe and performance high.

6.3.5 Main Functions of MRAM in Beyond-CMOS Nanoelectronics

MRAM plays an essential role in the "more-than-Moore" strategy, which aims to broaden the functionalities and energy efficiency of electronics by incorporating non-charge-based devices into the CMOS framework.

MRAM serves as a versatile memory solution by offering an appealing combination of SRAM's rapid speed, DRAM's high storage capacity, and flash memory's ability to retain data without power, along with exceptional endurance ($>10^{15}$ write cycles), positioning it as a potential "universal memory."

MRAM bridges the gap in the memory hierarchy by merging the quickness of SRAM, the density of DRAM, and the non-volatility of flash memory. This distinctive combination makes it suited to replace various tiers of the memory hierarchy, such as L1/L2 caches (using high-speed SOT-MRAM variants) and main memory, streamlining system architecture and enhancing overall performance.

The ability to offer hybrid integration is a significant characteristic of MRAM. The MRAM cells, particularly their central element, the MTJ, can be produced using standard CMOS processes, facilitating hybrid architectures that integrate magnetic and CMOS components. This convergence enables the creation of specialized, multi-functional devices and systems-on-chip (SoCs).

MRAM technology is being actively researched for innovative computing paradigms, like IMC and neuromorphic (brain-inspired) electronics, where both data storage and processing take place within the memory network itself, thereby alleviating data transfer issues (the von Neumann bottleneck), reducing power usage for AI tasks and leading to highly parallel, energy-efficient AI chips.

The provisions of "normally off" and "instant-on computing" are noteworthy attractions of MRAM. The inherent non-volatility of MRAM supports "normally off" computing, enabling systems to promptly save their state during power interruptions and swiftly resume when the power is restored (instant-on) without dependency on external batteries or lengthy boot sequences. This significantly lowers standby power consumption.

MRAM tackles various limitations associated with conventional CMOS security features. The natural physical variations in its MTJs are harnessed to develop robust physical unclonable functions (PUFs) for applications in hardware security, supplying unique, hard-to-replicate device identifiers.

MRAM provides robustness in harsh conditions. The data storage mechanism of MRAM, which uses magnetic fields rather than charge, renders it highly resistant to radiation, extreme temperatures, and other environmental stressors compared to charge-based devices. This quality makes it an essential technology for mission-critical uses in aerospace and industrial environments, such as satellites and flight control systems, and military applications.

MRAM guarantees enhanced performance and efficiency. Advanced MRAM variants like Spin-Orbit Torque (SOT-MRAM) achieve sub-nanosecond switching speeds and remarkably low switching energy levels, making them appropriate replacements for volatile L1/L2 cache memory while enhancing overall system performance and energy efficiency. The natural geometric variations found in MRAM cells can be utilized to develop MRAM-based Physically Unclonable Functions (PUFs), offering strong, on-chip security features for systems like IoT devices.

6.3.6 Technological Progress of MRAM

Continued research is focused on creating advanced MRAM technologies that address the constraints of existing designs. Spin-transfer torque MRAM (STT-MRAM) is currently the commercial benchmark. STT-MRAM is produced on cutting-edge CMOS nodes, offering high densities, endurance, and data retention. Nevertheless, its shared read/write path can lead to reliability issues at the most advanced nodes. Spin-orbit torque MRAM (SOT-MRAM) is the upcoming SOT-MRAM that separates the read and write paths, presenting the prospect of sub-nanosecond switching speeds, even lower power usage, and outstanding endurance, which could make it a future competitive substitute for L1/L2 cache memories.

Example 6.2 A spin-transfer torque magnetic tunnel junction (STT-MTJ) device is being developed for memory applications that go beyond CMOS technology and are non-volatile. The device's reliability is influenced by its thermal stability factor Δ, which helps prevent unwanted magnetic state changes caused by thermal variations. The equation for determining the thermal stability factor is

$$\Delta = \frac{E_b}{k_B T} \quad (6.5)$$

where E_b represents the energy barrier for magnetization reversal (measured in Joules, J), k_B denotes the Boltzmann constant (approx 1.38×10^{-23} J/K), and T stands for absolute temperature (in Kelvin, K). A common goal for ensuring dependable data retention in the industry is to achieve $\Delta \geq 60$ at room temperature. If an MRAM device operates at room temperature ($T = 300$ K) and requires a thermal stability factor of $\Delta = 60$, calculate the minimum energy barrier (E_b) needed for the magnetic tunnel junction in electron-volts (eV). Solution: Given: $\Delta = 60$, $T = 300$ K, $k_B \approx 1.38 \times 10^{-23}$ J/K. We know that $1\text{eV} \approx 1.602 \times 10^{-19}$ J. Rearranging the given Δ-formula to isolate the energy barrier E_b:

$$E_b = \Delta k_B T \quad (6.6)$$

Inserting the provided values into the equation:

$$E_b = 60 \times 1.38 \times 10^{-23}\ \text{J/K} \times 300\ \text{K} \quad (6.7)$$

Calculating the energy barrier in Joules:

$$E_b = 24{,}840 \times 10^{-23}\ \text{J} = 2.4840 \times 10^{-19}\ \text{J} \quad (6.8)$$

Converting the energy barrier from Joules (J) to electron-volts (eV):

$$E_b = \frac{2.4840 \times 10^{-19}\ \text{J}}{1.602 \times 10^{-19}\ \text{J/eV}} = 1.55056\ \text{eV} \quad (6.9)$$

The minimum energy barrier (E_b) required for reliable data retention at room temperature is roughly 1.55 eV. A widely recognized numerical concept in MRAM related to beyond-CMOS nanoelectronics is the minimum energy barrier (E_b), which is essential for assessing the dependability and consistency of data storage in an MRAM bit over time.

Example 6.3 A particular STT-MRAM bit cell is engineered to function in two distinct modes:

Non-Volatile Mode: This mode necessitates a high thermal stability factor Δ of 40 to guarantee a ten-year data retention period, but it results in a significant write energy consumption of 5 pJ (picojoules) for each write operation.

Low-Retention (Cache) Mode: By easing the non-volatility requirements, the thermal stability factor is lowered to 18, which helps to reduce energy usage. The write energy in this mode is only 1.5 pJ per write operation.

A computing framework is established to implement this MRAM using a multi-retention approach. Over a designated time frame, 1000 write operations are executed. If 80% of

these write operations are allocated to the low-retention mode and the remaining 20% to the non-volatile mode, what is the overall energy expended by all write operations?
Solution: To determine the total energy used, we compute the energy for each mode individually and then combine them. Then we calculate the number of writes for each mode: Low-retention writes: $1000 \times 0.80 = 800$ writes. Non-volatile writes: $1000 \times 0.20 = 200$ writes.
Next, computing the energy consumed for each mode:

$$\text{Energy for low} - \text{retention writes} = 800 \text{ writes} \times 1.5 \text{ pJ/write} = 1200 \text{ pJ} \quad (6.10)$$

$$\text{Energy for non} - \text{volatile writes} = 200 \text{ writes} \times 5 \text{ pJ/write} = 1000 \text{ pJ} \quad (6.11)$$

Calculating the overall energy consumed:

$$\begin{aligned}\text{Total energy} &= \text{Energy for low} - \text{retention writes} + \text{Energy for non} - \text{volatile writes}\\ &= 1200 \text{ pJ} + 1000 \text{ pJ} = 2200 \text{ pJ} = \frac{2200 \text{ nJ}}{1000} = 2.2 \text{ nJ}\end{aligned} \quad (6.12)$$

The total energy expended by all write operations amounts to 2200 pJ (picojoules) or 2.2 nJ (nanojoules).

6.4 PHASE-CHANGE MEMORY

6.4.1 Need of PCM

PCM is essential in the sphere of nanoelectronics beyond CMOS, providing a high-density, high-performance NVM solution that narrows the performance divide between traditional memory (DRAM) and storage (Flash). It also serves as a pivotal enabling technology for emerging non-von Neumann computing models, such as neuromorphic and IMC (Jeyasingh et al. 2014, Raoux et al. 2014, Syed et al. 2025).

6.4.2 PCM Cell Structure

Imagine a prototypical PCM cell as a tiny sandwich, with a special chalcogenide material (often $Ge_2Sb_2Te_5$ or GST) nestled between two electrodes (Figure 6.5). The top electrode (TE), usually made of titanium nitride (TiN), acts as a broad gateway for current, while the bottom electrode (BE) is a narrow path that focuses energy to generate intense, localized heat. At the heart of the cell, the phase-change layer (PCL) is the active storage medium that stores data by reversibly shifting between glassy, high-resistance (logic 0) and crystalline, low-resistance (logic 1) forms, all thanks to carefully controlled bursts of heat via joule heating. Surrounding insulation confines the heat to locations where it is needed, boosting efficiency. Each cell is individually addressed by a transistor (MOSFET or BJT), ensuring precise control. To change the cell's state, electrical pulses trigger rapid localized Joule heating: a quick, powerful pulse melts the material. It is then "quenched" (cooled rapidly) and locked into a disordered state for RESET (amorphization). A longer, gentler pulse

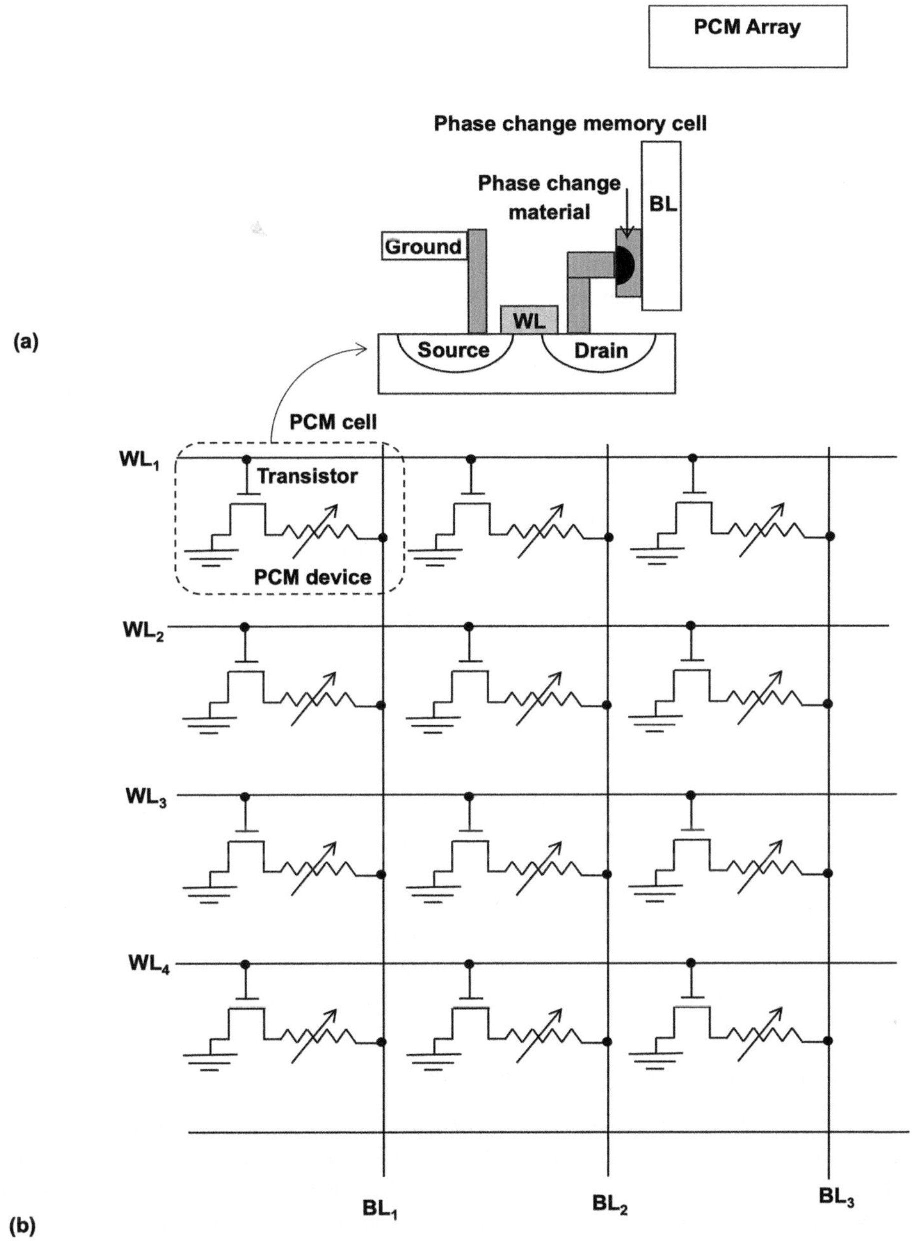

FIGURE 6.5 Phase-change memory: (a) Phase-change memory cell, and (b) array. In (a), the end of PCM cell containing the phase-change material is joined to the bit line (BL) and its opposite end to the drain of the transistor. The source of the transistor is grounded while the gate electrode is joined to the word line (WL). In (b) the PCM array is organized with word lines WL_1, WL_2, WL_3, and WL4 along the horizontal direction; and bit lines BL_1, BL_2, and BL_3 in the vertical direction.

heats the material above its crystallization temperature but below its melting point, allowing atoms to settle into an ordered crystal for SET state (crystallization). Reading the cell is gentle—a faint voltage checks resistance without disturbing the current phase. of the delicate structure.

6.4.3 PCM Array Architecture

PCM arrays are built like intricate city grids, typically in hierarchical structures with layers of organization to handle powerful write currents and pack in more memory. Each cell pairs a storage element with a selector device, such as a MOSFET or diode, to ensure that only the correct cell is selected. WLs connect to the gates of the selectors, acting as streets that activate entire rows, while BLs connect to the phase-change material to deliver the vital current needed to program or read each cell.

6.4.4 Working of the PCM Array

The operation of PCM relies on its mastery of heat for controlling thermal energy (Joule heating) to switch material phases. By channeling strong currents from the bottom to the top electrode, the cell's core heats up, coaxing the material to switch between phases. This transformation happens right where the heater meets the PCL, i.e., the PCL/heater interface, driven by a surge of energy, whereupon the phase transition is formed according to the resulting Joule heating. To read the stored data, a gentle current is sent through the cell, revealing its state without disturbing the memory.

(i) SET (Logic 1): A long, moderate current pulse is applied to heat the material to its crystallization temperature ~300°C–600°C. Consequently, the atoms rearrange into a highly conductive ordered lattice.
(ii) RESET (Logic 0): A short, high-intensity current pulse is sent to melt the material at a temperature ~600°C. Thereafter, rapid "quenching" (cooling) freezes the atoms of the material, yielding a disordered, HRS.
(iii) READ: A very low voltage pulse measures the resistance of the material without producing a sufficiently large quantity of heat to change the phase.

6.4.5 Function of PCM as a Memory Technology

Within traditional computing systems, PCM stands out as a primary option to enhance or replace present memory technologies because of its distinct array of features. Non-volatility of PCM is very useful because it maintains data even when power is lost, in contrast to volatile memory such as DRAM. High density and scalability of PCM provide a much higher memory density compared to Flash memory and show good scalability down to sub-20 nm process nodes and beyond. In terms of performance, PCM delivers rapid access times, nearing those of DRAM while also providing the persistence characteristic of storage, making it suitable for "storage class memory" (SCM). Owing to their CMOS compatibility, PCM cells can be integrated into the BEOL manufacturing process, facilitating simpler integration with advanced CMOS logic circuits.

6.4.6 Functions of PCM in Emerging Architectures

In addition to its role as stand-alone memory, PCM's intrinsic properties are immensely beneficial for innovative computing paradigms aimed at addressing the "von Neumann bottleneck" (the energy and time expenses associated with transferring data between separate processing and memory units):

PCM devices facilitate computing directly in memory and enable neuromorphic computing. These devices can function not just to store information but also to carry out computational tasks, such as analog matrix-vector multiplication, directly within the memory array itself. The capability to program individual PCM devices into a variety of stable intermediate resistance states (multi-level cell capability) allows them to replicate synaptic behavior in brain-inspired neural networks.

To provide functional diversification, the materials science behind PCMs is being researched in the field of integrated photonics, where the significant optical contrast between phases facilitates high-speed, energy-efficient optical switching and non-volatile photonic memory cells.

All in all, PCM's capacity to merge the best qualities of various memory types with innovative, non-volatile switching mechanisms positions it as a leader in energy-efficient data storage and for paving the way for the next generation of computing architectures.

Example 6.4 A single phase-change memory (PCM) cell is set to its high-resistance, amorphous state through the "RESET" operation, which involves the application of a brief, high-voltage electrical pulse. Given the following parameters for the programming pulse: Voltage (V): 4 V, Average Current (I_{Avg}): 0.5 mA, Pulse Width (t): 50 ns. Calculate the energy consumed during this RESET operation.

Solution: The energy (E) used during an electrical pulse can be determined using the formula

$$E = V \times I \times t \tag{6.13}$$

where V indicates voltage, I signifies current, and t represents time (pulse width). First, we identify the provided values: $V = 4$ V, $I_{\text{Avg}} = 0.5 \text{ mA} = 0.5 \times 10^{-3}$ A, $t = 50 \text{ ns} = 50 \times 10^{-9}$ seconds. We then apply the energy formula. Thus,

$$E = V \times I \times t = 4 \text{ V} \times 0.5 \times 10^{-3} \text{ A} \times 50 \times 10^{-9} \text{ seconds} \tag{6.14}$$

Next, we calculate the result, leading to

$$E = 4 \times 0.5 \times 50 \times 10^{-3} \times 10^{-9} \text{ Ws} = 100 \times 10^{-12} \text{ J/s} = 100 \text{ pJ} \tag{6.15}$$

The energy consumed during the RESET operation is 100 pJ (picojoules). The low energy consumption of these nanoscale devices (in the picojoule range) is a significant factor that makes them ideal candidates for energy-efficient, beyond-CMOS memory and in-memory computing applications.

6.5 FERROELECTRIC RANDOM-ACCESS MEMORY

6.5.1 Impact of FeRAM

FeRAM is set to significantly influence beyond-CMOS nanoelectronics, shifting traditional data storage into new territories by facilitating crucial emerging technologies such as IMC, neuromorphic systems, and non-volatile logic. The identification of ferroelectric properties in highly scalable, CMOS-compatible hafnium-based oxides has reignited interest and created a pathway for their integration into advanced nanoscale devices (Ishiwara 2012, Adnaan et al. 2025, Hwang et al. 2025).

6.5.2 FeFET Cell Structure

In a FeFET, the usual gate oxide of a MOSFET is replaced by a ferroelectric layer, often made from HfO_2 (Figure 6.6). This special layer, nestled between the gate and the channel, sometimes with a thin interfacial layer, acts like a dynamic capacitor in series with the underlying MOSFET's own capacitor. The device still has a source, drain, and gate, but now the ferroelectric material can switch its polarization to control channel conductivity, directly steering how easily current flows through the channel.

6.5.3 Programming an FEFET

To program a FeFET, a positive gate-source voltage ($V_{GS} > +V_T$) flips the ferroelectric layer to a positive state, storing a logic "1" and lowering the threshold voltage for an "ON" state (logic "1," low V_T). A negative voltage ($V_{GS} < -V_T$) does the opposite, setting a logic "0" and raising the threshold for an "OFF" state (logic "0," high V_T). The magic lies in the ferroelectric layer's ability to remember its polarization. For memory retention, the remanent polarization of the ferroelectric layer keeps the channel in the high or low V_T state. So, data is preserved even when the power is off. When reading, a gentle voltage (intermediate V_{GS}) is applied, and the amount of drain current (I_D) flowing through the cell indicates whether it holds a "1" or a "0" (high I_D for "1," low I_D for "0").

6.5.4 FeFET Array Architecture

In a 1F-1T memory array, each cell pairs a ferroelectric transistor (F) with a current-limiting partner transistor (T). This extra transistor reduces the impact of drain current (I_D) variations. It acts like a traffic cop, keeping the FeFET's current in check and smoothing out the wild swings between ON and OFF states. By capping current in the low-threshold state, the design effectively addresses the high-current variation (high OFF/ON current ratio) of stand-alone FeFETs to make FeFET arrays more reliable and better suited for demanding tasks such as AI acceleration. Compared to standard FeFET memory, this approach tames variability and reduces the risk of read or write errors. It thus improves upon standard 1T FeFET memory by mitigating the high variability and read/write disturbances inherent to 1T-only FeFET memory.

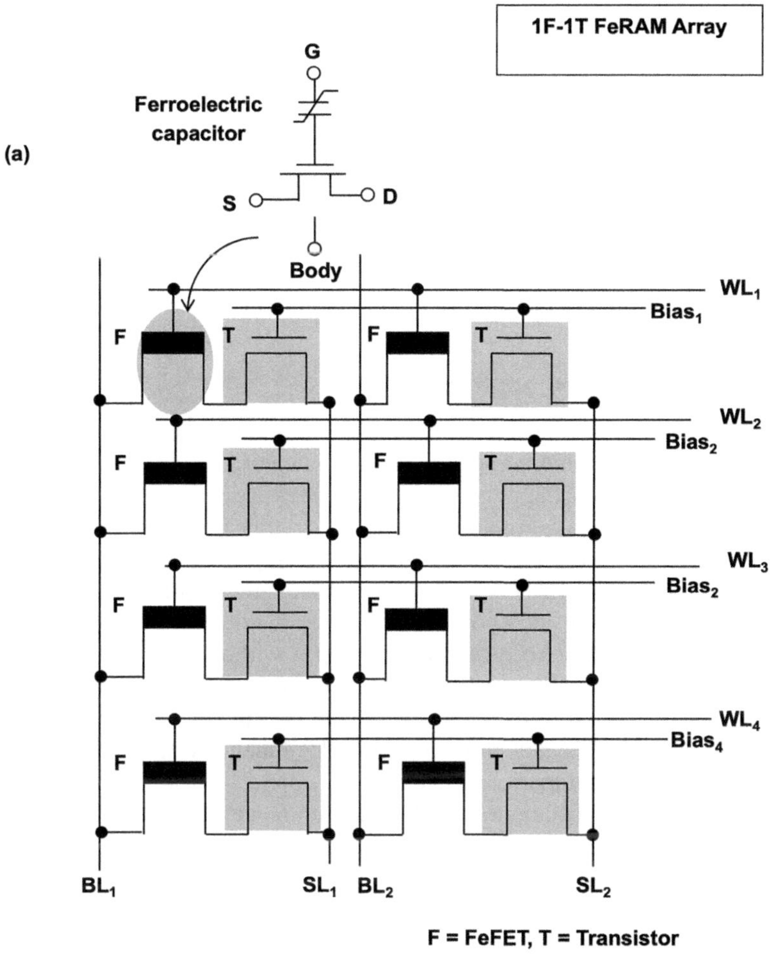

FIGURE 6.6 1F-1T FeRAM Array: (a) equivalent circuit of the FeFET consisting of a ferroelectric capacitor in series with a MOSFET, with its source, drain, gate, and body terminals marked. (b) Array organization in which each cell comprises an FEFET F and a regular transistor T. Source of F is joined to a bit line (BL), and its drain is connected to the source of T. The drain of T is tied to the source line (SL). In (b), word lines are WL_1, WL_2, WL_3, and WL_4; bit lines are BL_1 and BL_2; and source lines are SL_1 and SL_2.

6.5.5 Working of the FeFET Array

In each 1F–1T cell, the FeFET stores information by shifting its threshold voltage according to the direction of its ferroelectric polarization. WLs run along each row, connecting to the gates, while select lines (SLs) help pinpoint the exact cell to access. The current measured on the BLs then reveals the stored data, acting as a digital fingerprint for each cell.

Write Operation: To store data, a specific voltage is sent to the gate transistor's gate terminal to access the chosen cell, flipping its polarization and setting the threshold voltage high or low to represent the stored data as either a binary "0" or a "1."

Read Operation: To read data, the cell is accessed through the current-limiting transistor by applying the right gate voltages. The resulting current flowing through the cell is measured to reveal the cell's resistance and thus the stored value. Thanks to the extra transistor, the read current remains steady within acceptable limits and reliable, even as FeFETs differ from one cell to another, owing to manufacturing variations in the FeFET itself.

6.5.6 Chief Functions and Applications

FeRAM serves as a valuable component for "computing-in-memory" (CIM) systems and neuromorphic computing, which integrate memory and processing to address the traditional "memory wall" issue that arises between processors and separate memory units. Ferroelectric field-effect transistors (FeFETs) can act as artificial synapses in neuromorphic systems, utilizing their adjustable intermediate polarization states to replicate synaptic weight alterations in a brain-inspired architecture. FeRAM can be incorporated into non-volatile flip-flops and other logic circuits, allowing devices to maintain their states without the need for continuous power, enabling immediate power-on functions for ultra-efficient systems, such as those used in the Internet of Things (IoT). In addition to energy-efficient, non-volatile logic, FeRAM is well-suited for embedded memory applications, such as in microcontrollers for automotive systems and smart cards, due to its rapid write speeds, low energy consumption, and high endurance. Over and above, the capability of content-addressable memory (CAM) is a valuable characteristic of FeRAM. The FeFET-based CAM cells provide non-volatility and a reduced on-chip footprint when compared to current SRAM-based CMOS alternatives, making them beneficial for rapid search functions. For advanced memory architectures, traditional FeRAM encounters density restrictions. To address this issue for high-density uses and navigate these circumstances, new architectures and material enhancements are being investigated.

Ongoing research is focused on the vertical stacking of FeRAM components, including FTJs and capacitors, to elevate integration density and surpass physical scaling constraints. Besides standard computational tasks, ferroelectrics are undergoing evaluation for unique physics-based computing platforms, like Ising machines designed to tackle complex optimization challenges. Ising machines are dedicated hardware developed to address intricate optimization challenges by emulating the Ising model, a concept in physics involving interacting "spins" (similar to tiny magnets) that naturally reach low-energy states symbolizing solutions. These machines utilize physical systems such as optics or electronics rather than conventional digital computing methods. This represents a revolutionary approach, leveraging physics itself for computation, and holds promise for providing robust new tools to solve computationally challenging problems that can overwhelm even the most advanced supercomputers available today.

All things considered, FeRAM's exceptional characteristics of non-volatility, energy efficiency, rapid operation, and durability position it as a vital technology for pushing nanoelectronics beyond conventional CMOS limitations, especially in the creation of more efficient and intelligent computing systems.

Example 6.5 A ferroelectric RAM (FeRAM) memory device that uses HfO_2-based ferroelectric films boasts an exceptionally high write endurance of 1×10^{12} cycles. This device is employed in a system that writes data to the same memory cell continuously at an average rate of 10 MHz (megahertz). How long, in years, can this particular FeRAM cell be continuously operated (written to) before it is anticipated to reach its maximum specified endurance limit and possibly fail?

Solution: To determine this, we first need to calculate the total time in seconds and then convert that duration into years. First, we calculate total write cycles: The maximum endurance is already provided as 1×10^{12} cycles. Next, we compute the total time in seconds: The write frequency is 10 MHz, equivalent to 10×10^6 cycles/s 10^7 cycles/s.

$$\begin{aligned}\text{Time (in seconds)} &= \frac{\text{Total cycles}}{\text{Cycles/second}} = \frac{1 \times 10^{12}\ \text{cycles}}{10^7\ \text{cycles/s}} = 1 \times 10^5\ \text{seconds} \\ &= 100{,}000\ \text{seconds}\end{aligned} \tag{6.16}$$

Now, we convert the time to years: There are approximately $365 \times 24 \times 60 \times 60 =$ 31,536,000 seconds in a year (60 seconds/min, 60 minutes/h, 24 hours/day, and 365 days/year).

$$\text{Time (in years)} = \frac{\text{Time (in seconds)}}{\text{Seconds per year}} = \frac{100{,}000\ \text{s}}{31{,}536{,}000} \approx 0.00317\ \text{years} \tag{6.17}$$

The FeRAM cell can be continuously operated for roughly 100,000 seconds or about 0.00317 years (approximately 1 day and 4 hours) before reaching its specified maximum endurance limit.

Note: Although FeRAM exhibits higher endurance compared to flash memory, continuous operation at such elevated frequencies over prolonged periods rapidly depletes its cycle life. Real-world applications typically utilize FeRAM for specific tasks where this endurance is an advantage over the device's operational lifetime, rather than for constant high-frequency writes to a single cell.

6.6 DISCUSSION AND CONCLUSIONS

ReRAM, MRAM, PCM, and FeRAM are the frontrunners in NVM technologies that aim to supplant Flash in embedded systems, providing benefits such as quicker speeds, reduced power consumption, and superior endurance compared to older NVMs. MRAM and ReRAM are well-suited for advanced logic scaling, FeRAM offers exceptional endurance and data retention ideal for applications like smart cards and logging, while PCM holds promise for high density and IMC. However, each technology presents distinct trade-offs in terms of speed, power, cost, and density, catering to various applications ranging from caches to IoT devices. Table 6.1

TABLE 6.1
Comparison of Emerging Memories

Sl. No.	Technology Name	Acronym	Primary Mechanism	Strengths	Status/Usage
1	Resistive random-access memory	ReRAM	Formation of conductive filaments	High density, compatibility with CMOS, low energy usage	Used as embedded non-volatile memory for security features (PUFs) and artificial intelligence applications at the edge
2	Magnetoresistive random-access memory	MRAM	State of magnetic tunnel junction (MTJ)	Speed comparable to SRAM, high endurance, low energy consumption	Replacing SRAM in system-on-chips (SoCs) and microcontrollers (MCUs)
3	Phase-change memory	PCM	Phase transition of chalcogenide glass	Scalable, capable of multi-bit storage	Characterized as a niche memory/storage solution
4	Ferroelectric random-access memory	FeRAM	Polarization occurring in ferroelectric materials	Rapid write speeds, exceptional endurance, low power consumption	Applying in industrial sensors, transportation, and low-density independent chips

compares these technologies. The silver lining is that these technologies are progressively evolving from theoretical ideas to targeted embedded applications across AI, automotive, and IoT industries.

Non-volatile memories, discussed in this chapter, occupy central positions in tomorrow's computing breakthroughs. Traditional systems keep processing and memory separate, creating a stubborn memory wall that wastes precious time and energy by forcing data to shuttle back and forth. With NVM elements, analog multiply-and-accumulate operations can be performed directly in the memory array, supercharging AI and neural network performance. These advances help break the von Neumann bottleneck by enabling compute-in-memory operations. When non-volatility is woven into logic gates, systems cut down data transfer overhead and reveal bold, efficient new architectures. The next chapter will deal with the relevant computing architectures for the post-CMOS age.

REFERENCES

Adnaan M., S. A. Chamazcoti, E. Karimov, M. G. Bardon, F. Catthoor and J. van Houdt 2025 Benchmarking of FERAM-based memory system by optimizing ferroelectric device model, *IEEE Journal on Exploratory Solid-State Computational Devices and Circuits*, 11: 99–106, https://doi.org/10.1109/JXCDC.2025.3618883

Handy J. 2018 Emerging memories today: Understanding bit selectors, The Memory Guy Blog, https://thememoryguy.com/emerging-memories-today-understanding-bit-selectors/

Hellenbrand M., I. Teck and J. L. MacManus-Driscoll 2024 Progress of emerging non-volatile memory technologies in industry, *MRS Communications*, 14: 1099–1112, https://doi.org/10.1557/s43579-024-00660-2

Hu J.-M., Z. Li, L.-Q. Chen and C.-W. Nan 2011 High-density magnetoresistive random access memory operating at ultralow voltage at room temperature, *Nature Communications*, 2: 553, https://doi.org/10.1038/ncomms1564

Hwang H., S. Youn and H. Kim 2025 Recent advances in ferroelectric materials, devices, and in-memory computing applications, *Nano Convergence*, 12: 55, https://doi.org/10.1186/s40580-025-00520-2

Ielmini D. and G. Pedretti 2025 Resistive switching random-access memory (RRAM): Applications and requirements for memory and computing, *Chemical Reviews*, 125(12): 5584–5625, https://doi.org/10.1021/acs.chemrev.4c00845

Ishiwara H. 2012 Ferroelectric random-access memories, *Journal of Nanoscience and Nanotechnology*, 12(10): 7619–7627, https://doi.org/10.1166/jnn.2012.6651

Jeyasingh R., E. C. Ahn, S. B. Eryilmaz, S. Fong and H.-S. P. Wong 2014 Chapter 5: Phase change memory, In: Chen A., J. Hutchby, V. Zhirnov and G. Bourianoff (Eds.), *Emerging Nanoelectronic Devices*, Wiley Online Library, John Wiley & Sons, Hoboken, NJ, USA, pp. 78–109, https://doi.org/10.1002/9781118958254.ch05

Katehi L., S.-I. Yi, Y. C. Lin, S. Banerjee, Q. Xia and J. J. Yang 2025 Emerging nonvolatile memory technologies in the future of microelectronics, *ACS Omega*, 10(27): 28492–28498, https://doi.org/10.1021/acsomega.5c01414

Nguyen V.D., S. Rao, K. Wostyn and S. Couet 2024 Recent progress in spin-orbit torque magnetic random-access memory. *npj Spintronics* 2: 48, https://doi.org/10.1038/s44306-024-00044-1

Raoux S., F. Xiong, M. Wuttig and E. Pop 2014 Phase change materials and phase change memory, *MRS Bulletin*, 39: 703–710, https://doi.org/10.1557/mrs.2014.139

Sharma J., S. Akashe and G. M. Upadhyay 2026 Investigating non-volatile memory architectures for neuromorphic systems with CMOS spike-timed synaptic plasticity, *Discover Computing*, 29: 2, https://doi.org/10.1007/s10791-025-09780-2

Syed G. S., M. L. Gallo and A. Sebastian 2025 Phase-change memory for in-memory computing, *Chemical Reviews* 125 (11): 5163–5194, https://doi.org/10.1021/acs.chemrev.4c00670

Yu S. 2016 *Resistive Random Access Memory (RRAM): From Devices to Array Architectures, Synthesis Lectures on Emerging Engineering Technologies*, Springer Nature Switzerland AG, Cham, pp. 71.

Zahoor F., T. Z. Azni Zulkifli and F. A. Khanday 2020 Resistive random access memory (RRAM): An overview of materials, switching mechanism, performance, multilevel cell (MLC) storage, modeling, and applications, *Nanoscale Research Letters*, 15: 90, https://doi.org/10.1186/s11671-020-03299-9

Zhu J.-G. 2008 Magnetoresistive random access memory: The path to competitiveness and scalability, *Proceedings of the IEEE*, 96(11): 1786–1798, https://doi.org/10.1109/JPROC.2008.2004313

7 Computing Architectures and Paradigms

7.1 INTRODUCTION

Beyond-CMOS encompasses not only innovative materials but also their organization for data processing. Researchers are exploring various "state variables" that go beyond the traditional voltage and current. Instead of simply substituting transistors, researchers are crafting new architectures specifically designed to utilize the distinct advantages (such as non-volatility and parallel processing) offered by these new devices. This comprehensive approach promotes more efficient hardware tailored for specific tasks, transcending standard Boolean logic and the von Neumann bottleneck (Beyond von Neumann 2020). The computing architectures within Beyond-CMOS nanoelectronics signify a transformation in data-processing methods; rather than relying exclusively on electron charge, these architectures leverage alternative physical properties to enhance energy efficiency and overall performance. It is anticipated that Beyond-CMOS technologies will first complement rather than outright replace CMOS. This results in hybrid architectures where conventional silicon handles general-purpose functions while Beyond-CMOS accelerators optimize specialized workloads such as AI, security, or signal processing (Basu et al. 2019, Finocchio et al. 2024).

In continuation of our deliberations on non-volatile memories in the preceding chapter, we try to assess the importance of memory technologies in the upcoming computing architectures. Memory technologies have stepped out of the shadows of passive storage, taking center stage as the industry pivots from compute-centric to memory-centric architectures.

7.2 IN-MEMORY COMPUTING

In-memory computing (IMC) plays a vital role in "beyond-CMOS" nanoelectronics as a key strategy to address the challenges posed by the conventional von Neumann architecture, especially the "memory wall" issue, in which data transfer between the processor and separate memory units consumes substantial time and energy (Mannocci et al. 2023, Athle and Borg 2024, Zhang et al. 2026).

7.2.1 Unveiling In-Memory Architecture from the Viewpoint of the von Neumann Model

We try to uncover the shared traits and distinguishing features of the two architectures. Figure 7.1 highlights a persuasive contrast between von Neumann and in-memory architectures, revealing their strengths and differences. The main components

DOI: 10.1201/9781003753469-7

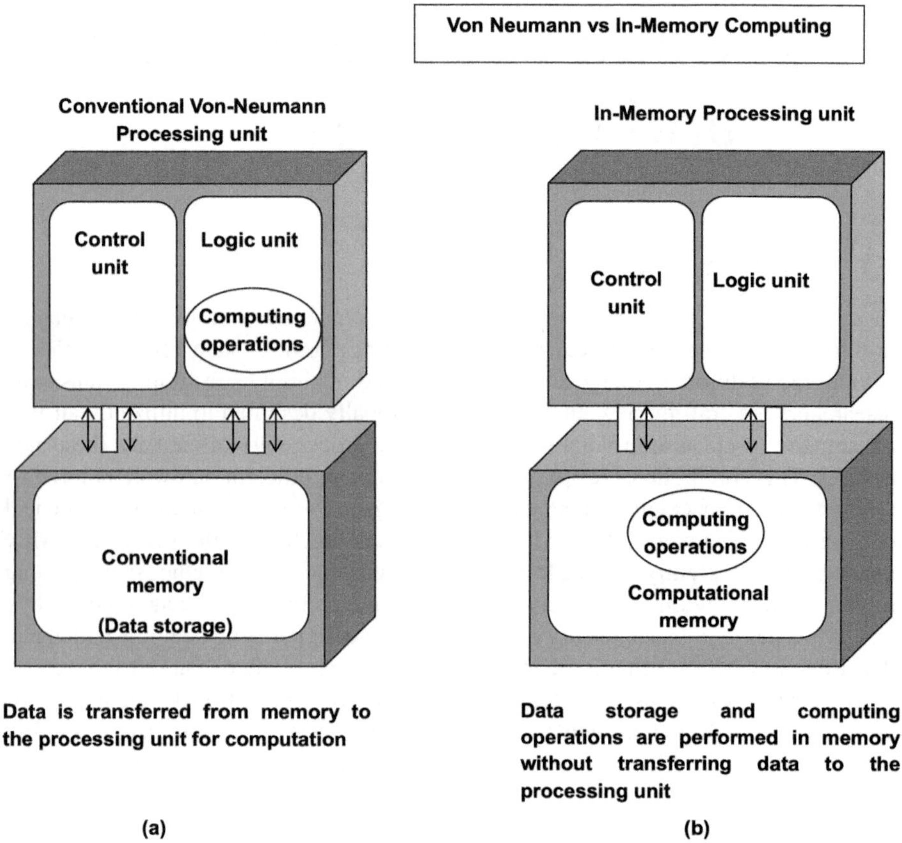

FIGURE 7.1 Von Neumann vs. in-memory computing. (a) The von Neumann processing unit features a control and a logic unit that relies on traditional memory for storing data, shuttling information back and forth for each computation. (b) The in-memory processing unit merges data storage and computation within the same memory, allowing operations to happen directly where the data resides, eliminating the need for constant data transfers.

of a von Neumann processing unit are the control unit and the logic unit (Figure 7.1 (a)). The processing unit works in collaboration with a conventional memory unit. The computing operations are performed in the logic unit in association with the conventional memory unit, which is used for data storage. Data must be transferred from memory to the processing unit for computation, and results must be returned to memory after computation, incurring undue expenditure of time and decelerating the work of the processor heavily. For the purpose of making a comparative analysis, we look at the parts of an in-memory processing unit. The principal parts of an in-memory processing unit (Figure 7.1(b)) are the control unit and logic unit, as in a von Neumann processing unit (Figure 7.1(a)). These work in collaboration with the memory unit. But the memory unit has a different role here because it has the dual functions of computational work and data storage, i.e., the computing operations are performed directly in the computational memory unit, which is also used for data storage. So, data transfer from memory to the processing unit is not necessary for computation, as was essential for the von Neumann case.

7.2.2 Structural Unit Cell of an In-Memory Computer, and Organization of the Array

The "structural unit cell" of an in-memory computer (IMC) is the basic memory component, along with the related circuitry that both stores information and executes logic functions directly within the memory array, typically arranged as a crossbar grid. It is essentially the memory bit cell that has been enhanced or explicitly designed to execute both data storage and computation. This cell is usually configured in a high-density crossbar array format. Ultimately, the structural unit cell in an IMC is designed to eliminate the conventional division between memory storage and processing, enabling the physical properties of the memory device itself to perform computations (e.g., by applying Ohm's law and Kirchhoff's current law for analog MVM). The components of an IMC unit cell vary depending on the underlying memory technology (such as ReRAM, SRAM, PCM, FeFET memory) (Figure 7.2). However, they typically encompass the following:

Storage Component: The primary element that retains the data bit(s) (e.g., a capacitor and transistor in DRAM, a 6T latch in SRAM, or a resistive-switching element in ReRAM). In IMC, the storage element typically needs to support multi-level states to execute multi-bit functions. At its core, an NVM device is commonly utilized, such as resistive random-access memory (ReRAM), PCM, or a FeFET. These devices are selected for their ability to store multiple bits of data (multi-level cells) and their physical characteristics (e.g., conductance states) that can be exploited for analog computations. ReRAM devices function as artificial synapses in neural networks, leveraging their variable resistance states to store synaptic weights and perform computations with high energy efficiency. PCM devices leverage phase transitions to alter resistance, offering an additional method for data storage and in-memory computations. FeFETs and capacitors (FeCaps) are employed to retain multi-bit information and execute multiply-and-accumulate (MAC) operations within the memory cell, thereby reducing data variability challenges. Magnetic Devices (MRAM, STT-MRAM, SOT-MRAM) are spintronics-based devices that utilize electron spin to decrease power consumption and provide dense, non-volatile storage for future computing needs. Electrochemical random-access memory (ECRAM) devices are under investigation for enhancing AI training by employing fully parallel tensor-product operations in crossbar arrays.

Access/Select Transistors: The memory component is generally combined with one or more transistors (e.g., a 1T-1R cell or a 6T/8T SRAM cell) that serve as access or selection transistors, managing data retrieval. These transistors select the specific cell for read or write operations and isolate it from external circuits when not active—mitigating sneak-path currents and improving operational stability.

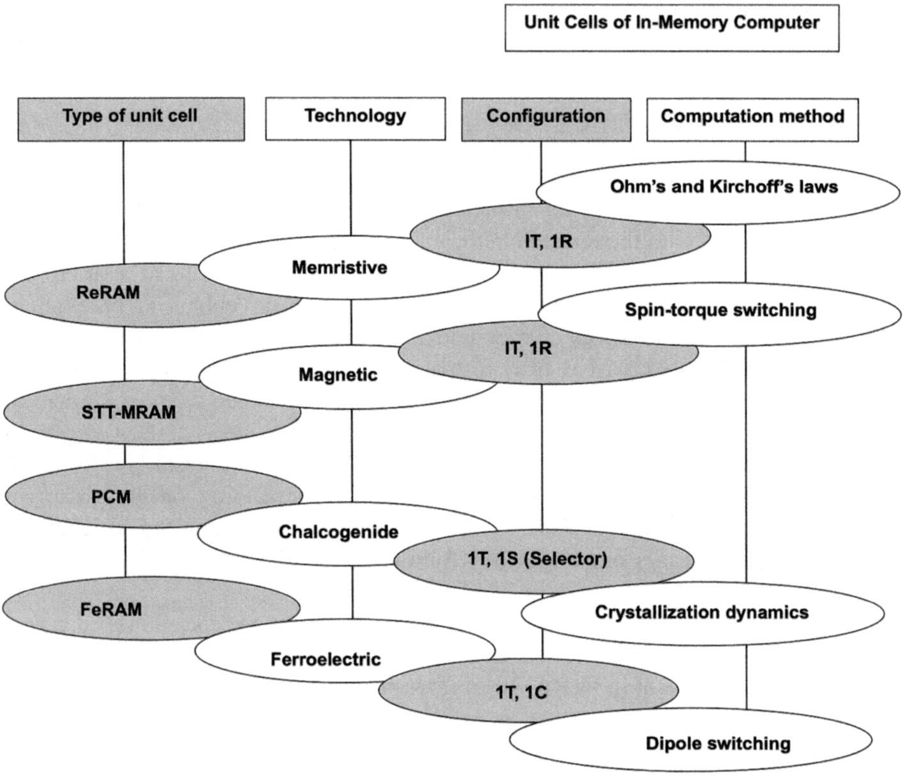

FIGURE 7.2 Unit cell options of an in-memory computer. The unit cell is based on one of the several types of non-volatile memory, such as ReRAM, STT-MRAM, PCM, and FeFET memory. Respective technologies (memristive, magnetic, chalcogenide, and ferroelectric), configurations (1T, 1R; IT, 1R; 1T, 1S; and 1T, 1C), and computation methods (Ohm's and Kirchhoff's laws, spin-torque switching, crystallization dynamics, and dipole switching) are also mentioned.

Interconnects: The cells interconnect in a grid (array) through horizontal and vertical lines (word lines and bit lines). In analog IMC, computations frequently use fundamental circuit principles, such as Ohm's law and Kirchhoff's current law, to combine currents flowing through the array and facilitate operations such as matrix-vector multiplication (MVM).

Typical IMC unit cell illustrations are:

SRAM-Based: Often incorporates modified standard cells like 6T or 8T SRAM cells (six or eight transistors). Additional transistors (read ports) and control lines are integrated to facilitate logic operations (such as XOR, AND/OR) directly within the memory array.

Resistive RAM (ReRAM)/Memristor-Based: A standard unit cell for analog MVM comprises a resistive-switching component (the memory element) and an access transistor (1T-1R cell). The component's resistance corresponds to the synaptic weight in a neural network, and applying voltage pulses enables the array to perform a weighted summation of inputs in the current domain.

FeFET-Based: The proposed 1FeFET-1R cell employs a multi-level FeFET to store multi-bit data and carry out multiply-accumulate (MAC) operations with high precision and energy efficiency.

Array Organization and Peripheral Circuitry: The unit cells are arranged in a tightly packed, grid-like crossbar array, comprising rows (word lines) and columns (bit lines). This setup facilitates extensive parallel processing, as voltage pulses applied to the word lines and current sensing on the bit lines enable concurrent MVM operations for uses such as neural networks. Memory arrays/crossbars serve as the hub where computations take place. Beyond-CMOS technologies, particularly emerging NVM devices, are integrated into high-density crossbar arrays to enable parallel operations, such as analog vector-matrix multiplication (VMM).

Although the primary computation occurs within the memory array, the IMC unit cell is augmented by the surrounding supporting circuits. These peripheral circuits may include analog-to-digital converters (ADCs), digital-to-analog converters (DACs), voltage/current sources, and sense amplifiers to support data input, output, and necessary conversions for data processing and the retrieval of computational results.

7.2.3 Core Components and Functional Processes of an In-Memory Computing (IMC) System

The core components typically include a unified memory array for data processing, input/output memory devices for data storage, and data-processing components such as ADCs and DACs.

The memory array is an essential part of the IMC system, where data is retained, and calculations are executed on-site, eliminating the need to transfer large amounts of data to a separate central processing unit (CPU). It capitalizes on multiple memory technologies, such as the following:

(i) Charge-based devices, including SRAM, DRAM, Flash, and FeCAP, can be modified for IMC. However, they sometimes necessitate additional components and might require moving data to a nearby cache before processing. While the purpose of IMC is to eliminate the traditional disk I/O, some architectures still incorporate high-speed SRAM cache for interim data storage and quicker access before transferring the data to the IMC macro for computation. SRAM (Static Random-Access Memory) is used as cache memory for its speed and compatibility with standard CMOS processes. DRAM is investigated for its high storage density.

(ii) Emerging Non-Volatile Memories (eNVMs) include ReRAM, PCM, and Spin-Transfer Torque Magnetic Random-Access Memory (STT-MRAM), which are especially advantageous for analog IMC (AIMC) because they can store weights with high precision and perform MVM effectively. The memory devices, including ReRAM, FeFET, and SRAM, are engineered to execute computations directly within the memory array, thereby avoiding data transfers between the processor and memory. They frequently use the "in-situ" (IS) compute model, executing calculations directly at the data's storage location.

A digital processor is the component that manages computational functions that are not appropriate for in-memory execution, such as batch normalization and the application of non-linear activation functions in neural networks.

Input and output memory are usually scratchpad memories, e.g., local SRAM buffers used to temporarily hold input data (activations) before they are sent to the crossbar array and to gather the results of the computation (output activations).

Dataflow controller is the element that oversees the data and instruction flow within the IMC core, orchestrating data retrieval, in-memory computation, and the transmission of results to other processing layers or components. ADC converts digital input data into analog signals that the analog IMC array can process. DAC converts the results of analog computation back to a digital format for use by the rest of the system.

From a functional viewpoint, IMC encompasses three primary processes:

Data Loading: Data sets are transferred from diverse sources into high-speed memory.

Data Processing: Calculations are performed directly in memory, often using in-memory analytics tools.

Data Storage: The results of the processing are either retained in memory or saved to disk for long-term archiving.

7.2.4 Role of In-Memory Computing in Nanoelectronics

Alleviating Data Transfer Issues, Reducing the Memory Wall, and Energy Usage: The physical separation of processor and memory in traditional systems (von Neumann architecture) necessitates continuous data transfers, which are both time-consuming and energy-intensive. IMC relieves this burden by performing operations directly within memory arrays and eliminating the need for frequent, energy-intensive data transfers to a separate processing unit. This significantly enhances energy efficiency and minimizes latency.

Supporting Innovative Computing Models: While beyond-CMOS devices may not match the performance of traditional silicon field-effect transistors (FETs) for standard Boolean logic, their exceptional physical attributes (e.g., analog resistance modulation) are ideally suited for non-von Neumann and brain-inspired computing frameworks, such as the following:

(i) Analog AI Hardware: Memristors and PCM can retain synaptic weights through analog means (multiple resistance levels), enabling high-density, efficient in-memory MVM, an essential operation in neural networks.

(ii) Neuromorphic Computing: IMC enables the integration of memory and processing units, simulating the organization of human brain synapses, making it ideal for machine learning and pattern recognition applications. IMC emerges as the most promising solution for developing brain-inspired systems that integrate sensing, memory, and computation in an event-driven, parallel manner. It is particularly efficient for AI applications such as neural networks, where tasks like MVM can be performed effectively in analog memory arrays.

(iii) Probabilistic Computing: Certain emerging devices exhibit inherent stochastic properties that can be exploited to efficiently implement probabilistic computing methods for tackling complex tasks such as combinatorial optimization.

Utilizing Emerging Nanoscale Devices: Research beyond CMOS explores innovative memory and logic elements such as memristors, ReRAM, PCM, and magnetic tunnel junctions (MTJs) that are highly scalable and compatible with current fabrication processes. These devices typically exhibit unique traits (e.g., analog resistance tuning, non-volatility, stochastic behavior) that lend themselves well to efficient implementations of IMC, particularly for neuromorphic and probabilistic computing.

Promoting Monolithic 3D Integration: IMC architectures that frequently employ emerging non-volatile memories (NVMs) advocate monolithic 3D integration, which can boost device density and shorten interconnect lengths.

High Integration Density and Scalability: Numerous new memory devices align with standard CMOS fabrication techniques and can be seamlessly integrated into monolithic 3D configurations. This facilitates ultra-dense packing of computing and memory components onto a single chip, addressing the limitations posed by dimensional scaling (Moore's Law).

Enhanced Functionality: Beyond-CMOS nanoelectronics support "functional scaling" by integrating sensing, memory, and computing capabilities into a single device. For instance, a single ferroelectric transistor can perform both logic and IMC tasks, enabling advanced "smart edge" technologies.

Enhancing Energy Efficiency and Performance: By reducing data movement, IMC significantly reduces energy per operation, a critical objective for extending computing capabilities as CMOS scaling nears its limits. This strategy holds the promise of faster computation for applications that require real-time data analysis and decision-making.

Ultimately, IMC leverages the unique physical properties of beyond-CMOS nanoelectronic devices to produce specialized, highly effective hardware accelerators that complement traditional computing systems, especially for data-intensive applications such as AI. Devices beyond CMOS, such as memristors, PCM, and MTJs, possess distinctive properties that facilitate the effective implementation of the IMC paradigm.

Example 7.1 Examine a straightforward 2×2 ReRAM array designed to calculate the product of a voltage vector $\mathbf{V}$ and a conductance matrix $\mathbf{G}$ for an in-memory computing architecture. The conductance matrix $\mathbf{G}$ is characterized by specific values.

$$\mathbf{G} = \begin{pmatrix} G_{11} & G_{12} \\ G_{21} & G_{22} \end{pmatrix} = \begin{pmatrix} 10\ \mu\text{S} & 20\ \mu\text{S} \\ 30\ \mu\text{S} & 40\ \mu\text{S} \end{pmatrix} \tag{7.1}$$

An input voltage vector **V** is supplied.

$$\mathbf{V} = \begin{pmatrix} V_1 \\ V_2 \end{pmatrix} = \begin{pmatrix} 0.5\text{ V} \\ 0.5\text{ V} \end{pmatrix} \tag{7.2}$$

Determine the output currents I_1 and I_2 in milliamperes (mA).
Solution: A resistive random-access memory (ReRAM) crossbar array, which is essential for in-memory computing (IMC), is utilized to execute an analog matrix-vector multiplication (MVM). The current found in each column of the array results from the aggregate of currents from every cell, according to the formula:

$$I_j = \sum_i G_{ij} V_i \tag{7.3}$$

where I_j represents the output current in the j-th column. V_i denotes the input voltage applied to the i-th row. G_{ij} refers to the conductance (which is the inverse of resistance) of the memory cell located at the intersection of the i-th row and j-th column.
Computing the current I_1 for the first column.

$$\begin{aligned} I_1 &= G_{11}V_1 + G_{21}V_2 = 10\ \mu\text{S} \times 0.5\text{ V} + 30\ \mu\text{S} \times 0.5\text{ V} = 5\ \mu\text{A} + 15\ \mu\text{A} = 20\ \mu\text{A} \\ &= 20 \times 10^{-3}\text{ mA} = 0.02\text{ mA} \end{aligned} \tag{7.4}$$

Computing the current I_2 for the second column.

$$\begin{aligned} I_1 &= G_{12}V_1 + G_{22}V_2 = 20\ \mu\text{S} \times 0.5\text{ V} + 40\ \mu\text{S} \times 0.5\text{ V} = 10\ \mu\text{A} + 20\ \mu\text{A} = 30\ \mu\text{A} \\ &= 30 \times 10^{-3}\text{ mA} = 0.03\text{ mA} \end{aligned} \tag{7.5}$$

The resulting output currents are $I_1 = 0.02$ mA and $I_2 = 0.03$ mA.

Example 7.2 In-memory computing utilizing non-volatile memory (NVM) devices frequently employs a crossbar array for executing analog multiply-accumulate (MAC) operations, which are essential for neural network processing. Within this design, synaptic weights are represented as the conductance (G) of the memory cells (such as memristors or ReRAM). Voltages (V) corresponding to the input activation values are applied to the word lines, while the output current (I) is accumulated along the bit lines, adhering to Ohm's Law $(I = G \times V)$ and Kirchhoff's Current Law.
Consider a small 4 × 4 ReRAM crossbar array designed for in-memory computing. The programmed conductance values (G), measured in microSiemens (μS), in the memory cells for one row are: $G_1 = 10$ μS, $G_2 = 20$ μS, $G_3 = 5$ μS, $G_4 = 15$ μS. The input voltage vector applied to the respective word lines includes: $V_1 = 0.2$ V, $V_2 = 0.1$ V, $V_3 = 0.4$ V, $V_4 = 0.3$ V. Find the total accumulated current (I_{Sum}) on the output bit line.
Solution: To determine the total accumulated current (I_{Sum}) on the output bit line, we follow the steps given below. First, we calculate the individual currents produced by each cell using Ohm's Law $(I = G \times V)$. Then, we sum up the individual currents along the bit line to establish the total accumulated current (I_{Sum}).

Individual currents are as follows:

$$I_1 = G_1 \times V_1 = 10\ \mu\text{S} \times 0.2\ \text{V} = 2\ \mu\text{A} \tag{7.6}$$

$$I_2 = G_2 \times V_2 = 20\ \mu\text{S} \times 0.1\ \text{V} = 2\ \mu\text{A} \tag{7.7}$$

$$I_3 = G_3 \times V_3 = 5\ \mu\text{S} \times 0.4\ \text{V} = 2\ \mu\text{A} \tag{7.8}$$

$$I_4 = G_4 \times V_{41} = 15\ \mu\text{S} \times 0.3\ \text{V} = 4.5\ \mu\text{A} \tag{7.9}$$

Finally, the total accumulated current is computed as:

$$I_{\text{Sum}} = I_1 + I_2 + I_3 + I_4 = 2\ \mu\text{A} + 2\ \mu\text{A} + 2\ \mu\text{A} + 4.5\ \mu\text{A} = 10.5\ \mu\text{A} \tag{7.10}$$

The accumulated current on the output bit line amounts to 10.5 microamperes (μA). This analog current is subsequently converted into a digital value by an analog-to-digital converter (ADC) in the associated circuitry.

7.3 NEUROMORPHIC COMPUTING

Neuromorphic computing represents a transformative shift in nanoelectronics beyond CMOS, enabling energy-efficient, high-density, and highly parallel computing systems that address the shortcomings of the conventional von Neumann architecture (Xiao et al. 2024, Marrows et al. 2024, Chowdhury et al. 2025, Greatorex et al. 2025).

7.3.1 Examining Neuromorphic Architecture in the Context of the von Neumann Model

Figure 7.3 delves into a side-by-side exploration of von Neumann and neuromorphic architectures, highlighting their unique approaches to computing. As previously mentioned, the heart of a von Neumann processing unit lies in its control and logic units, which work together to execute instructions (Figure 7.3(a)). Input and output data units perform their respective functions, while the processing unit communicates with memory via the data bus. On the other hand, the principal parts of a neuromorphic processing unit are the input and output neuron layers, which are connected through synaptic connections to hidden neuron layers (Figure 7.3(b)). Input and output data units are also shown. The data is fed from the input data unit to the input neuron layer, from where it flows through synaptic connections to the first hidden layer. From the first hidden neuron layer, the data passes through synaptic connections to the second hidden neuron layer. From the second hidden neuron layer onward, the data moves through synaptic connections to the output neuron layer, which finally delivers the computed results for the given input data to the output data unit.

Imagine the classic von Neumann architecture: a system where memory and processing are kept apart, forcing data to travel back and forth and creating a bottleneck that slows things down and wastes energy. In contrast, neuromorphic computing draws inspiration from the brain, weaving memory and computation together within synaptic connections. Here, layers of artificial neurons—input, hidden,

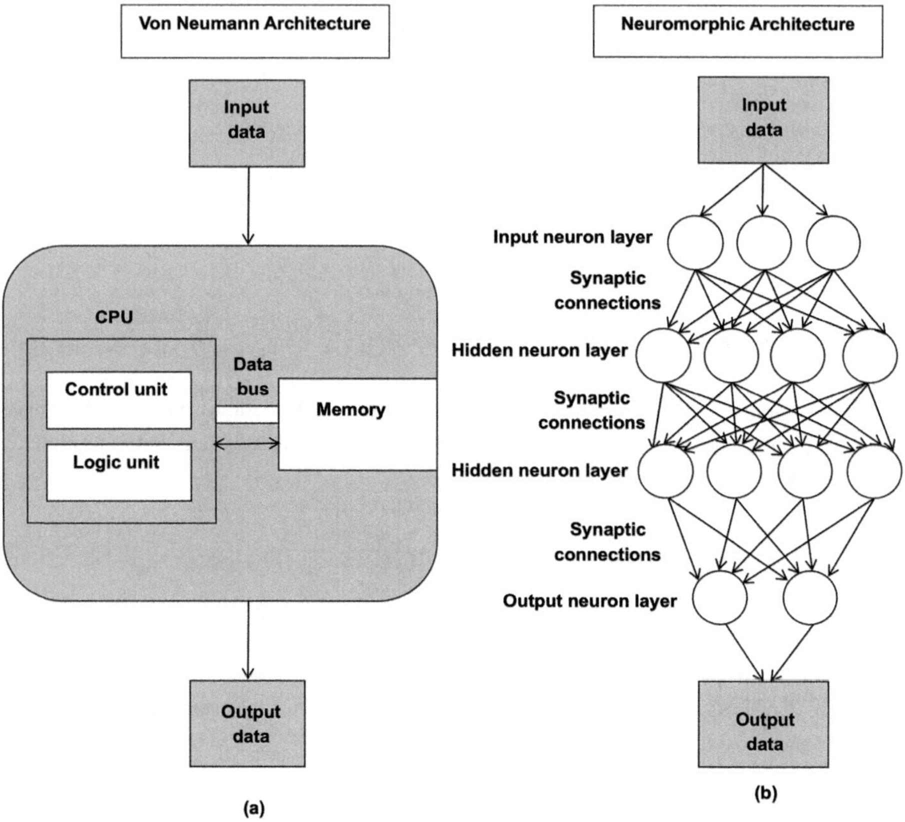

FIGURE 7.3 Von Neumann and neuromorphic architectures. (a) In the von Neumann architecture, a control unit and a logic unit work together, exchanging information with memory along a data bus. Input and output data paths are indicated. (b) In contrast, neuromorphic architecture mimics the brain: input data enters the input neuron layer, travels through synaptic connections to the first hidden neuron layer, then to the second hidden neuron layer, and finally reaches the output neuron layer, which produces the result.

and output—work in parallel and respond to events as they happen, boosting both efficiency and adaptability. The latest 2D spintronic neuromorphic systems take this even further, unlocking new levels of scalability and energy savings.

7.3.2 The Unit Cell of a Neuromorphic Computing System

The "unit cell" is a specialized hardware design that emulates the core elements of a biological neural system: the neuron (a computational component) and the synapse (a storage and learning component). In contrast to the conventional von Neumann architecture, which separates processing and memory spatially, the unit cell in neuromorphic hardware combines these functions to facilitate parallel, energy-efficient, and event-driven processing, similar to the brain's operation (Figure 7.4).

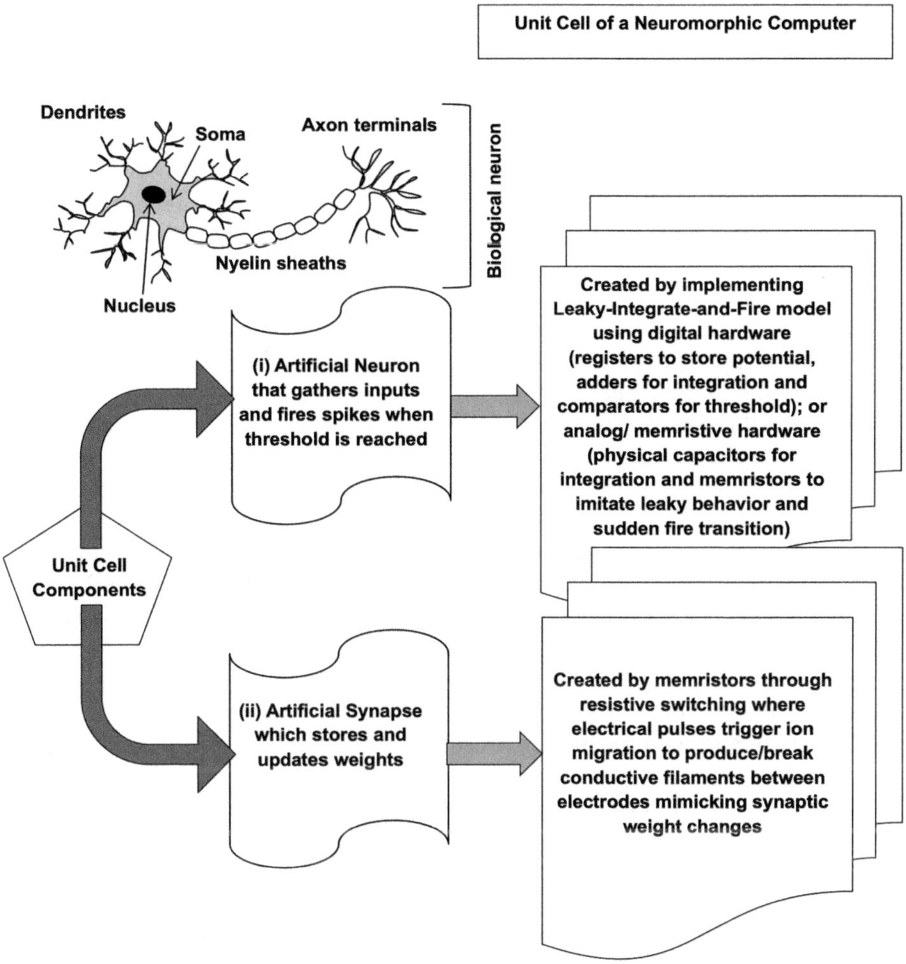

FIGURE 7.4 Unit cell components/options of a neuromorphic computer computational unit. Artificial neurons serve as signal processors, integrating inputs from multiple synapses and generating outputs when specific thresholds are met, following the leaky integrate-and-fire principle. Artificial synapses manage information storage and computational operations by regulating the strength of transmitted signals, known as synaptic weights. Memristors are used to implement their functions.

The main elements of a neuromorphic unit cell are crafted to mimic the essential functions of biological neurons and synapses.

Computational Unit (Artificial Neuron): This section processes incoming signals (spikes) and produces an output spike when a specific threshold is exceeded, often using an "integrate-and-fire" model to simulate biological neuronal behavior.

Information Storage Unit (Artificial Synapse): This component retains the connection strength, or "weight," linking neurons. Importantly, these weights are typically stored using NVM technologies that can adapt dynamically, a characteristic known as synaptic plasticity, which is critical for learning.

Standard hardware implementations of the unit cell comprise the following:

A variety of materials and devices are utilized to construct these unit cells, moving away from conventional silicon CMOS-based transistors toward more specialized elements:

Memristors: These devices can alter their resistance depending on the voltage history applied to them, making them well-suited for replicating the adaptable, multi-state nature of synapses.

Resistive Random-Access Memory (ReRAM): ReRAM devices, including particular configurations such as a nanosheet 4T2R (4 transistors, 2 resistors) cell, are frequently researched for their capacity to provide multi-level conductance states.

Ferroelectric Transistors (FeFETs) and Memory (FeRAM): These integrate ferroelectric materials within the gate stack of a transistor or in a 1T1C (one transistor, one capacitor) cell configuration, using polarization to control current flow and store information.

Magnetic RAM (MRAM): Multi-state MRAM cells, particularly utilizing spin-transfer torque (STT-MRAM), can create quantized weights for the network and are a focus in some hardware designs.

These material-based devices enable computation and memory to be co-located or integrated into a single component, effectively addressing the "von Neumann bottleneck" prevalent in traditional computing architectures.

7.3.2.1 Memristors in Neuromorphic Computing

Memristors are regarded as ideal components for neuromorphic computing due to their ability to modulate resistance, effectively mimicking the functions of biological synapses and neurons and facilitating critical brain activities such as learning, memory, and signal processing.

Artificial Synapses: Memristors are essential for creating artificial synapses because they can alter their conductance (synaptic weight) based on the history of the voltage or current applied, a characteristic known as plasticity. They can reflect critical biological learning processes:

Short-Term and Long-Term Potentiation/Depression (STP/LTP): Memristors can exhibit both temporary and permanent changes in synaptic weight, mimicking the brain's short-term and long-term memory capabilities.

Spike-Timing-Dependent Plasticity (STDP): A memristor's conductance changes can depend on the timing relationship between input (presynaptic) and output (postsynaptic) spikes, which is a crucial mechanism for learning in the brain.

Spike-Rate-Dependent Plasticity (SRDP): Memristors can modify synaptic strength in response to spike frequency, increasing the flexibility of neural circuits.

Artificial Neurons: Certain volatile memristors can mimic neuronal activity, such as the leaky integrate-and-fire (LIF) model. They accumulate incoming signals (charge) until a threshold is reached, then generate an action potential before resetting. A single memristor can replace more complex transistor-based neuron circuits, significantly reducing power consumption and size.

In-Memory Computing (IMC): Memristor-based crossbar arrays combine memory and processing, enabling calculations (exceptionally efficient analog MVM, a fundamental task in neural networks) to happen directly within the memory unit, eliminating the "von Neumann bottleneck" prevalent in traditional computing architectures.

Advantages of memristors are as follows:

Energy Efficiency: Memristive systems can conduct operations with substantially less energy than traditional CPUs and GPUs, with some artificial neurons functioning at levels as low as femtojoules per spike, which is comparable to biological neurons.

High Integration Density: Their uncomplicated, two-terminal design permits dense three-dimensional stacking in crossbar arrays, allowing the creation of systems with a vast number of artificial synapses in a compact physical space.

Scalability: The nanoscale dimensions and compatibility with standard CMOS processes enable the design of extensive, integrated neuromorphic chips.

Bio-Realistic Dynamics: The inherent physics of ion movement in memristors provides a natural hardware foundation for the intricate, non-linear dynamics of biological neural networks, which are challenging to replicate with conventional silicon-based components.

A few use cases of memristors are as follows:

Pattern Recognition and Classification: Memristor-based neural networks have proven effective in tasks such as recognizing handwritten digits (e.g., from the MNIST dataset), achieving remarkable accuracy.

Edge Computing and Robotics: Their low energy consumption and real-time processing abilities render them suitable for AI applications in edge devices, including robotic tactile perception and autonomous driving systems, where quick adaptation to unpredictable surroundings is essential.

Sequential Data Processing: Systems designed with short-term memory memristors can effectively handle sequential and temporal data, such as training and generating biomedical sequences, including antimicrobial peptides.

7.3.2.2 Materials Utilized in the Construction of Memristors

Memristors designed for neuromorphic computing are generally made using a metal-insulator-metal (MIM) "sandwich" structure, primarily incorporating binary metal oxides, 2D materials, and perovskites as the insulating layers, along with various metal electrodes. Principal material categories are:

Binary Metal Oxides: These materials are the most prevalent and widely used due to their reliability and compatibility with conventional CMOS fabrication techniques. Examples include titanium oxide (TiO_x), hafnium oxide (HfO_2), tantalum oxide (TaO_x), silicon oxide (SiO_2), aluminum oxide (Al_2O_3), nickel oxide (NiO), copper oxide (CuO), and zinc oxide (ZnO). The resistive-switching behavior in these materials typically relies on the creation and breakdown of conductive filaments made of oxygen vacancies.

Two-Dimensional (2D) Materials: These materials exhibit distinctive properties, including atomic-scale thickness, flexibility, and tunable electrical properties, making them attractive for compact, low-power devices. Examples include graphene, graphene oxide (GO), molybdenum disulfide (MoS_2), tungsten disulfide (WS_2), hexagonal boron nitride (h-BN), and black phosphorus (BP). Their role can be as switching layers, electrodes, or components of heterojunctions, employing mechanisms such as defect migration, phase shifts, and charge trapping.

Perovskite Materials: Recognized for their outstanding optoelectronic traits and mixed ion/electron conductivity, perovskites are being investigated for enhanced neuromorphic applications, including optical signal processing. Examples include methylammonium lead iodide ($MAPbI_3$), a lead-free double perovskite ($Cs_2AgBiBr_6$), and ($LaFeO_3$). Their unique crystal configurations enable a diverse range of physical and chemical properties to be obtained through doping and structural modifications.

Organic Materials and Polymers: These substances are beneficial due to their low cost, flexibility, and straightforward preparation, making them suitable for flexible and wearable technology. Examples include organic polymers, such as PEDOT: PSS, fluoropolymers, and biomaterials, such as DNA and egg proteins.

Electrode Materials: These materials not only facilitate current conduction but can also play a crucial role in the resistive-switching process. Inert/noble metals include platinum (Pt) and gold (Au). Active metals—including those that participate in filament development—are copper (Cu), silver (Ag), tantalum (Ta), titanium (Ti), and nickel (Ni). Other materials compatible with CMOS include titanium nitride (TiN), tantalum nitride (TaN), indium tin oxide (ITO), and graphene.

7.3.2.3 Phase-Change Memory in Neuromorphic Computing

PCM stands out as a prominent technology for neuromorphic computing because it can replicate biological synapses, facilitating energy-efficient computing architectures that do not rely on the von Neumann model.

Artificial Synapses: PCM is primarily utilized as a synthetic synapse. By applying a sequence of electrical pulses, the electrical conductance of a PCM device can be accurately and gradually modified, akin to the way biological synapses strengthen or weaken over time (synaptic plasticity).

In-Memory Computing (IMC): PCM supports IMC systems by integrating memory and processing units at the exact location, addressing the "von Neumann bottleneck" (the delay and energy waste incurred by transferring data between separate memory and processing units). Fundamental tasks in neural networks, such as MVM, can be efficiently implemented in PCM arrays.

Spiking Neural Networks (SNNs): Hardware implementations of SNNs use PCM devices that process information via spikes, mimicking natural brain function. These devices can store synaptic weights and gather signals from presynaptic neurons.

Rapid Adaptation and Learning: Neuromorphic hardware using PCM has demonstrated "learning-to-learn" capabilities, enabling AI models to swiftly adapt to new tasks with minimal data and computational resources, mirroring human learning behaviors.

Photonic Neuromorphic Computing: PCM is also being investigated for photonic systems, leveraging its optical properties to enable all-photonic neuromorphic circuits with high bandwidth and reduced energy consumption.

The benefits of PCM are non-volatility (data retention without power), high scalability, and density multi-level storage (storing multiple bits per cell), fast switching speed (nanosecond range), and low power consumption per operation.

Obstacles to PCM include resistance drift (changes in conductance over time), endurance limits (a finite number of write cycles), variability (inter- and intra-device inconsistencies), and thermal crosstalk in dense arrays.

Researchers are actively investigating materials and circuit designs to address these challenges, such as employing innovative materials like monolayer Sb_2Te_3 or applying intelligent compensation techniques to counter resistance drift, which is essential for maintaining long-term accuracy in applications like image classification.

In addition to PCM, other emerging non-volatile memory (eNVM) technologies under extensive study for neuromorphic computing include ReRAM, magnetic random-access memory (MRAM), and FeFETs. These technologies function as artificial synapses due to their non-volatility and ability to retain multiple conductance levels.

7.3.3 Parts of a Neuromorphic Computer

7.3.3.1 Hardware

Neuromorphic computers are specialized hardware designed to replicate biological neural networks, primarily using artificial neurons, synapses, and integrated neuromorphic chips.

Artificial Neurons: These hardware elements function as fundamental processing units, resembling biological neurons. They generate electrical "spikes" (action potentials) in response to incoming signals that exceed a specific threshold, a mechanism known as the LIF model. The timing and frequency of these spikes carry information, which distinguishes them from the binary framework of conventional computers. While standard analog or digital

CMOS circuits can implement neurons, post-CMOS nanoelectronics provide more efficient options, especially in achieving low power consumption and specific dynamic responses.

Artificial Synapses: In a neuromorphic computer, the main components reliant on post-CMOS nanoelectronics are primarily the synapses, with neurons being of lesser importance. These evolving technologies are essential for achieving the high-density, low-power consumption, and in-memory processing required to replicate brain functions efficiently. These connections between artificial neurons regulate signal transmission, much as biological synapses do. They possess "weights" (conductance values) that can be altered dynamically based on activity, enabling the system to "learn" and adapt over time, a characteristic known as synaptic plasticity.

Synapses represent the most crucial aspect of post-CMOS applications. Conventional CMOS designs for synapses require substantial silicon area to store synaptic weights, limiting scalability and power efficiency. Post-CMOS nanoelectronics present a solution through innovative NVM devices integrated directly on top of the silicon circuitry (Back-End-Of-Line integration).

Neuromorphic Chips/Processors: These integrated circuits are specifically devised to accommodate extensive networks of artificial neurons and synapses. They are optimized for parallel, event-driven processing (only active neurons and synapses consume energy), facilitating high efficiency and reduced power usage. Notable examples include Intel's Loihi and IBM's TrueNorth chips.

7.3.3.2 Supporting Hardware and Materials

The physical realization of neurons and synapses typically employs various technologies:

Memristors: Short for "memory resistors," these components are resistive-switching devices with two or three terminals and are the leading post-CMOS technology for simulating artificial synapses. They are crucial NVM elements whose resistance can be dynamically adjusted based on prior voltage applications, effectively emulating synaptic behavior and enabling IMC. They modify their electrical characteristics based on previous signals, allowing them to retain multiple weight states and mimic synaptic plasticity (learning and memory) in a compact, non-volatile manner.

Transistors: Neuromorphic systems can be constructed using existing CMOS technology, although this may pose challenges for scaling complex networks. Researchers are also investigating innovative transistor types, such as ferroelectric transistors, to serve as artificial synapses. Nanomaterial-based transistors, notably anti-ambipolar (Gaussian) transistors made from 2D materials, can provide the necessary positive and negative feedback for spiking and refractory behaviors characteristic of biological neurons, often requiring simpler circuitry than complex multi-transistor CMOS arrangements.

Spintronic Devices: Technologies such as MTJs with low-barrier nanomagnets are being examined to create extremely low-power "probabilistic bits" (p-bits), which can be used to build stochastic neurons.

Emerging Materials: To address scalability and power-efficiency limitations, researchers are examining materials such as PCM, spintronic memories (e.g., MTJs), and various 2D materials to develop more efficient and denser artificial neurons and synapses. PCM devices are currently being studied for their ability to store multiple data states and enable IMC.

7.3.3.3 Software and Architecture

In addition to physical components, neuromorphic computing includes the following:

Spiking Neural Networks (SNNs): These foundational computational models operate on hardware, processing information via discrete spikes rather than continuous data streams, aligning more closely with biological brain function.

Event-Driven Processing: This architectural concept means computation happens only when triggered by a spike (an "event"), resulting in exceptionally efficient power usage compared to traditional continuously clocked processors.

Plasticity Mechanisms: Software and hardware systems that permit the network to modify synaptic weights and neuron thresholds over time in response to experience, facilitating learning and adaptation.

7.3.3.4 Overall System Integration

In addition to individual components, post-CMOS nanoelectronics facilitate architectural enhancements by enabling monolithic 3D integration. Stacking layers of artificial neurons and synapses vertically significantly boosts device density, minimizes interconnect lengths, and alleviates the "von Neumann bottleneck" (the inefficiency of moving data between distinct processing and memory units), a common issue in traditional computers. Ultimately, post-CMOS nanoelectronics are vital for the tangible realization of the essential computational and memory elements of a neuromorphic system, enabling the convergence of memory and processing functions at a single location, akin to the brain's architecture.

7.3.4 Contributions of Neuromorphic Computing to Nanoelectronics

Alleviating the Von Neumann Bottleneck: Conventional computing divides memory and processing units, resulting in significant data transfers that are both time- and energy-inefficient. Neuromorphic computing combines memory and processing (in-memory computing), removing this bottleneck and allowing for quicker, more efficient data management.

Brain-Inspired Architecture: In contrast to traditional computers that distinctively separate memory and processing (the von Neumann bottleneck), neuromorphic computing integrates these roles, minimizing the necessity for continuous data transfers and decreasing power usage.

Significant Energy Efficiency Improvements: Drawing inspiration from the human brain's 20-watt operation, neuromorphic systems utilize event-driven (spiking) computation, consuming power only when a signal (spike) occurs. This approach provides substantial enhancements in energy efficiency compared to traditional processors.

Facilitating New Device Materials and Structures: As silicon-based CMOS technology approaches physical scaling limits (below 3 nm), neuromorphic computing encourages the exploration and integration of innovative materials and architectures.

Novel devices: It employs innovative components, such as memristors and spintronic devices, to function as artificial synapses and neurons, which can be produced using beyond-CMOS materials and methods.

Memristors and Other Emerging Components: Neuromorphic systems rely heavily on devices such as memristors, PCM, and ferroelectric components, which simulate biological synapses and neurons. These devices store and process information by changing resistance or other properties in an analog manner, enabling more compact, functional designs than standard transistors.

3D Integration: The distinct attributes of new materials, such as atomically thin two-dimensional (2D) materials (e.g., graphene, MoS_2), enable vertical stacking of functional layers in monolithic 3D (M3D) integration. This greatly enhances integration density and shortens interconnect distances, resulting in higher bandwidth and lower power consumption.

Hardware Realization of AI and Machine Learning: Neuromorphic chips are specifically engineered to implement artificial neural networks (ANNs) and SNNs in hardware, excelling in tasks such as pattern recognition, image processing, and autonomous navigation. This permits real-time, on-chip learning and adaptability that is hard to achieve with traditional systems.

AI Enhancement: By emulating the brain, neuromorphic hardware is inherently suited for tasks the brain excels at, such as pattern recognition, decision-making, and learning. This positions it as a catalyst for advancing AI applications.

Hybrid Systems and Future Integration: Neuromorphic computing is driving the growth of hybrid systems that merge established CMOS logic with newer materials and innovative nanoelectronic devices for memory and specialized processing, facilitating a gradual transition, as exemplified by CMOS-compatible neuromorphic chips, and setting the stage for future systems that optimize both energy and area efficiency. This practical approach helps bridge the gap between current manufacturing capabilities and future "beyond-CMOS" objectives.

Energy Efficiency: This integrated architecture, along with brain-inspired algorithms such as SNNs, enables extremely energy-efficient computation, making it ideal for devices with limited resources, such as those used in edge computing and biosignal processing.

Addressing Moore's Law Constraints: As a non-Von Neumann approach, neuromorphic computing is a crucial strategy for advancing performance beyond the scaling limitations of conventional silicon-based transistors, making it a promising option for the post-Moore era.

Briefly, neuromorphic computing transcends being merely an alternative application; it embodies a fundamental transformation in design principles that harnesses advanced nanoelectronics to enable more powerful, efficient, and brain-inspired computing systems.

Example 7.3 A standard CMOS (Complementary Metal-Oxide-Semiconductor) AI accelerator requires 6000 J of energy to execute a particular image classification task. In contrast, a new neuromorphic chip prototype, which employs nanoelectronic memristive devices in a beyond-CMOS framework, accomplishes the same task using just 2527 J. Determine the percentage of energy conserved by employing the neuromorphic chip as opposed to the conventional CMOS system.
Solution: We use the following formula:

$$\text{Energy saved (\%)} = \frac{\begin{array}{c}\text{Energy consumed by CMOS chip}-\\ \text{Energy consumed by neuromorphic chip}\end{array}}{\text{Energy consumed by CMOS chip}} \times 100\% \tag{7.11}$$

Let's identify the provided values: Energy consumed by CMOS $(E_{\text{CMOS}}) = 6000$ J and energy consumed by neuromorphic chip $(E_{\text{Neuro}}) = 2527$ J. Now, calculate the total energy difference:

$$\text{Energy difference} = E_{\text{CMOS}} - E_{\text{Neuro}} = 6000\text{ J} - 2527\text{ J} = 3473\text{ J} \tag{7.12}$$

Next, we calculate the percentage of energy saved:

$$\text{Energy saved (\%)} = \frac{3473\text{ J}}{6000\text{ J}} \times 100\% = 57.88\% \tag{7.13}$$

Thus, the neuromorphic chip conserves approximately 57.88% of the energy needed by the traditional CMOS system for the same task. This demonstrates the substantial energy efficiency improvements that neuromorphic architectures and innovative nanoelectronic materials present when compared to established computing models.

Example 7.4 A particular neuromorphic processor utilizing nanoelectronic resistive random-access memory (ReRAM) devices for its synapses requires 361 femtojoules (fJ) of energy for one read operation. In contrast, a conventional CMOS-based processor performing a similar operation uses 10 picojoules (pJ). Determine how many times more energy efficient the neuromorphic processor is compared to the traditional CMOS processor for this task.

Solution: To compare the energy consumption of the two types of processors, we need to standardize the units (e.g., convert both to femtojoules, knowing that 1 pJ equals 1000 fJ). The energy used by the neuromorphic processor E_{Neuro} is 361 fJ, while the energy for the CMOS processor E_{CMOS} amounts to 10 pJ, which converts to 10 × 1000 fJ, or 10,000 fJ. Next, we calculate the ratio of the energy consumption of the CMOS to that of the neuromorphic system:

$$\text{Efficiency ratio} = \frac{E_{\text{CMOS}}}{E_{\text{Neuro}}} = \frac{10{,}000\text{ fJ}}{361\text{ fJ}} = 27.7 \tag{7.14}$$

The neuromorphic processor demonstrates roughly 27.7 times greater energy efficiency than the traditional CMOS processor for the specified operation. This significant enhancement in energy efficiency is a key objective of neuromorphic computing for advancements in beyond-CMOS nanoelectronics.

7.4 QUANTUM COMPUTING

Quantum computing serves a dual purpose in the evolution of "beyond CMOS" nanoelectronics. It serves as both a technology that can be physically realized using advanced nanoelectronic systems and a significant application area that drives demand for novel nanomaterials and architectures (Zhang et al. 2019, Gonzalez-Zalba et al. 2021, Nikandish et al. 2021, Bhat et al. 2022, Zito 2025).

7.4.1 Quantum Computing as a Beyond-CMOS Implementation Approach

Quantum computing represents a crucial technological avenue for sustaining performance gains that become unfeasible with traditional CMOS scaling. The nanoelectronics sector is actively investigating how to build quantum computers using the semiconductor industry's existing infrastructure. Researchers are creating qubits using adapted CMOS processes and materials such as silicon spin qubits. This CMOS-compatible qubit strategy aims to capitalize on the ongoing multitrillion-dollar global investment in silicon manufacturing to achieve scalable quantum processing, effectively developing a "quantum computer on a chip."

To address power and performance challenges, quantum-mechanical devices, such as quantum-dot cellular automata (QCA), hold the promise of extremely low power consumption, addressing a significant constraint that conventional CMOS faces due to physical scaling limits and heat dissipation problems.

The future anticipates hybrid systems in which classical CMOS electronics coexist with quantum processors on a single chip, especially in cryogenic (very low-temperature) settings. The CMOS components would handle control and readout circuitry. In contrast, quantum components perform complex computations, enabling real-time processing, minimizing latency, and enhancing scalability for applications such as AI and drug discovery.

7.4.2 Quantum Computing as a Plausible Target Application Catalyzing Advancements in Nanoelectronics

The aim is to apply the same manufacturing principles that enabled conventional microprocessors to create large arrays of qubits on a single chip, thereby facilitating large-scale quantum computers that incorporate control and readout circuitry alongside the qubits, with integration of classical electronics. A pivotal step is merging conventional and quantum electronics, which is essential for constructing a comprehensive quantum system.

The requirements of quantum computing exceed the capabilities of traditional CMOS technology, particularly concerning atomic-scale operations and extremely low power consumption, thereby fostering innovation in alternative nanoelectronic methodologies. Progress in beyond-CMOS nanoelectronics can lead to high-density qubit arrays, promoting denser, more efficient quantum systems and potentially resulting in smaller, more powerful quantum computers with a high qubit-to-area ratio.

Constructing functional qubits (quantum bits) requires manipulating and controlling matter at the atomic level, where quantum-mechanical phenomena such as superposition and entanglement become significant. Nanotechnology provides the essential tools to create devices with this level of atomic-scale accuracy. Spin qubits confined in quantum dots (QDs) can serve as qubits, requiring meticulous nanoscale engineering.

The pursuit of stable, scalable qubits has led to investigations into innovative beyond-CMOS device concepts and nanomaterials. Research beyond CMOS emphasizes new materials, such as 2D materials (e.g., transition-metal dichalcogenides), and spintronics, due to their potential for developing smaller, more efficient devices suitable for quantum information processing. The merging of two-dimensional (2D) materials with three-dimensional electronic materials could deliver new functionalities for logic switches and memory devices that surpass those of silicon technology. Superconducting qubits and circuits, which use concepts such as the Josephson junction (JJ), represent another major implementation approach in quantum nanoelectronics.

7.4.3 Structural Unit Cell of a Quantum Computer

The design of the structural unit cell in a quantum computer is determined by engineering choices that vary with the specific hardware architecture and the error-correction codes used (Figure 7.5).

Essential elements of a quantum unit cell maintained in a supercool ambience are as follows:

Physical Architecture: The unit cell is a part of a larger quantum chip or processor, typically contained in a controlled environment, such as a dilution refrigerator for superconducting qubits, to maintain the extremely low operating temperatures required.

There is a wide diversity in the design of a unit cell. The precise configuration of a unit cell largely depends on the type of quantum hardware and the computational model being pursued:

Qubits: These are the fundamental elements of any quantum computer, and therefore, its unit cell is the qubit (quantum bit), which represents the basic unit of quantum information. Physical qubits serve as the processing units. Examples include superconducting circuits, trapped ions, or electron spins within QDs.

Superconducting Qubits: Organizations like IBM implement a "heavy-hex" lattice topology, where the unit-cell design supports specific connectivity patterns across the chip.

Spin Qubits: In silicon-based spin systems, a unit cell can consist of two electron spins with a controllable exchange interaction, aiming for scalability through standardized semiconductor fabrication techniques.

Topological Quantum Computing: This method employs a 3D lattice-cell design to create qubits that are more resistant to noise by encoding information in the physical system's geometry, requiring complex unit-cell structures.

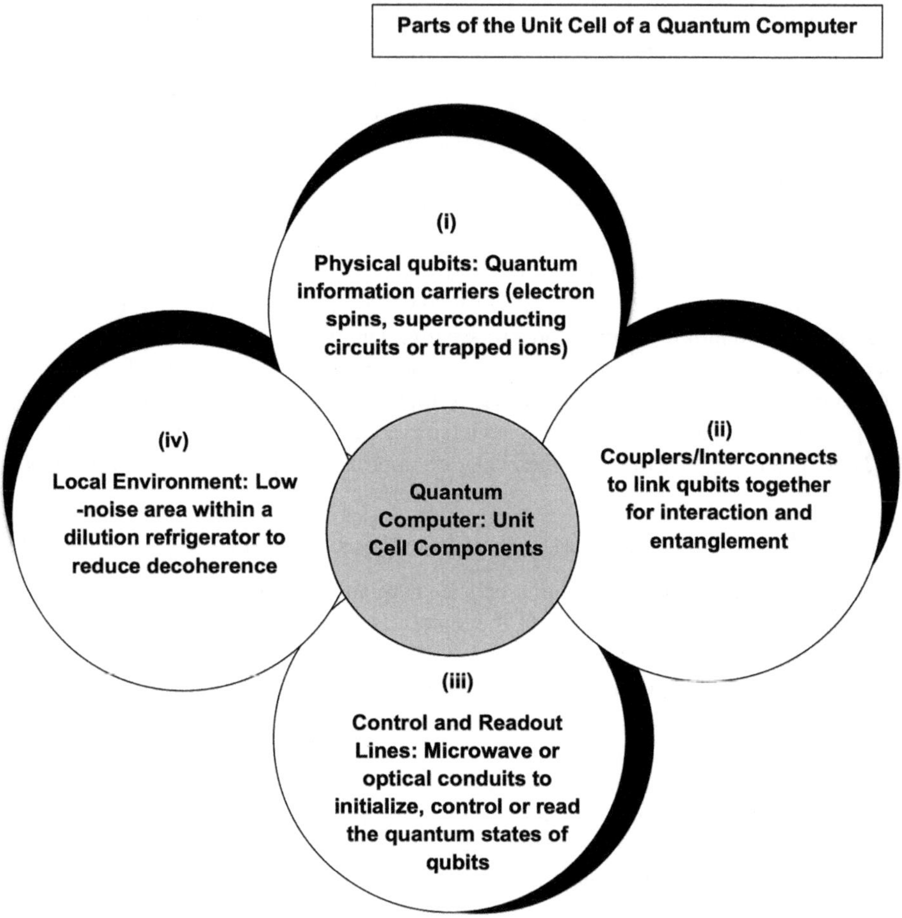

FIGURE 7.5 Unit cell components/options of a quantum computer: a physical architecture built from qubits—whether spin-based, superconducting, trapped ions, or QCA—interconnected with precise coupling, guided by advanced control and readout systems, all functioning in an ultra-cold environment to ensure flawless performance.

Quantum-Dot Cellular Automata (QCA): In this alternative computing model, the fundamental unit (cell) consists of four QDs, with two electrons encoding binary information based on their positions (polarization).

Connectivity/Coupling: The unit cell outlines how qubits are physically linked to adjacent qubits. For instance, specific architectures use a unit cell consisting of four qubits and a central coupling circuit to achieve full connectivity within the module, which can then be arranged into a larger array.

Control and Readout: Each unit cell incorporates the essential components (wiring and control electronics) for initializing, executing quantum logic gates, and measuring qubit states.

Ultimately, the structural unit cell serves as a design framework that enables engineers to replicate numerous small, well-defined modules to construct a larger, fault-tolerant quantum processor, presenting a significant challenge on the journey toward practical large-scale quantum computing.

7.4.4 Parts of a Quantum Computer

Post-CMOS nanoelectronics play a vital role in advancing large-scale, reliable quantum computers by influencing qubit, control, and measurement circuits, as well as interconnects, particularly those operating at cryogenic temperatures. A quantum computer consists of an integrated arrangement of distinct hardware and software elements.

7.4.4.1 Adoption of a Classical-cum-Quantum Computing Structure

A quantum computer consists of a classical processing unit (CPU) and a quantum processing unit (QPU) as depicted in Figure 7.6. In this classical-cum-quantum arrangement, the CPU is the medium of communication of the end user with the application interface and software. Its main components are the applications/software unit, quantum data encoder, and measurement output unit. The quantum data encoder receives the input data from the applications/software interface and sends it to the quantum RAM of the QPU located inside the QPU. The measurement output unit receives data from the results output buffer of the QPU and passes the classical bits to the application interface. Thus, the CPU plays the roles of encoding input data and receiving output data as classical bits.

The principal parts of QPU are a quantum RAM, quantum logic gates, classical measurement interface, and result readout buffer. The quantum RAM receives encoded data from the quantum data encoder of CPU. Quantum logic gates in the QPU work through to-and fro data transfer from quantum RAM to process the encoded data. The classical measurement interface in QPU records the results of quantum computation. The readout output buffer of QPU delivers the results of the quantum calculations to the measurement output unit of CPU in the form of classical bits.

7.4.4.2 Partitioning the Computer into Separate Housing Portions for Quantum and Classical Processing

As the parts of a quantum computer devoted to classical and quantum-mechanical operations work at different temperatures, a quantum stack is designed keeping in mind their separate temperature requirements. The typical housing of a quantum computer comprises an ultracool space for quantum operations and room-temperature ambience for classical operations. Therefore, the outer enclosure of a quantum computer consists of a quantum portion and a classical portion, as shown in Figure 7.7. The qubits sit at the bottom of the stack in the quantum plane, along with associated electronics needed to help with the processing of information from raw qubits. The qubits work under cryogenic control as they require an extremely cool environment, which is stringently maintained within close tolerance limits for satisfactory operation. After quantum processing in ultracool conditions, there is a

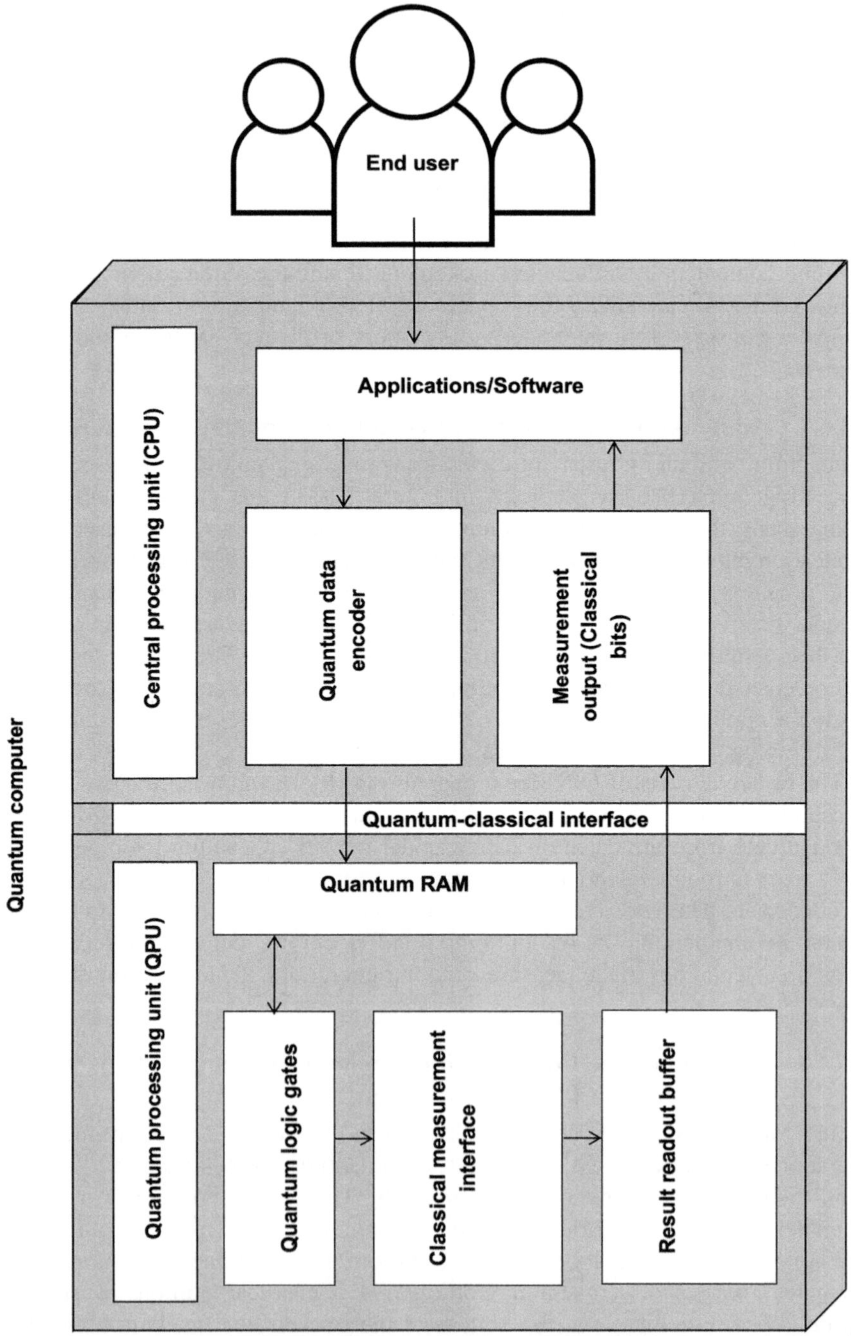

FIGURE 7.6 A quantum computer is shown as made up of a central processing unit (CPU) and a quantum processing unit (QPU). Within the CPU, we find the applications and software unit, the quantum data encoder, and the measurement output unit, while the QPU is equipped with a quantum RAM, quantum logic gates, a classical measurement interface, and a result readout buffer.

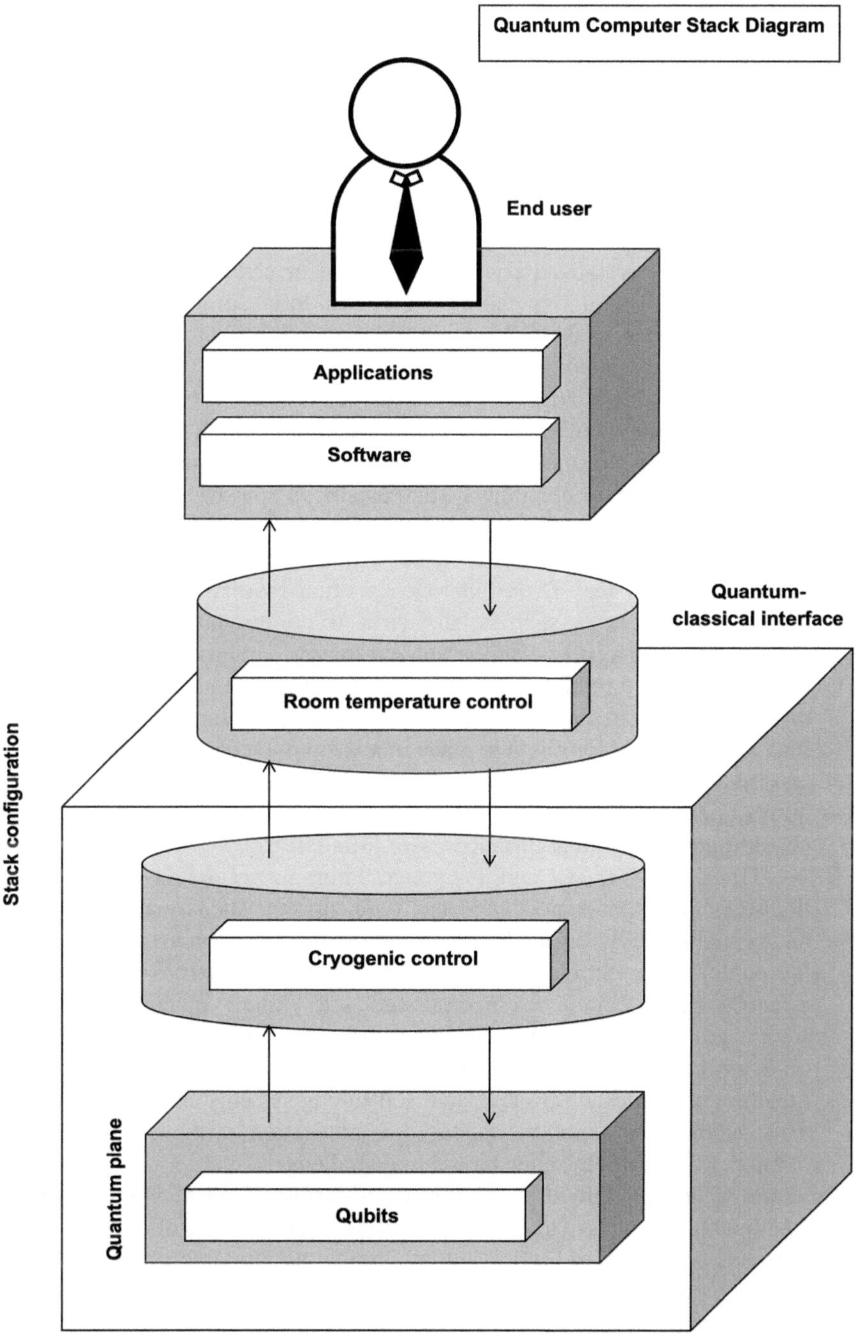

FIGURE 7.7 The quantum and classical segments of a quantum computer. In the quantum segment lie the qubits and accompanying electronics processing circuitry maintained under cryogenic control. This segment lies at the quantum plane. In the classical segment are located the applications interface and software. These accessories are placed at the classical plane. The quantum and classical planes meet at the quantum-classical interface.

transition at the quantum-classical interface toward room-temperature operation. The applications and software units of the CPU for which room-temperature conditions are adequate serve as the end user's facilities for working with the quantum computer while sitting in a comfortable cabin.

7.4.4.3 Role of Beyond-CMOS Nanoelectronics

We re-emphasize that the primary component of a quantum computer is the QPU, which houses the qubits, a control system for manipulating the qubits, a measurement system, and a specialized cryogenic system (for certain types of qubits) that provides the necessary operating conditions. The essential components of a quantum computer that rely on post-CMOS nanoelectronics include:

Qubits (Quantum Bits): These are the fundamental units of information in the realm of quantum computing. Unlike classical bits, which can be either 0 or 1, qubits can be in a superposition of both states at the same time. The core elements of quantum computers, such as silicon spin-based or QD qubits, depend on sophisticated nanofabrication techniques that extend beyond traditional CMOS technology to achieve the required precision and control over quantum states. These approaches often involve fabricating nanometer-scale structures, such as nanowires, to confine individual electrons. Various physical systems can be utilized to create qubits.

QDs are nanoscale particles employed as qubits. Superconducting circuits are tiny circuits that require cooling to extremely low temperatures. Trapped ions are individual ions held in place in a vacuum by electromagnetic fields and manipulated with lasers. Photons are light particles that are utilized as qubits due to their ability to transport quantum states (such as polarization or path) and cover great distances with little noise.

Novel Transistors/Devices and Integrated Photonics: Beyond conventional field-effect transistors, new nanoelectronic concepts such as single-electron and tunneling FETs are under investigation. These advanced devices offer the potential for ultra-low power consumption while harnessing quantum effects, which could be incorporated into future quantum systems. In specific quantum-computing frameworks, particularly silicon photonic quantum computers, post-CMOS technologies are employed to integrate quantum photonics, electronics, and self-tuning systems onto a single chip. This integration supports optical communication within the quantum computer, which helps alleviate scaling challenges.

Quantum Gates and Circuits: Quantum gates perform fundamental operations (the quantum counterpart of logic gates) to alter the state of qubits. These gates are combined to form quantum circuits that run algorithms.

Cryogenic Control System (Dilution Refrigerator) and Measurement Circuitry: This encompasses classical computers and specialized electronics that produce precise signals (such as microwave pulses or laser beams) to govern qubits and determine their final states after computation. For systems that utilize superconducting qubits, this assembly is a sophisticated, multi-stage cooling system that reduces the temperature of the quantum chip to levels close to absolute zero (a few millikelvins) to maintain its quantum properties and minimize thermal noise-induced errors. To oversee thousands or even millions of qubits, the control electronics must be placed close to the quantum processor and operate at extremely low, "cryogenic" temperatures (just a few Kelvin or lower). Post-CMOS nanoelectronics, particularly cryo-CMOS and new superconducting electronics, are being designed to produce integrated circuits that can operate reliably and efficiently in this cold environment, thus minimizing wiring complexity and energy consumption.

Interconnects, Cabling, and Wiring: The physical connections (wiring) that link the quantum processor to room-temperature classical computers create a significant scaling challenge. Post-CMOS nanoelectronics enable the creation of integrated, high-density quantum interconnects, potentially using optical interconnects or innovative superconducting materials to transmit large volumes of control and data signals with minimal thermal output. High-speed, cryogenic cables designed for specialized applications are required to convey signals between the room-temperature control electronics and the cold quantum chip within the refrigerator.

Quantum Error Correction (QEC): While it is not a tangible component, QEC is a crucial technique that processes data to reduce the noise and errors naturally present in quantum computations.

7.4.4.4 Using Beyond-CMOS Nanoelectronics to Build Quantum Computers

Beyond-CMOS nanoelectronics supply the necessary foundational technologies and materials to build the physical components of quantum computers by facilitating the precise manipulation and control of quantum bits (qubits). These technologies connect classical semiconductor manufacturing with the specific needs of quantum mechanics.

A prominent application of beyond-CMOS nanoelectronics in quantum computing is qubit creation. Nanofabrication methods are crucial for producing the minuscule structures that act as qubits. To fabricate silicon QDs, researchers use nanometer-level precision to position individual dopant atoms within silicon, forming single-electron transistors that function as QDs, which are essential for silicon-based qubits. This method aims to be compatible with existing silicon large-scale integration (LSI) circuits, which are vital for scalable manufacturing.

Atomic-Scale Components: Beyond-CMOS techniques enable the creation of atomic-scale electronic components, such as those constructed from silicon dangling bonds (DBs) on a hydrogen-terminated silicon surface, which can act as building blocks for QCA and quantum-computing circuits.

Superconducting Qubits: Nanoscale design is applied to develop superconducting circuits, a critical technology for leading quantum-computing initiatives by companies such as Google Quantum AI, IBM, and AWS.

Integration of Control Electronics: Nanotechnology contributes to the development of cryo-CMOS transistors, classical semiconductor devices designed to operate at the extremely low temperatures required by various quantum-computing systems. This advancement reduces the need for bulky, costly external control electronics, enhancing the efficiency and scalability of quantum computers.

Novel Materials: The exploration of materials beyond conventional CMOS has driven researchers to investigate innovative materials and physical principles (e.g., spintronics and magnetic skyrmions) for advanced logic gates, memory devices, and qubits. The potential for employing two-dimensional materials on ferroelectrics is also being explored to discover new physical effects that could enhance quantum applications.

Thus, beyond-CMOS nanoelectronics presents the essential toolkit to address the scaling and performance challenges of classical computing, facilitating the careful and accurate assembly of the fundamental hardware (qubits and control systems) required for an effective, large-scale quantum computer.

7.4.4.5 Methods for Qubit Creation Beyond Conventional CMOS

Innovative approaches to nanoelectronics, such as silicon spin QDs, are currently being explored to develop scalable qubits. This method capitalizes on established semiconductor fabrication techniques to embed quantum capabilities into conventional microelectronics frameworks, paving the way for large-scale, fault-tolerant quantum computing.

Silicon Spin Qubits in QDs: Rather than relying on rare materials or entirely new technologies, a primary strategy is to adapt the designs of standard silicon transistors to operate as quantum devices. Researchers create QDs within the silicon channel of a modified transistor framework. These QDs can trap individual electrons, whose spin states are then utilized as physical qubits. This approach is highly compatible with current industrial 300 mm wafer fabrication processes, offering a significant advantage for scalability.

Nanowires: Nanowires, which are extremely narrow structures typically only a few atoms wide, exhibit distinctive quantum properties that can be harnessed for information processing. They are being investigated as an innovative method for qubit development and are a crucial element of quantum nanotechnology.

Superconducting Qubits with CMOS: Superconducting qubits have been effectively created using industrial CMOS pilot lines. This technique combines superconducting material islands with standard manufacturing methods, yielding good coherence times and providing an alternative approach for large-scale fabrication.

Atomic-Scale Silicon Dangling Bonds: Another technique involves precisely manipulating silicon DBs on a hydrogen-passivated silicon surface to create artificial molecules and QCA circuits. These structures can function as QDs and serve as foundational elements for “beyond CMOS” architectures.

7.4.4.6 Benefits of This Approach

The approach offers scalability. By utilizing existing, high-volume semiconductor manufacturing infrastructure (such as 300 mm wafer fabs), it becomes feasible to integrate billions of components and potentially achieve the millions of physical qubits necessary for practical, fault-tolerant quantum computing. Further, the approach is well-suited to integration, as the qubits can be combined with classical CMOS control electronics on the same chip or nearby, streamlining the complex wiring and control systems essential for a large-scale quantum computer operating at cryogenic temperatures. Cost-effectiveness is another advantage of the approach. Leveraging decades of semiconductor research and established fabrication techniques can reduce costs and accelerate the development of quantum hardware.

This intersection of nanotechnology and quantum mechanics illustrates how “beyond CMOS” nanoelectronics play a vital role in addressing the challenges of traditional computing and facilitating the creation of viable quantum computers.

7.4.5 Nanoelectronic Fabrication Techniques for Quantum Computers

Fabrication strategies can be split into top-down approaches, which focus on shaping larger bulk materials, and bottom-up approaches, which concentrate on assembling components from singular atoms or molecules.

Top-down methods (compatible with industry) leverage the well-established semiconductor industry infrastructure (fabs) to produce nanoscale features with high accuracy and the potential for mass manufacturing. Lithography is an essential technique for circuit patterning. Electron beam lithography (EBL) employs a concentrated electron beam to directly inscribe patterns onto a substrate with exceptional resolution, primarily utilized for prototyping and research due to its accuracy. Optical lithography employs light to transfer patterns from a mask onto a photosensitive material, a staple of industrial processes and now being adapted for extensive quantum chip fabrication on 300 mm wafers. Nanoimprint lithography (NIL) involves mechanically deforming a resist using a pre-patterned mold.

Several deposition and etching techniques are applied. ALD is a repetitive process that permits the precise application of ultrathin, conformal films, providing exceptional control at the atomic scale. CVD is a technique in which gaseous precursors chemically react on a heated substrate to form a thin film. The selective removal of substrate material to achieve desired shapes is done by etching, often using chemical or physical techniques such as low-energy ion beam etching (IBE).

Atomic-precision advanced manufacturing (APAM) is an exceptionally specialized approach for customizing silicon nanoelectronics with complete 3D atomic precision, effective for prototyping and systematically studying multi-qubit devices.

Bottom-up methods involve constructing structures atom-by-atom or molecule-by-molecule, essential for producing the QDs and nanowires that constitute the qubits. Molecular beam epitaxy (MBE) is a method for generating high-quality crystalline films with exact control over composition and thickness, used in the development of quantum wells and QDs. During self-assembly, nanostructures form spontaneously through interactions among individual components, a process that may support the integration of the millions of qubits required for scalable quantum computing. In wet-chemical synthesis, colloidal QDs are generated in a liquid medium, an economical and scalable fabrication process suitable for large-area deposition.

7.4.5.1 Production of Silicon Quantum Dots

This mainly involves two primary methodologies: atomically precise manufacturing utilizing a scanning tunneling microscope (STM), and the adaptation of existing industrial 300 mm CMOS fabrication lines.

Atomically precise manufacturing (beyond conventional CMOS) techniques extend beyond the capabilities of traditional lithography to fabricate QDs with atomic precision, representing a "beyond-CMOS" electronic strategy. In this method, a scanning tunneling microscope (STM) selectively eliminates individual hydrogen atoms from a hydrogen-passivated silicon (001) surface. In the resulting configuration, the DBs function as QDs. These atomic-scale QDs can be meticulously organized to create tailored structures such as QCA circuits, which function at room temperature and consume ultra-low power.

The strategy using industrial CMOS manufacturing leverages existing high-throughput semiconductor fabrication facilities (fabs) to produce silicon QD devices, meeting the essential scalability requirement for quantum computing. While leveraging CMOS infrastructure, the intent is to create quantum-computing components, serving as a "beyond-CMOS" application of the technology. All-optical lithography is utilized in 300 mm semiconductor manufacturing facilities to attain high yield and uniformity in nanoscale gate designs. Standard techniques include CVD, plasma dry etching, and ALD to form gate stacks and dielectric layers. By process enhancement, Specific engineering of the silicon/silicon dioxide interface and gate stack materials is implemented to achieve reduced charge noise levels, which are vital for qubit performance. As a result, spin qubits are generated in QDs situated at the Si/SiO_2 interface, functioning within a metal-oxide-semiconductor (MOS) stack configuration. These methods enable combining the QD layer with classical CMOS control electronics on the same chip, facilitating multiplexed readout and control at cryogenic temperatures.

These two methodologies showcase distinct approaches to moving beyond traditional microelectronics: one through atomic precision, and the other through the industrial scalability of existing manufacturing techniques for a new era of electronics.

7.4.5.2 Superconducting Qubits

Superconducting qubits are fabricated in specialized facilities using advanced nanotechnology methods, employing techniques similar to those used in the semiconductor industry. The production process comprises multiple stages to fabricate complex circuits on a substrate, with a primary focus on the JJ, an essential element. High-quality silicon or sapphire wafers serve as the substrate. Typical choices of superconducting materials include aluminum (Al), niobium (Nb), and, more recently, tantalum (Ta). A JJ is central to the qubit's design. It is a weak connection between two superconductors, generally comprising a thin insulator layer (like aluminum oxide, AlO_x) sandwiched between two layers of superconducting metal (e.g., $Al/AlO_x/Al$). The non-linear inductance of this junction generates the quantum energy levels that create the qubit.

The production process demands an extremely controlled cleanroom setting to reduce defects that might lead to decoherence (loss of quantum state). The qubit and circuit schematic are initially designed and simulated using specialized software to enhance parameters such as coherence time and gate fidelity. Thin films of superconducting materials are deposited on the substrate using methods such as sputtering or evaporation. Cutting-edge lithography (either electron beam or optical) is employed to create patterns on the circuit at the micro- and nanometer scales. This specifies the architecture of capacitive pads, resonators, and electrical connections. Unnecessary material is removed using techniques such as reactive ion etching (RIE) or wet etching. A lift-off technique is frequently applied in laboratory environments, while industrial methods utilize approaches consistent with standard semiconductor fabrication.

Josephson-junction formation is a vital phase. A prevalent technique involves depositing the first electrode, undertaking controlled oxidation to construct the insulating barrier, and then depositing the second electrode in situ (without breaking vacuum). This guarantees a pristine, high-quality interface. By layering and planarization, multiple layers of metal and insulator films can be stacked to form intricate 3D structures and interconnections. For integration and packaging, the finished quantum chip is mounted on a specially designed microwave-compatible carrier and connected via wire bonding.

Cryogenic cooling, testing, and calibration: For functionality, the packaged chip must be cooled to extremely low temperatures, usually below 15 millikelvin (mK), with a dilution refrigerator. This is crucial for sustaining the stable quantum states needed for computation. The qubits are rigorously tested within the cryostat to evaluate performance metrics, including coherence times and gate fidelities, using microwave pulses for control and readout.

Current research aims to utilize high-volume, high-precision industrial semiconductor (CMOS) fabrication lines to enhance yield, consistency, and scalability, with recent advancements achieving high-coherence qubits on 300 mm silicon wafers.

7.4.5.3 Production of Qubits for Quantum-Dot Cellular Automata

QCA can be used in two ways: for standard digital circuits or for quantum-computing circuits (coherent QCA). The kind of "qubit" produced and the fabrication method vary based on the application being pursued.

In classical digital QCA circuits (binary logic), information is represented by the positions of two extra electrons within a cell housing four (or more) QDs. These stable charge configurations symbolize binary states ("0" and "1" or polarizations $P = +1$ and $P = -1$), rather than authentic quantum-mechanical qubits that exhibit superposition.

Initial experimental QCA devices have been created using metal dots and operated at cryogenically low temperatures. The present research investigates the molecular implementation of QCA, in which redox-active molecular sites serve as QDs. This method shows potential for room-temperature operation, rapid speeds, and high density but remains in the research stage. Typical strategies for fabricating semiconductor QDs based on semiconductor nanostructures involve colloidal chemistry and epitaxial growth (e.g., molecular-beam epitaxy or MOCVD) using materials such as cadmium selenide (CdSe), zinc sulfide (ZnS), and others.

For coherent quantum computing circuits used in genuine quantum computing, QCA cells must operate as true qubits capable of existing in superpositions and being entangled. The quantum state of these qubits evolves in response to interactions with neighboring cells and external electrical signals. Spin qubits in semiconductor QDs are a key focus for this purpose. These are formed when electrons (or "holes") are confined within a potential well in a semiconductor material. Production methods for these devices utilize advanced Very-Large-Scale Integration (VLSI) technology, akin to modern chip manufacturing, and employ established silicon-based processes. The QDs are created by nanoscale patterning and epitaxial growth on a substrate. External gate voltages regulate the charge and tunneling rates inside the dots to facilitate single- and multi-qubit operations.

Thus, the production of QCA involves nanoscale fabrication methods, ranging from metal deposition to sophisticated semiconductor techniques and molecular self-assembly, with the specific approach contingent upon whether the goal is a classical binary logic cell or a coherent quantum spin qubit.

7.4.6 Utilization of Quantum Computers in Research and Development of Nanoelectronics

The development of quantum computers will significantly advance materials science and nanoelectronics research. This application is viewed as one of the most promising near-term "killer applications" for quantum computing. Quantum computers can model intricate molecules and innovative materials, such as those for next-generation batteries and advanced semiconductor technology. Traditional computers face challenges with these simulations due to the massive increase in computational power required to represent quantum systems accurately. By investigating material properties at the quantum level, researchers can speed up the identification and development of new nanoelectronic devices, including 2D quantum materials for future transistors. For optimizing nanomanufacturing, quantum algorithms can be used to tackle complex optimization tasks in the production of nanoscale components, especially those involving self-assembly or precise atomic arrangements.

Essentially, quantum computing serves as both a potential replacement for CMOS technology and a robust tool to enhance the research and development of nanoelectronics beyond CMOS.

7.4.7 Trends of Quantum Computing

Quantum computing, alongside other alternatives to traditional CMOS such as spintronics and neuromorphic systems, signifies a transition to new paradigms of information processing as the scaling of conventional CMOS approaches fundamental physical limitations. Prospects for the future indicate that creating functional, large-scale quantum computers is an ambitious long-term objective, with practical and fault-tolerant applications likely still decades away (some forecasts suggest that the first viable applications may emerge around 2035–2040). In the short term (during the Noisy Intermediate-Scale Quantum, or NISQ phase), researchers are concentrating on developing systems with a limited number of qubits (50–100) to investigate problems that may provide a computational edge, often requiring support from classical computing and error-correction methods.

Example 7.5 A particular design of silicon spin qubit, significant for future nanoelectronics beyond CMOS, has a T_2 coherence time measured at 1 millisecond (1 ms). If a fundamental quantum operation (quantum gate) requires 1 microsecond (1 μs) to execute, how many consecutive operations could theoretically be conducted within the coherence time of that individual qubit?
Solution: To determine the number of operations, we simply divide the total coherence time by the time taken for one operation. First, we need to convert both values to the same time unit: Coherence time: $T_2 = 1\text{ ms} = 1000\ \mu\text{s}$. Operation time: $T_{\text{Op}} = 1\ \mu\text{s}$

$$\text{Number of operations} = \frac{\text{Coherence time}}{\text{Operation time}} = \frac{T_2}{T_{\text{op}}} = \frac{1000\ \mu\text{s}}{1\ \mu\text{s}} = 1000 \qquad (7.15)$$

In theory, 1000 consecutive quantum operations can be executed before the qubit is apt to lose its quantum state due to decoherence. This straightforward calculation underscores a fundamental challenge in quantum computing: the requirement for operations to be significantly quicker than the coherence time to facilitate complex algorithms and error correction.

Example 7.6 A proposed quantum computer architecture beyond traditional CMOS utilizes a quantum error correction (QEC) code in which each logical qubit is represented by 100 physical qubits to achieve fault tolerance (resulting in a 100:1 ratio of physical to logical qubits). If a given algorithm necessitates the use of 50 logical qubits for successful execution, how many physical qubits in total are required in this design?
Solution: The total quantity of physical qubits is determined by multiplying the number of logical qubits by the ratio of physical to logical qubits. Number of logical qubits required: 50. Ratio of physical to logical qubits: 100 physical qubits for each logical qubit.

$$\text{Total physical qubits} = \text{Logical qubits} \times \text{Ratio} = 50 \times 100 = 5000 \qquad (7.16)$$

For this computation within the described architecture, 5000 physical qubits are necessary. This significant overhead presents a major challenge in the development of scalable quantum computers. This was a simple numerical question concerning quantum-computing architecture and error correction.

7.5 DISCUSSION AND CONCLUSIONS

Post-CMOS nanoelectronics offer crucial components, such as innovative memory solutions and materials, that support the fundamental functions of next-generation

TABLE 7.1
Comparison Chart of In-Memory, Neuromorphic, and Quantum Computing

Sl. No.	Computing Paradigm	Components Dependent on Post-CMOS Nanoelectronics
1	In-memory computing (IMC)	Non-volatile memory (NVM) arrays: IMC requires compact, non-volatile memory technologies directly integrated with processing units to mitigate the von Neumann bottleneck. Significant NVM devices include resistive random-access memory (ReRAM), phase-change memory (PCM), ferroelectric random-access memory (FeRAM), spin-transfer torque magnetic random-access memory (STT-MRAM), and ferroelectric field-effect transistors (FeFETs).
2	Neuromorphic computing	Artificial synapses replicate biological synapses by modulating analog conductance. Non-natural neurons are components that implement non-linear activation functions. Memristors are used to set synaptic weights due to their non-volatility and ability to retain analog information, e.g., in memristive crossbar arrays. The spintronic devices are employed to produce magnetic memristors and spintronic neural networks. For 2D materials, silicon nanosheets and other innovative materials are used to create distinctive synaptic devices and neuromorphic structures.
3	Quantum computing	Qubits (quantum bits) are the fundamental units of quantum information, derived from quantum-mechanical systems that go beyond conventional transistors. Examples include superconducting circuits/SQUIDs (superconducting quantum interference devices), quantum dots (e.g., used in semiconductor-based quantum computers), trapped ions, nitrogen-vacancy centers, photonic elements (for quantum photonic chips), and single-electron transistors (SETs) for controlling individual electrons within quantum dots.

computing models, facilitating more efficient, faster, and versatile in-memory, neuromorphic, and quantum computing. Table 7.1 highlights the essential components for each type of computing that rely on technologies extending beyond conventional CMOS manufacturing.

Apart from these, several architectures are focused on reducing energy consumption while increasing functional density and computational speed. Some of these are listed in the extensive Table 7.2.

Hitherto, we have been extensively covering the materials, logic and memory devices, and computing architectures. We now set our sights on the fields of nanofabrication, integration, and packaging. Mastery of these areas is essential for transforming designs into real, functional devices. In the next chapter, we take up nanofabrication.

TABLE 7.2
Holistic Comparison of Advanced Computing Architectures for Beyond-CMOS Nanoelectronics

Sl. No.	Architecture	Core Mechanism/Technology	Primary Advantages	Applications
1	Analog in-memory computing	ReRAM or PCM integrated into logic, magnetic tunnel junctions (MTJ)	Eliminates von Neumann bottleneck and breaks the "memory wall" by eliminating data movement	Deep learning, vector-matrix multiplication (VMM) acceleration
2	Neuromorphic computing	Spiking neural networks (SNNs), memristive crossbars, and analog synaptic devices	Massive parallelism, brain-like energy efficiency, extreme energy efficiency for AI, in-memory computing	AI inference, pattern recognition, sensory processing
3	Spintronic/ magnonic	Spin waves, spin-transfer torque (STT), spin-orbit torque (SOT), spin-transfer torque (STT), magnons, and magnetic tunnel junctions (MTJs)	Low-power, non-volatility assures non-volatile logic and storage, data processing with spin, extremely low power switching	Low-power logic, ultra-dense non-volatile memory
4	Reservoir computing	Physical/material substrates, e.g., nanowire networks, 2D materials	Explores inherent physical dynamics for computation	Temporal pattern processing, chaotic time-series prediction
5	Unconventional charge-based	Steep-slope devices (TFETs), carbon nanotube FETs (CNTFETs), and 2D materials	Low voltage operation, ultra-low power dissipation	Mobile electronics, IoT, high-performance edge computing
6	Nanomechanical computing	NEMS (nano-electro-mechanical systems)	Extremely low leakage; potential for high-temperature operation	High-temperature applications
7	Molecular electronics	Molecular switches, self-assembly	Ultimate density scaling, high functionality at molecular level	Ultra-high-density memory: molecular crossbar memories, creating non-volatile memory cells
8	Photonic/optical computing	Light-based signal processing	Very high speed, low-power interconnects and logic	Accelerating matrix multiplication for AI training, low-latency LiDAR for autonomous vehicles, solving complex combinatorial optimization problems, telecommunications and data centers
9	Stochastic computing	Probabilistic logic, random number generators	Efficient, fault-tolerant processing	AI and probabilistic algorithms
10	Quantum computing	Superconducting qubits, topological insulators, and quantum gates.	Exponential speedup for specific algorithms	Optimization, cryptography

REFERENCES

Athle R. and M. Borg 2024 Ferroelectric tunnel junction memristors for in-memory computing accelerators, *Advanced Intelligent Systems*, 6(3): 2300554, https://doi.org/10.1002/aisy.202300554

Basu S., R. E. Bryant, G. De Micheli, T. Theis and L. Whitman 2019 Nonsilicon, non-von Neumann computing—Part I, *Proceedings of the IEEE*, 107(1): 11–18, https://doi.org/10.1109/JPROC.2018.2884780

Beyond von Neumann 2020 *Nature Nanotechnology*, 15: 507, https://doi.org/10.1038/s41565-020-0738-x

Bhat H. A., F. A. Khanday, B. K. Kaushik, F. Bashir and K. A. Shah 2022 Quantum computing: Fundamentals, implementations and applications, *IEEE Open Journal of Nanotechnology*, 3: 61–77, https://doi.org/10.1109/OJNANO.2022.3178545

Chowdhury S. S., D. Sharma, A. Kosta and K. Roy 2025 Neuromorphic computing for robotic vision: algorithms to hardware advances, *Communications Engineering*, 4: 152, https://doi.org/10.1038/s44172-025-00492-5

Finocchio G., J. A. C. Incorvia, J. S. Friedman, Q. Yang, A. Giordano, J. Grollier, H. Yang, F. Ciubotaru, A. V. Chumak and A. J. Naeemi 2024 Roadmap for unconventional computing with nanotechnology, *Nano Futures*, 8(1): 012001, https://doi.org/10.1088/2399-1984/ad299a

Gonzalez-Zalba M. F., S. de Franceschi, E. Charbon, T. Meunier, M. Vinet and A. S. Dzurak, 2021 Scaling silicon-based quantum computing using CMOS technology, *Nature Electronics*, 4(12): 872–884, https://doi.org/10.1038/s41928-021-00681-y

Greatorex H., O. Richter, M. Mastella, M. Cotteret, P. Klein, M. Fabre, A. Rubino, W. Soares Girão, J. Chen, M. Ziegler, L. Bégon-Lours, G. Indiveri and E. Chicca 2025 A neuromorphic processor with on-chip learning for beyond-CMOS device integration, *Nature Communications*, 16(1): 6424, https://doi.org/10.1038/s41467-025-61576-6

Mannocci P., M. Farronato, N. Lepri, L. Cattaneo, A. Glukhov, Z. Sun and D. Ielmini 2023 In memory computing with emerging memory devices: Status and outlook, *APL Machine Learning*, 1: 010902, https://doi.org/10.1063/5.0136403

Marrows C.H., J. Barker, T. A. Moore and T. Moorsom 2024 Neuromorphic computing with spintronics, *npj Spintronics*, 2: 12, https://doi.org/10.1038/s44306-024-00019-2

Nikandish R., E. Blokhina, D. Leipold and R. B. Staszewski 2021 Semiconductor Quantum Computing: Toward a CMOS quantum computer on chip, *IEEE Nanotechnology Magazine*, 15(6): 8–20, https://doi.org/10.1109/MNANO.2021.3113216

Xiao Y., C. Gao, J. Jin, W. Sun, B. Wang, Y. Bao, C. Liu, W. Huang, H. Zeng and Y. Yu 2024 Recent progress in neuromorphic computing from memristive devices to neuromorphic chips, *Advanced Devices & Instrumentation*, 5: 0044, https://doi.org/10.34133/adi.0044

Zhang G., X. Fan, Z. Wang, P. Li, Z. Zhang, B. Yu, D. Gao, D. Loke, S. Zhong, Q. Wan and Y. Zhang 2026 Self-rectifying memristors for beyond-CMOS computing: Mechanisms, materials, and integration prospects. *Nanomicro Letters*, 18(1): 188, https://doi.org/10.1007/s40820-025-02035-1

Zhang X., H.-O. Li, G. Cao, M. Xiao, G.-C. Guo and G.-P. Guo 2019 Semiconductor quantum computation, *National Science Review*, 6(1): 32–54, https://doi.org/10.1093/nsr/nwy153

Zito D. 2025 Qubit-size low-power cryogenic CMOS ICs for monolithic quantum processors, *Microprocessors and Microsystems*, 118: 105192, https://doi.org/10.1016/j.micpro.2025.105192

8 Nanofabrication Techniques

8.1 INTRODUCTION

Nanofabrication is the foundation of nanoelectronics, making the impossible possible at the tiniest scales. Nanofabrication serves as the essential technology that enables the construction, manipulation, and control of materials and devices at the atomic and molecular scale. If we could not shape and control materials at the nanometer scale, building advanced nanoelectronic devices like transistors, dense memory, and sensors would remain out of reach. Nanofabrication transforms the promise of nanotechnology into reality, enabling the mass production of electronic devices that connect the mysteries of quantum mechanics with the technology we use every day. Nanofabrication methods fall into two contrasting camps: the top-down approach, where one begins with something grand and carve sit down, and the bottom-up approach, where one assembles greatness from the tiniest pieces. We can imagine the top-down method as a sculptor facing a hulking block of marble, chipping away bit by bit until a masterpiece emerges. In contrast, the bottom-up method is like a bricklayer, patiently stacking minuscule bricks, each one essential, until a majestic castle stands tall (Figure 8.1).

Beyond-CMOS nanoelectronics increasingly depends on merging top-down precision with bottom-up scalability to construct intricate structures, such as 2D-material heterojunctions and 3D nanostructures (Biswas et al. 2012). Top-down fabrication techniques carve out nanostructures from bulk materials by removing material from a larger piece. In contrast, bottom-up approaches create structures by assembling atoms, molecules, or nanoparticles into functional configurations (Lu and Lieber 2007).

8.2 TOP-DOWN TECHNIQUES

8.2.1 LITHOGRAPHY

8.2.1.1 Electron Beam Lithography

EBL is an essential nanofabrication method utilized to produce extremely small patterns that exceed the resolution capabilities of conventional optical lithography (Tseng et al. 2003, Sharma et al. 2022). An EBL system is shown in Figure 8.2, with a majestic castle that stands tall (Figure 8.1).

EBL employs a focused electron beam for accurate pattern creation to directly "write" designs onto a substrate coated with an electron-sensitive resist. This technique enables the fabrication of features down to a few nanometers, which is crucial for advancing next-generation devices.

DOI: 10.1201/9781003753469-8

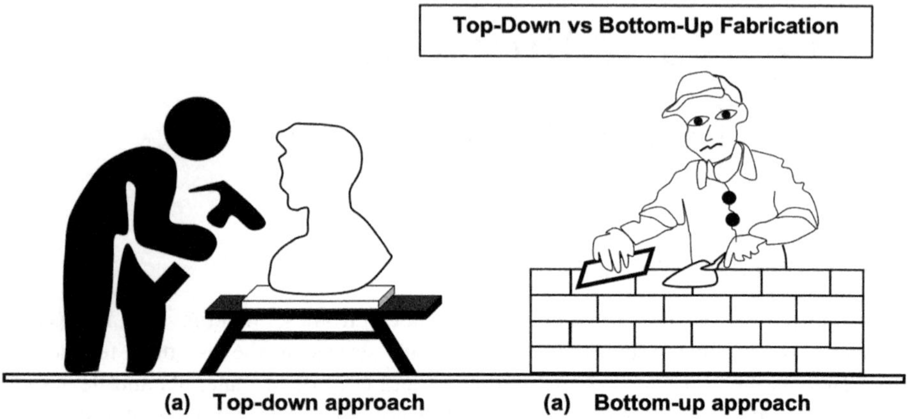

FIGURE 8.1 Simple everyday analogies of top-down and bottom-up fabrication, such as a sculptor carving a statue by shaping a large piece of marble and a mason laying down bricks to build a house. Top-down fabrication resembles an artist chiseling a masterpiece from marble, carving away excess to reveal delicate details. In contrast, bottom-up fabrication mirrors the careful stacking of bricks, piecing together atoms or molecules to construct something grand from the ground up.

The significance of EBL in beyond-CMOS technology resides in creating advanced architectures. EBL serves as the foremost tool for research and development as well as for prototyping new beyond-CMOS devices, including single-electron transistors (SETs), for which precise nanometer-sized regions (quantum dots) are necessary for electron tunneling; nanowire devices by shaping silicon or III–V semiconductor nanowires for highly scaled logic and sensing applications; and spintronics and quantum devices by constructing the complex structures required for spin-wave guides or superconducting qubits. EBL is also instrumental in heterogeneous integration, enabling the incorporation of novel materials such as graphene, carbon nanotubes, and ferroelectrics onto established silicon CMOS platforms to develop "hybrid" systems.

Unlike typical photolithography, EBL operates without the need for costly physical masks. Non-requirement of masks offers considerable flexibility for experimental designs. EBL attains high resolution. It can reach sub-10 nm resolution, which is essential for the "beyond-CMOS" era where device sizes approach atomic dimensions. Although EBL delivers superior resolution, its primary drawback for mass production is its slow serial writing speed (low throughput). Drawing patterns pixel by pixel, it operates much more slowly than parallel optical techniques, limiting its application to low-volume production and research rather than large-scale manufacturing. To overcome this challenge, it is often paired with other methods, such as NIL, in which EBL is used to fabricate high-resolution stamps (masters) for faster subsequent replication.

Example 8.1 You are utilizing an EBL system to design a single-electron transistor (SET) structure intended for a beyond-CMOS application. The pattern necessitates a

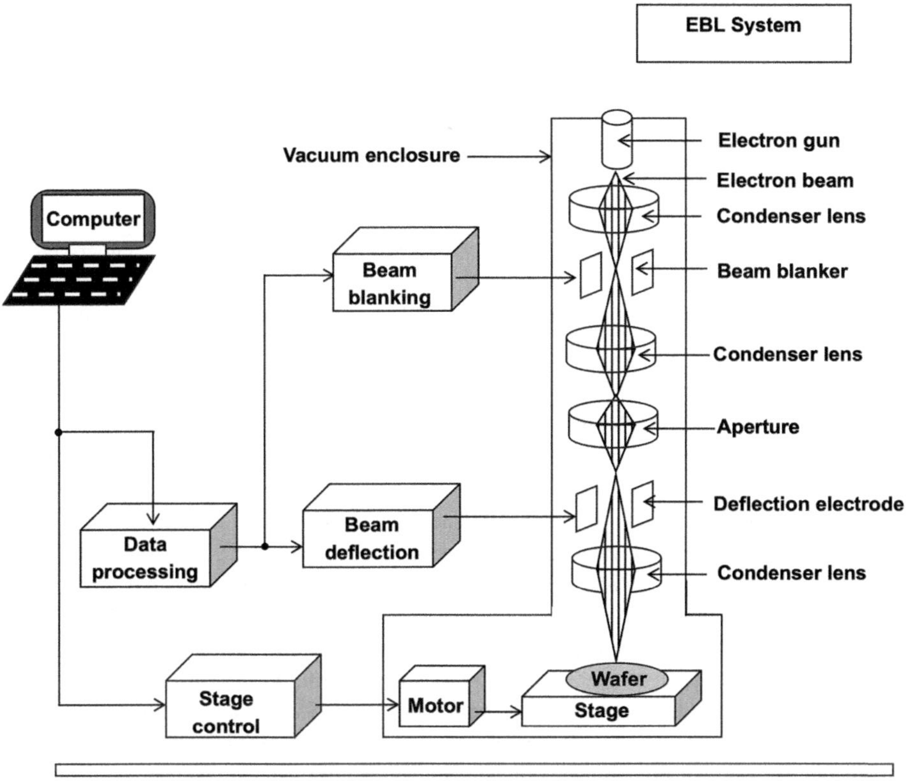

FIGURE 8.2 Electron beam lithography system. The system features a high-vacuum chamber with an electron gun and an electromagnetic column for focusing and scanning. The electron gun produces high-energy electrons (10–100 keV) using either a thermionic (tungsten/LaB_6) or field-emission source. The electron column includes magnetic or electrostatic lenses, apertures, and deflectors to focus, steer, and shape the beam. Condenser lenses adjust beam intensity and focus, while the deflection system uses electromagnetic coils or electrostatic plates to scan the beam across the sample. The beam blanker rapidly switches the beam on and off to control exposure. A computer, through control software, manages beam deflection, blanking pulses, and the motorized stage. It exposes patterns with nanometer precision onto an electron-sensitive, resist-coated substrate, such as PMMA, whose solubility changes upon exposure. The stage, guided by a laser interferometer, ensures precise positioning for large-area, multi-field, or stitching-intensive exposures.

specific exposure dose. Given parameters: Desired dose $D: 300\ \mu C/cm^2$, Beam current $I: 5000$ pA. Total area to be exposed $A: 0.01 cm^2$. Determine the minimum exposure time (t) needed to expose the designated area. What does your result highlight?

Solution: The basic equation connecting dose (D), area (A), current (I), and time (t) is

$$D = \frac{It}{A} \tag{8.1}$$

which is reorganized to find the time:

$$t = \frac{DA}{I} = \frac{300 \times 10^{-6}\ \mathrm{C/cm^2} \times 0.01\ \mathrm{cm^2}}{5000 \times 10^{-12}\ \mathrm{C/s}} = \frac{300 \times 10^{-6} \times 0.01\ \mathrm{C}}{5000 \times 10^{-12}\ \mathrm{C/s}} = \frac{3 \times 10^{-6}}{5 \times 10^{-9}}\ \text{seconds}$$
$$= \frac{3000}{5}\ \text{seconds} = 600\ \text{seconds} = 10\ \text{minutes} \quad (8.2)$$

The required exposure duration is 600 seconds (10 minutes). The outcome underscores a significant drawback of EBL in high-volume production: even small areas can take considerable time to expose, which is why EBL is mainly used for research and mask fabrication.

8.2.1.2 Extreme Ultraviolet Lithography

EUVL is the leading patterning technology in contemporary semiconductor manufacturing and is a key pathway to advancing "Beyond CMOS" nanoelectronics (Banqiu and Kumar 2014, Johnson 2019, Giannopoulos et al. 2024). Using a wavelength of 13.5 nm, EUVL enables the industry to overcome the physical constraints of traditional DUV light, allowing the creation of features at the sub-10 nm level. Figure 8.3 illustrates the main components of an EUVL machine.

EUVL plays a diverse role in advancing nanoelectronics. It enables sub-3 nm scaling. EUVL is critical for the mass production of the 5, 3, and forthcoming 2 nm process nodes. It streamlines the manufacturing process by substituting intricate multi-patterning procedures for essential layers with a single-exposure approach, resulting in reduced costs and higher yields. The next phase, High-NA (Numerical Aperture) EUVL systems utilizing 0.55 NA scanners, is currently being implemented by leading companies. This technology has already demonstrated the capability to print 10 nm-dense lines and 9.5 nm random-logic structures, which are crucial for next-generation 2D and 3D device architectures. EUVL offers support for emerging architectures, including the precision required to manufacture beyond-CMOS devices such as GAA FETs, TFETs, and ultra-thin devices utilizing 2D materials like molybdenum disulfide or graphene.

EUVL faces various technological and financial hurdles, including significant complexity and expense. A single EUV scanner requires a substantial financial investment, particularly given the high cost of high-NA equipment. As feature sizes decrease, random fluctuations in photon count and in the number of resist molecules (stochastic effects) can lead to defects such as line-width roughness (LWR), adversely affecting device performance and reliability. Power efficiency is a concern because the 13.5 nm light is generated by targeting high-power lasers at tin droplets, a method that demands significant energy. Efforts are ongoing to identify more efficient light sources to enhance throughput and sustainability.

As EUVL nears its own diffraction limits, investigations are already underway into "Beyond EUV" alternatives, the next horizon. Wavelengths shorter than 13.5 nm (blue-X or soft X-rays) are being explored to further extend resolution capabilities. Advanced light sources use laser technologies that are ten times more efficient than current CO_2-based systems. Wide aperture EUV lithography is dawning as

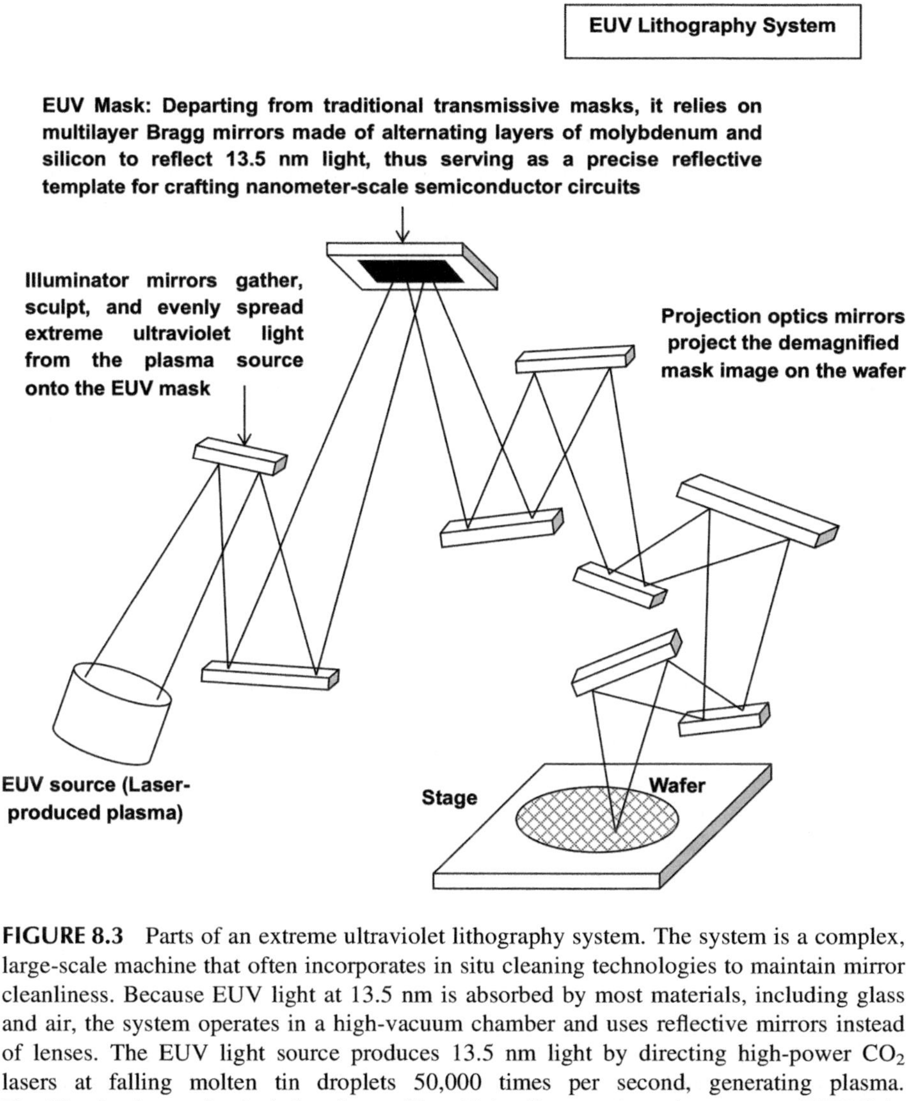

FIGURE 8.3 Parts of an extreme ultraviolet lithography system. The system is a complex, large-scale machine that often incorporates in situ cleaning technologies to maintain mirror cleanliness. Because EUV light at 13.5 nm is absorbed by most materials, including glass and air, the system operates in a high-vacuum chamber and uses reflective mirrors instead of lenses. The EUV light source produces 13.5 nm light by directing high-power CO_2 lasers at falling molten tin droplets 50,000 times per second, generating plasma. The illumination optics include a large ellipsoidal collector mirror that captures EUV light from the plasma and focuses it into a beam. This beam is shaped by multilayer mirrors made of molybdenum and silicon to ensure uniform illumination for the mask. A hydrogen buffer gas is often used near the collector to prevent tin debris from contaminating the mirrors. The reflective mask, which contains the circuit pattern, is designed to reflect rather than transmit light. Unlike traditional transparent masks, EUV masks are reflective. The projection optics consist of ultra-smooth, high-precision mirrors that project and reduce the pattern from the mask onto the wafer. The wafer stage is a highly precise, often dual-stage platform that holds and moves the silicon wafer with nanometer-level accuracy in a vacuum.

EUV lithography is moving into the sub-10 nm half-pitch era, powered by the next wave of high numerical aperture EUV scanners that promise to propel Moore's Law past the 2 nm frontier (van de Kerkhof et al. 2017). By boosting the optical system's numerical aperture from 0.33 to 0.55, these machines deliver crisper, more detailed transistor patterns for 2 nm and smaller logic chips in a single exposure, lowering the need for complicated multi-patterning. This leap is made possible by high-NA scanners equipped with anamorphic lenses that stretch magnification 8x in one direction and 4x in the other. Looking ahead, researchers are already setting their sights on "Hyper-NA" systems, aiming for 0.7–0.75 NA to break new ground.

Alternative patterning involves combining EUVL with methods such as directed self-assembly (DSA) or NIL to overcome purely optical limitations.

Example 8.2 A cutting-edge high-NA EUV lithography scanner operates at a wavelength λ of 13.5 nm and has a numerical aperture (NA) of 0.55. Given that the process is highly optimized with a low complexity factor k_1 value of 0.3, what is the minimum achievable critical dimension (CD) or half-pitch of the advanced high-NA EUV system?

Solution: The issue revolves around the Rayleigh criterion formula, which is essential for optical lithography:

$$\text{Critical dimension (CD)} = k_1 \times \frac{\lambda}{\text{NA}} \tag{8.3}$$

where CD represents the smallest feature size (resolution), also known as half-pitch. k_1 indicates the complexity factor of the process, usually ranging between 0.25 (the physical limit) and 0.8. The symbol λ denotes the wavelength of the exposure light. NA stands for the Numerical Aperture of the projection lens.

First, we note the provided values: $\lambda = 13.5$ nm, NA $= 0.55$, and $k_1 = 0.3$. Next, utilizing the Rayleigh criterion formula, inserting the variables in the equation, and proceeding with the calculation:

$$\text{Critical dimension (CD)} = 0.3 \times \frac{13.5\text{ nm}}{0.55} = 7.3636\text{ nm} \tag{8.4}$$

Thus, CD ≈ 7.36 nm.

The smallest possible critical dimension (half-pitch) achievable with this high-NA EUV system is approximately 7.36 nm (nanometers). Additionally, we can calculate the theoretical physical limit for this machine by applying the absolute minimum k_1 value of 0.25.

$$\text{CD}_{\text{Limit}} = 0.25 \times \frac{13.5\text{ nm}}{0.55} = 6.136\text{ nm} \approx 6.14\text{ nm} \tag{8.5}$$

8.2.1.3 Nanoimprint Lithography

NIL is a cost-effective, high-resolution patterning method that is crucial for producing next-generation devices that exceed conventional semiconductor capabilities (Resnick 2014, Barcelo and Li 2016, Lin and Guo 2025). It serves as a leading alternative to EUV lithography for fabricating features with sub-5 nm and sub-10 nm dimensions. The NIL process involves pressing a patterned mold (stamp) into a polymer resist on a substrate, yielding high-resolution device patterns (Figure 8.4).

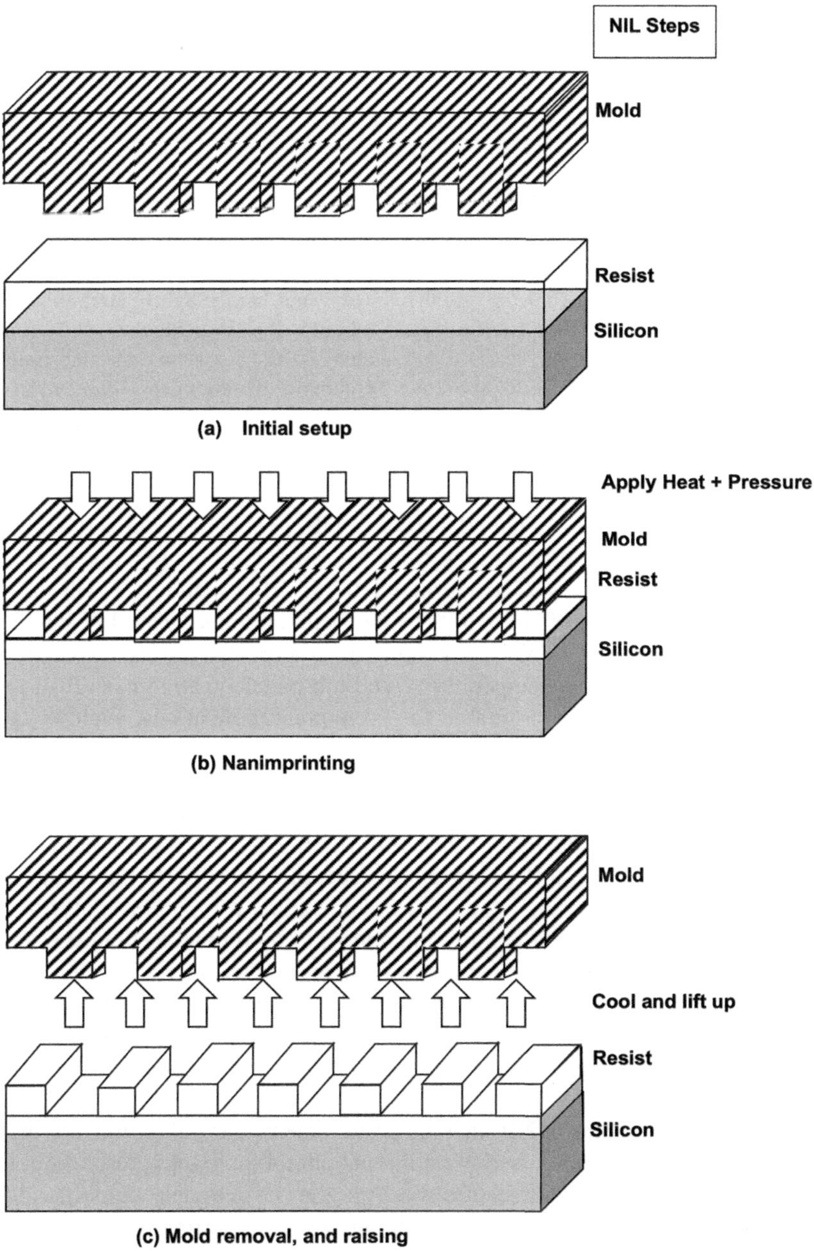

FIGURE 8.4 Principal steps of thermal NIL. (a) The process begins with the creation of a master mold featuring nanoscale circuit patterns, produced by high-precision methods such as electron-beam lithography (EBL) or photolithography. The mold is coated with an anti-sticking layer, and a thin thermoplastic polymer resist is spin-coated onto the silicon substrate.

(Continued)

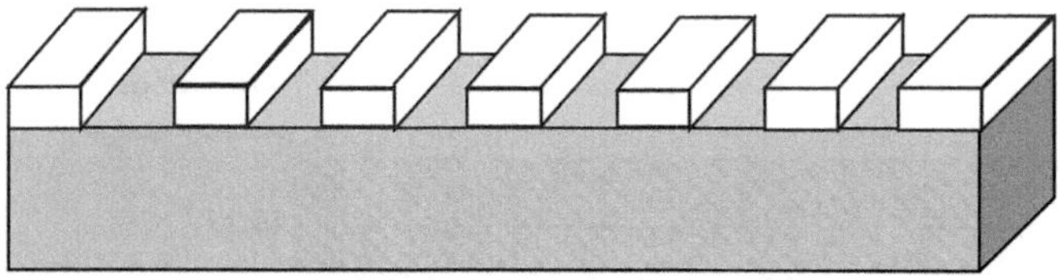

(d) Reactive ion etching

FIGURE 8.4 (Continued) (b) During nanoimprinting, the mold is pressed onto the substrate under controlled force, and the system is heated above the polymer's glass transition temperature to shape the resist, then cooled to solidify it. (c) After hardening, the mold is separated from the substrate, leaving the patterned resist behind. (d) The residual resist layer in the imprinted areas is removed by reactive ion etching (RIE), exposing the substrate. The remaining resist serves as a mask for subsequent etching or material deposition.

Let us explain the basics of the NIL process. NIL generates patterns by physically deforming a resist with a template (mold) instead of employing photons or electrons to alter chemical properties. The thermal NIL (T-NIL) method applies heat and pressure to embed a mold into a thermoplastic polymer. The UV-based NIL (UV-NIL) technique uses a UV-curable liquid photopolymer and a transparent mold, enabling processing at room temperature with high-precision alignment. Roll-to-Roll NIL is a continuous process suitable for large-area applications, such as flexible displays and solar cells.

The contributions of NIL to nanoelectronics need elaboration. NIL, often paired with Scanning Probe Lithography (SPL), is a top contender for high-throughput manufacturing of devices featuring sub-5 nm patterns. NIL facilitates the fabrication of quantum-effect devices. It provides the accuracy needed to fabricate single-electron and quantum-dot devices at room temperature, a capability not achievable with conventional optical lithography. In contrast to traditional techniques, NIL can directly pattern three-dimensional structures and multilayer stacks, which is essential for making 3D and stacked structures for developing advanced 3D integrated circuits and sophisticated photonic crystals. NIL is currently under consideration for large-scale production of next-generation cross-point memory, magnetic logic, and lab-on-a-chip sensors.

A few strategic benefits of NIL include cost efficiency, as NIL eliminates the need for the costly, intricate optics used in traditional photolithography; resolution, which is primarily confined to the feature size on the 1x template (mold), allowing for features as small as 1–2 nm; and scalability by supporting large-wafer (up to 300 mm) or continuous roll production, which is crucial for the commercial viability of beyond-CMOS technologies.

8.2.1.4 Block Copolymer Lithography

Block copolymer (BCP) lithography is a "bottom-up" approach with top-down goals to patterning that leverages the DSA of polymer chains to produce periodic nanostructures—such as lines, dots, or holes—at resolutions under 10 nm, which is typically challenging or costly to achieve using standard photolithography methods (Tallegas et al. 2013, Hu and Xiong 2022, Cheng et al. 2025).

BCP lithography functions through several mechanisms: self-assembly, in which two or more chemically different polymer blocks spontaneously segregate into organized structures (spheres, cylinders, lamellae) on a substrate; DSA, in which external influences (topographic patterns, chemical brushes) guide the BCPs to arrange themselves into designated, long-range ordered patterns; and pattern transfer, where one of the blocks is selectively eliminated, allowing the other to act as a mask for etching the substrate beneath, thereby transferring the nanoscale design.

BCP lithography plays a significant role in nanoelectronics. It enables resolution and density multiplication by enhancing the density of patterns produced by conventional lithography (such as EUV) by two to four times, enabling densities of 1 terabit per square inch. It aids in patterning 2D materials, such as graphene and MoS_2, into nanoscale structures, including nanoribbons and nanoporous sheets, which are crucial for high-performance phototransistors and multichannel FETs.

Among new architectures, BCP templates facilitate the creation of many non-CMOS structures. It is used to make nanowire crossbars: aligned arrays of nanowires (e.g., tungsten trioxide) for future generations of logic circuits and memory devices. Magnetic nanodots are accurately positioned on bit-patterned media used in hard disk drives for magnetic storage. High-density arrangements are used in NEMS to enable extremely sensitive sensors.

Implementation strategies for BCP lithography include graphoepitaxy, which uses physical topographical molds (trenches or holes) to direct the self-assembly of BCPs to specific sites; chemoepitaxy, which employs chemically patterned surfaces that have differing interfacial energies to control the phase segregation of the polymer domains, and sequential infiltration synthesis (SIS), a method in which BCP domains serve as "pockets" for the growth of metallic or metal-oxide structures directly within the polymer template.

Strategic advantages of BCP lithography include its cost-effectiveness: it enables sub-10 nm resolution without the high costs associated with multiple-masking techniques or low-throughput EBL; dimensional control by providing improved line-edge roughness (LER) and critical dimension uniformity (CDU) when compared to traditional photoresists at similar dimensions; and industrial compatibility because it can be engineered to be incorporated into current 300 mm semiconductor fabrication processes as "smart resists" or hard masks.

Example 8.3 A frequently utilized block copolymer for directed self-assembly is poly (styrene)-*b*-poly(methyl methacrylate) (PS-*b*-PMMA). The periodic structure (or domain spacing, L_0) of the self-assembled arrangement mainly varies with the total molecular weight of the polymer (M_n). If a particular PS-*b*-PMMA polymer with a number-average molecular weight M_n of 50 kg/mol creates a lamellar structure with a full period L_0 of 20 nm, what would be the anticipated period for another batch of the same polymer type with a molecular weight of 100 kg/mol? Assume that the Flory-Huggins interaction parameter χ and the composition remain unchanged for both polymers. The parameter χ measures how much energy is shifted when a polymer blends with a solvent or mixes with another polymer.

Solution: The domain spacing L_0 of a block copolymer is approximately proportional to its molecular weight M_n as $L_0 \propto M^{2/3}$. Given: Polymer 1 has a molecular weight of $M_{n,1} = 50$ kg/mol and a contour length $L_{0,1} = 20$ nm. Polymer 2 has a molecular weight of $M_{n,2} = 100$ kg/mol, but its contour length $L_{0,2} = ?$ is unknown. We can establish a ratio utilizing the scaling relationship:

$$\frac{L_{0,2}}{L_{0,1}} = \left(\frac{M_{n,2}}{M_{n,1}}\right)^{2/3} \tag{8.6}$$

Substituting the values and simplifying the ratio,

$$\frac{L_{0,2}}{20\text{ nm}} = \left(\frac{100\text{ kg/mol}}{50\text{ kg/mol}}\right)^{2/3} = (2)^{2/3} \approx 1.587 \tag{8.7}$$

$$\therefore L_{0,2} = 1.587 \times 20\text{ nm} = 31.74\text{ nm} \tag{8.8}$$

The anticipated length L_0 for the polymer with a molecular weight of 100 kg/mol would be around 31.74 nm.

8.2.2 Reactive Ion Etching

RIE is an essential subtractive manufacturing technique in nanoelectronics, employed to create patterns in materials that traditional silicon etching methods struggle to handle (Figure 8.5). As the industry gears up for upcoming technology nodes and unconventional architectures, RIE delivers the precise patterning required for nanostructures as size reductions approach fundamental limits (Wiśniewski et al. 2016, Huff 2021).

RIE is used in nanoelectronics to structure 2D materials; it plays a crucial role in the selective patterning of 2D materials such as graphene. Low-energy oxygen plasma RIE is a conventional process for establishing active regions in graphene-based gas sensors and FETs. RIE can also perform nanowire and trench isolation. In structures such as silicon-on-insulator (SOI), RIE is used to create deep trenches that provide electrical isolation for nanoscale islands and silicon "necks" in SETs. RIE facilitates the heterogeneous integration of III–V materials, RF MEMS, and other materials onto silicon CMOS platforms by enabling accurate etching of TSVs and selective material removal.

Several advanced RIE variations are crucial for overcoming "plasma-induced damage" that may compromise the functionality of fragile next-generation devices. Atomic-layer etching (ALE) is a precise variant of RIE that removes one atomic layer at a time, vital for sub-5nm nodes and GAA transistors, where accuracy is critical. Area-selective etching is an emerging method that concentrates on specific chemical domains on a substrate, minimizing the necessity for intricate lithographic masks in 3D integrated circuits. The low-power oxygen plasma process is applied explicitly to delicate 2D lattices to avoid lattice damage while ensuring clean edge termination.

Examples of emerging materials processed through RIE are 2D materials, e.g., for making graphene/TMDC FETs, RIE is used for selective etching and definition of edges; spintronic materials, e.g., in fabricating magnetic tunnel junctions, RIE is

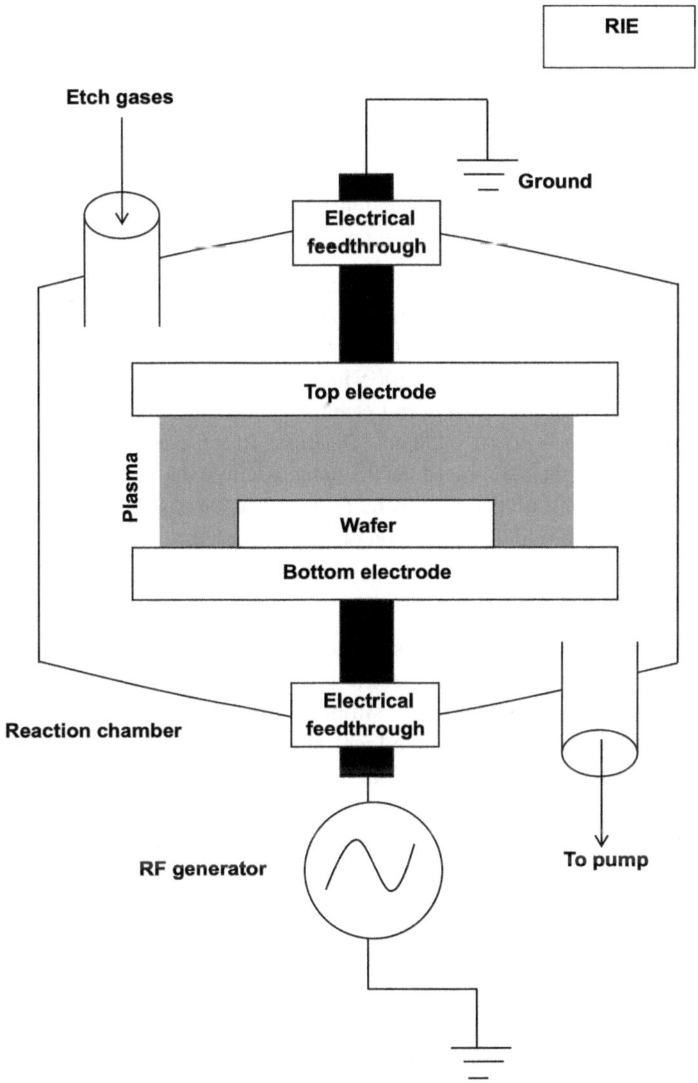

FIGURE 8.5 A reactive ion etching system. The reaction chamber is a sealed vacuum enclosure kept at a low pressure, typically from a few millitorr to several hundred millitorr, to ensure a controlled, clean environment. The substrate is placed on the bottom electrode (cathode), which is powered by radio frequency (RF) energy, while the top electrode (anode) is usually grounded. An RF generator, typically operating at 13.56 MHz, produces an electric field that ionizes gases, forming a plasma of ions and radicals. Reactive gases such as CF_4, SF_6, or Cl_2 are introduced for etching, and high-capacity pumps, including turbo pumps, maintain the required pressure and remove volatile byproducts.

applied for patterning of nanopillars for spin torque memory; and ferroelectrics, e.g., in FeFET non-volatile memory, RIE performs accurate etching of $Hf_xZr_{1-x}O_2$ films. RIE is also utilized to create nanowires for defining grids of nanowires under 30 nm for junctionless transistors.

Example 8.4 A research team is employing reactive ion etching (RIE) to create silicon nanowires for an innovative "beyond-CMOS" device. They must achieve a specific width for the nanowires with a high degree of accuracy. The target width for the silicon nanowire is 20 nm. Throughout the RIE process, a consistent amount of lateral etching (etching that occurs perpendicular to the desired vertical direction) results in the mask being undercut. This lateral etching is constantly measured as 15 nm on each side due to the conditions of the process. Determine the initial width that the photoresist mask pattern should be designed with to reach the target width of 20 nm for the nanowire.

Solution: It is necessary to consider the material removed by lateral etching from both sides of the pattern. The total loss from lateral etching is calculated as follows: Lateral etch per side multiplied by 2. Hence, total lateral etching loss equals 15 nm $\times$ 2 = 30 nm. Consequently, the required initial mask width will be the desired final width of the nanowire plus the total loss from lateral etching. In this case, the required initial mask width becomes 20 nm + 30 nm = 50 nm. Therefore, to achieve a final silicon nanowire width of 20 nm after the RIE procedure, the initial width of the photoresist mask pattern prior to etching should be designed to be 50 nm. This emphasizes the critical role of dimension control (CD) in the fabrication of nanoelectronics.

8.2.3 Focused-Ion-Beam Milling

Focused ion beam (FIB) milling is a fundamental method for fabrication and metrology in beyond-CMOS nanoelectronics (Han W. and S. Xiao 2013, Höflich et al. 2023). The milling mechanism is simple (Figure 8.6). A concentrated ion beam (typically Gallium, Ga^+) ejects (dislodges) atoms from the material's surface with accuracy, effectively removing material.

Applications: FIB milling has evolved from basic sample preparation to active device engineering and precise manufacturing. It is applied for prototyping new architectures. It enables the rapid production of sub-10 nm features without the need for masks, which is crucial for evaluating beyond-CMOS devices such as memristors, nanophotonic waveguides, and plasmonic devices. FIB milling is used for 3D nano-structuring. In contrast to standard planar CMOS processes, numerous beyond-CMOS designs incorporate vertical or 3D structures. FIB is employed for "nano-origami" and high-aspect-ratio 3D milling to develop intricate vertical interconnects or GAA configurations. In nanofluidics and lab-on-chip devices, FIB milling is used to fabricate nanochannels in materials such as lithium niobate for cutting-edge nanofluidic sensors that surpass traditional electronic logic. Beyond subtractive milling, FIB milling is used for direct material manipulation. It enables localized ion implantation to introduce defects (e.g., generating single-photon emitters for quantum computing) or to modify specific regions of a nanostructure. FIB milling enables advanced metrology and preparation methods, e.g., TEM lamella preparation. As devices progress beyond the 2 nm threshold, FIB plays a vital role in creating ultra-thin cross-sections (lamellae) for transmission electron microscopy (TEM) to examine atomic-level interfaces in 2D multi-fin FETs and other emerging

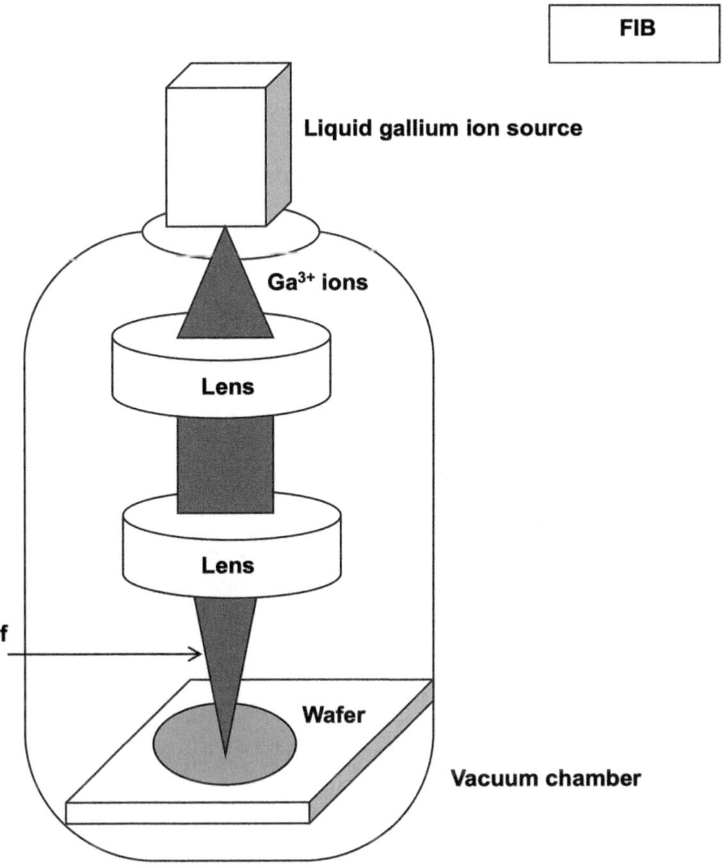

FIGURE 8.6 Focused ion beam system. The vacuum chamber provides an oil-free, high-vacuum environment ($<10^{-6}$ torr) to prevent beam scattering and contamination, both of which are essential for precise milling. The most common ion source is the liquid metal ion source (LMIS), which typically uses gallium (Ga^+) due to its low melting point and stability. Electrostatic condensers and objective lenses focus the beam to a spot often smaller than 10 nm. A high-precision, motorized *XY* stage with tilt and rotation capabilities holds and positions the sample.

technologies. For delicate organic or biological nanoelectronics, cryo-FIB milling prevents beam damage and thermal distortion, enabling high-resolution imaging of fragile materials in their "native" condition. In circuit editing and debugging, FIB is employed to redirect connections on sophisticated prototype chips to verify "beyond-CMOS" logic gates or to conduct backside deprocessing of high-density interconnects.

Current Challenges to FIB Milling: Redeposition of atoms is a problem. Sputtered atoms can redeposit elsewhere, impacting feature fidelity. Ion damage takes place. Gallium ions can be implanted into the substrate, forming amorphous regions or defects that alter material properties.

To prevent gallium contamination in sensitive semiconductor junctions, modern systems use Plasma FIB (PFIB) with inert gas ions (such as xenon) or multi-species ion sources to ensure Ga-free milling and material purity. Finding a trade-off or balance between milling speed and acceptable resolution remains a persistent concern. Throughput improvements are necessary. Conventional FIB operations are slow and tend to be serial. New methods, such as masking processes, can facilitate nanoscale machining that is up to 75 times quicker than traditional techniques. Research will focus on modeling and minimizing "FIB-induced surface artifacts" such as amorphization layers, which can distort the electrostatic potential of nanoscale devices.

8.3 BOTTOM-UP TECHNIQUES

8.3.1 Atomic-Layer Deposition

ALD is an essential technique for manufacturing beyond-CMOS nanoelectronics. Figure 8.7 shows the schematic diagram of an ALD system. The capability of ALD to achieve atomic-level control over thickness and provide excellent conformality on intricate 3D structures makes it vital for incorporating new materials and device designs needed for the future of computing (Hwang and Yoo 2014, Amashaev et al. 2025)

Uses of ALD in Beyond-CMOS Devices: ALD is used to deposit high-*k* gate dielectrics (such as Al_2O_3 and HfO_2) on 2D semiconductors, such as MoS_2 and graphene. This is crucial for ensuring effective gate control in atomically thin channels while reducing interface traps. The discovery of ferroelectric properties in ALD-grown doped HfO_2 has led to the development of ferroelectric FeFETs. These devices serve as non-volatile memory and logic components in low-power, neuro-inspired computing applications. ALD enables the precise construction of ultra-thin tunneling barriers and gate stacks in TFETs, aiming for a sub-60 mV/dec subthreshold swing, enabling extremely low-power operation. Additionally, ALD enables the integration of III–V materials (such as InGaAs and GaN) with high-*k* dielectrics, overcoming the "Fermi-level pinning" problems that have previously limited their use.

Advanced ALD Methods for Nanoelectronics: Area-selective ALD (AS-ALD) is a "bottom-up" technique that enables films to grow only on chosen surfaces, minimizing the necessity for complicated lithography and etching in ultra-scaled nodes. The plasma-enhanced ALD (PEALD) approach uses plasma to facilitate lower deposition temperatures, which is essential for delicate beyond-CMOS materials such as organics or specific 2D layers. 3D conformal coating is crucial for GAA nanosheets and 3D DRAM architectures, where high-aspect-ratio structures require uniform coating.

Ultimately, ALD transitions from merely reducing CMOS size to enabling entirely new nanoelectronic paradigms by manipulating materials and structures at the atomic scale, which is vital for the future of computing beyond current scaling trajectories.

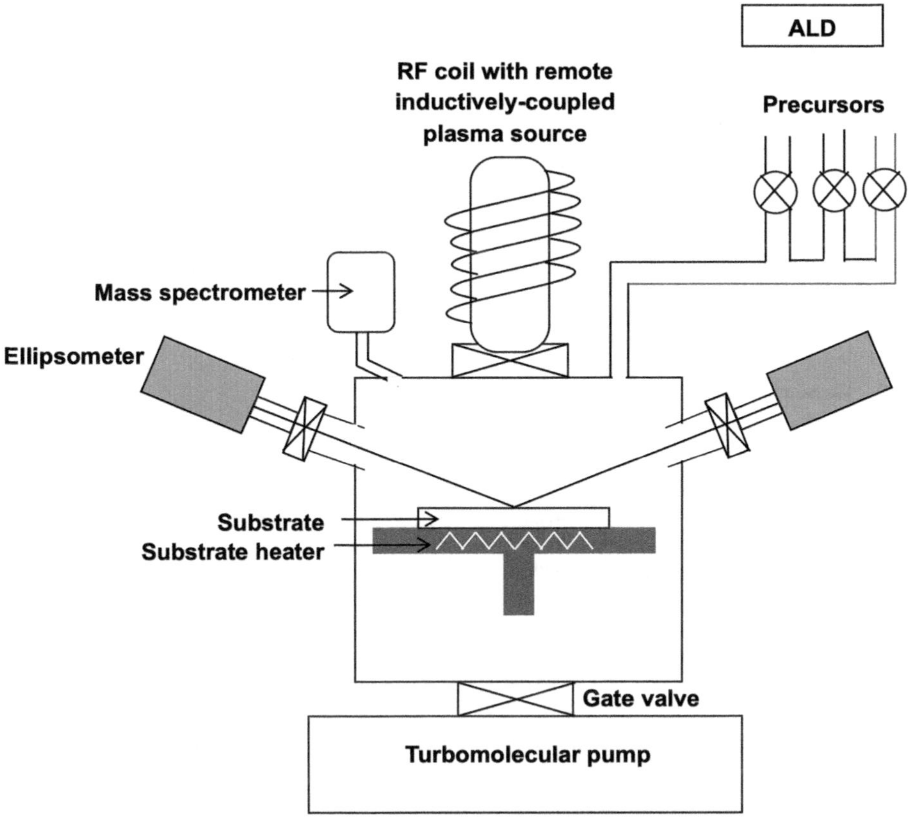

FIGURE 8.7 Schematic of an ALD system. Precursors, typically organometallics or halides, are stored in canisters or bubblers. Liquid precursors such as TMA ($Al(CH_3)_3$) are used for Al_2O_3 deposition, while gaseous precursors such as $TiCl_4$ are used for TiN deposition. The substrate heater or chuck maintains the deposition temperature, usually between 100°C and 300°C. A remote inductively coupled plasma source, often a helical copper coil around a dielectric chamber, is powered by a high-frequency (13.56 MHz) RF generator through an impedance-matching network. This setup generates a magnetic field to sustain the plasma, with the source located outside the main chamber. This configuration allows radicals to reach the substrate while minimizing exposure to charged particles and UV radiation, reducing surface damage. A turbomolecular pump creates ultra-high vacuum conditions using rapidly spinning blades. In situ ellipsometry and mass spectrometry provide comprehensive process monitoring: the mass spectrometer verifies precursor reactions and byproduct removal in the gas phase, while the ellipsometer confirms material growth on the substrate.

8.3.2 Molecular Beam Epitaxy

MBE serves as a crucial technology for nanoelectronics (Figure 8.8). MBE offers the atomic-scale accuracy necessary for the growth of high-purity, non-silicon materials and intricate nanostructures on silicon substrates (Chang et al. 2011, Henini 2012, Shen et al. 2024).

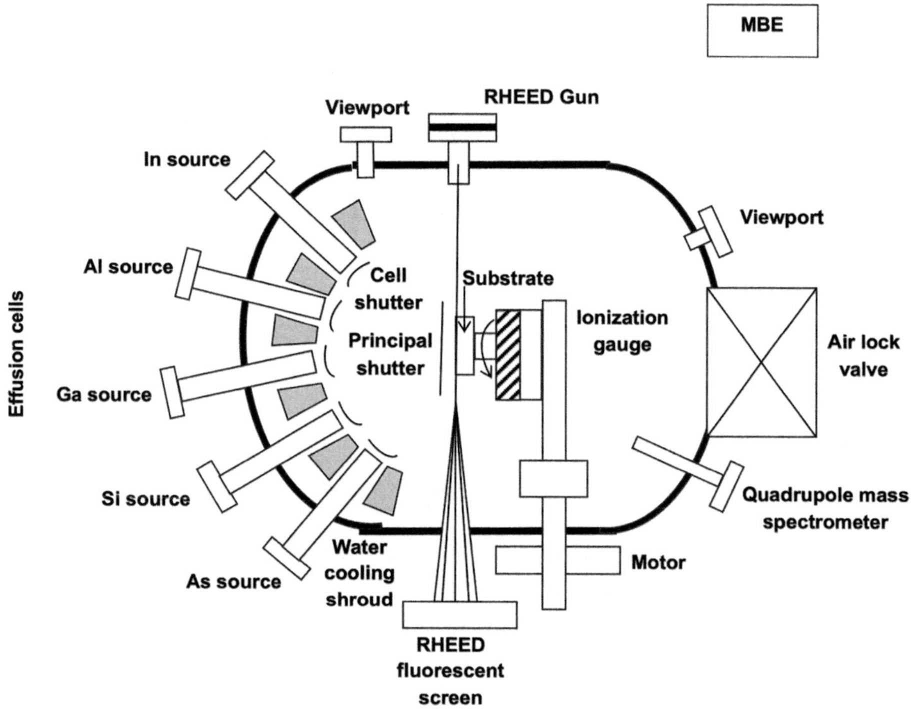

FIGURE 8.8 A molecular beam epitaxy (MBE) system. It is an ultra-high vacuum (UHV) apparatus for growing high-purity epitaxial thin films. Effusion cells, or Knudsen cells, are heated crucibles that melt or sublime source materials to generate atomic or molecular beams. These include standard, high-temperature, and valved cracker cells for specialized materials, such as Al, Ga, Si, and As. Water cooling shrouds, often water- or LN_2-cooled cryopanels, are essential for managing heat loads from hot cells (200°C–1400°C), stabilizing source temperatures, improving vacuum quality by trapping contaminants, and preventing crosstalk between adjacent cells. Electronically controlled shutters in front of each cell enable precise, atomic-layer control of film thickness. The viewport is a transparent window in the UHV chamber for visual inspection, beam alignment, and in situ monitoring of crystal growth. The ionization gauge monitors UHV conditions (down to 10^{-11} Torr) and measures atomic or molecular beam flux (Beam Equivalent Pressure, BEP). The airlock or load-lock valve allows sample transfer into or out of the UHV chamber without breaking vacuum, maintaining pressure integrity during exchanges. Reflection high-energy electron diffraction (RHEED) uses an electron gun and phosphor screen for real-time, in situ monitoring of surface structure, crystal quality, and growth rate. The quadrupole mass spectrometer (QMS) monitors beam flux, verifies flux stability, and detects background gases or contaminants.

Significant Functions of MBE Beyond CMOS: MBE facilitates the growth of semiconductor channels with high carrier mobility (such as InGaAs, Ge, and GaN) on silicon. These materials can achieve a lower equivalent oxide thickness (EOT) and superior interface quality with high-κ dielectrics. The MBE technique is employed to fabricate vertically aligned nanowires (NWs) and quantum dots (QDs) for future logic applications. For instance,

GaSb/InAs nanowires, grown via MBE on Si, are being explored for advanced logic feature dimensions. MBE is a key technique for producing "artificial" quantum materials by meticulously depositing oxides and multi-component materials, which are vital for advancing quantum computing and sensing. MBE produces the refined heterojunctions and doping profiles required for TFETs, which seek to achieve steeper subthreshold swings and enhanced energy efficiency compared to conventional MOSFETs at low supply voltages. MBE enables seamless integration of III–V semiconductors (such as GaAs or GaP) onto silicon wafers compatible with CMOS via methods like nanoheteroepitaxy (NHE). For NHE, gas-source MBE (GS-MBE) is applied to grow low-defect GaAs and GaP nanoislands on silicon nanotips, marking a significant advancement in integrated photonics. In advanced device engineering studies, MBE-grown hybrid axial core-shell nanowire photodetectors and self-assembled nanostructures have been designed for ultra-low-power electronics. Techniques involving NH_3-MBE mitigate current collapse in high-mobility transistors (HEMTs), thereby enhancing their reliability for high-power and automotive applications.

Thus, MBE offers the material versatility and nanoscale precision essential for elevating electronic performance beyond the capabilities of traditional silicon CMOS, facilitating new functionalities and increased integration densities.

Example 8.5 A molecular beam epitaxy (MBE) apparatus is utilized to deposit a thin film of gallium arsenide (GaAs) to form a quantum well structure. The target thickness for the GaAs film is set at 100 nm. This system functions with a beam flux density (J) of 1×10^{15} atoms per square centimeter per second (10^{15} atoms/cm^2 s), and the growth rate of GaAs under these conditions is roughly 1 monolayer (ML) per second. Each monolayer of GaAs has an average thickness of 0.3 nm. Determine the necessary duration for the growth process to reach the intended film thickness.

Solution: We start by calculating the total number of monolayers needed: Desired thickness = 100 nm; Thickness per monolayer = 0.3 nm/ML. Then, the Total number of monolayers (MLs) is

$$\text{Total MLs} = \frac{\text{Desired thickness}}{\text{Thickness per ML}} = \frac{100\ \text{nm}}{0.3\ \text{nm}} \approx 333.33\ \text{MLs} \tag{8.9}$$

Next, we calculate the overall growth duration in seconds: Growth rate = 1 ML/s; Total time (seconds) is given by

$$\begin{aligned}\text{Total time (seconds)} &= \text{Total MLs} \times \text{Growth rate} = 333.33\text{MLs} \times 1\text{ML/s} \\ &= 333.33\ \text{seconds}\end{aligned} \tag{8.10}$$

Finally, we convert the time into minutes:

$$\text{Total time (minutes)} = \frac{\text{Total time (seconds)}}{60} = \frac{333.33}{60} \approx 5.56\ \text{minutes} \tag{8.11}$$

The growth duration should be around 5.56 minutes to achieve a film thickness of 100 nm.

8.3.3 Metal-Organic Chemical Vapor Deposition

MOCVD has become a fundamental technology for the production of nanoelectronics, primarily by enabling wafer-scale fabrication of two-dimensional (2D) materials and innovative compound semiconductors (Yang 2014, Zhang et al. 2025). Recent developments have focused on developing CMOS-compatible MOCVD techniques, including low-temperature growth of 2D materials for BEOL integration, which is crucial for industrial applications and large-scale manufacturing. Leading equipment manufacturers are designing MOCVD systems for high-volume production of these next-generation devices. A typical MOCVD system is shown in Figure 8.9.

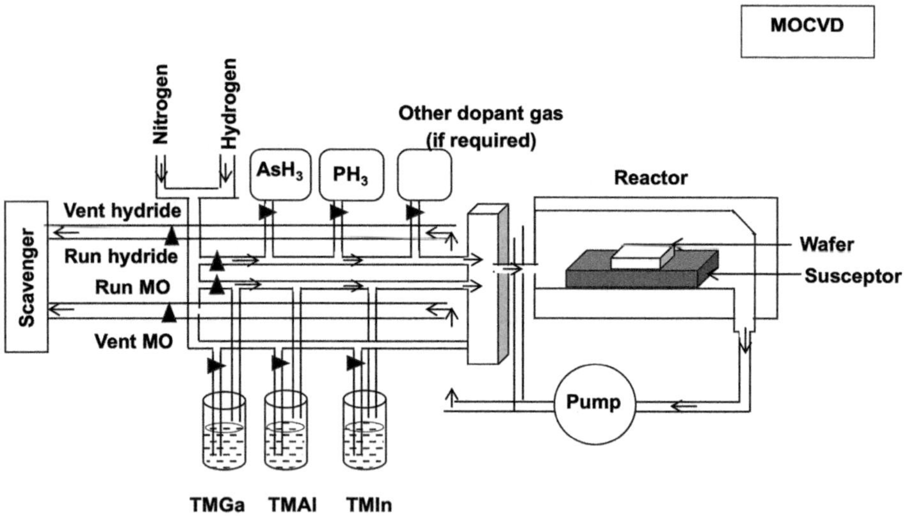

FIGURE 8.9 Metal-organic chemical vapor deposition system. The reactor, typically made of stainless steel or quartz, contains a susceptor (often SiC-coated graphite) to hold and heat the wafers. Run and Vent modes control the introduction of precursor gases, such as metal organics and hydrides, into the reactor chamber. This enables the formation of sharp interfaces in thin-film semiconductor layers. Vent Hydride Mode: When a hydride, such as arsine (AsH_3) or phosphine (PH_3), is in the "Vent" position, it bypasses the reactor chamber and flows directly to the exhaust or scrubber system. This stabilizes gas flow and pressure before introduction into the chamber. Run Hydride Mode: In the "Run" position, the valve directs the hydride into the reactor chamber for the epitaxial growth reaction. Vent MO Position: The metal-organic (MO) precursor flows through the manifold to the exhaust system, bypassing the reactor. This mode is used during idle periods, precursor stabilization, or recipe changes. Run MO Position: The valve directs the MO precursor into the main reactor chamber for deposition. TMGa ($Ga(CH_3)_3$), TMAl ($Al(CH_3)_3$), and TMIn ($In(CH_3)_3$) are the primary sources for group-III elements. A scavenger is a component or mechanism used to remove, neutralize, or trap hazardous, unwanted, or contaminating species from the process gas stream or reaction chamber.

Functions of MOCVD in Nanoelectronics: MOCVD delivers the accuracy and scalability needed to progress beyond conventional silicon-based architectures into next-generation device categories. MOCVD is the premier technique for 2D semiconductor production, enabling the synthesis of transition-metal dichalcogenides (TMDs) such as MoS_2 and WSe_2 on 300 mm silicon wafers. These fragile materials are vital for "More Moore" advancements, as they retain excellent carrier mobility even at thicknesses below 3 nm, where silicon performance starts to decline. MOCVD can grow MoS_2 at low thermal budgets, under 300°C. This allows 2D materials to be deposited directly onto the existing BEOL layers of silicon circuits without harming the underlying metal interconnects, enabling 3D monolithic integration. MOCVD plays a crucial role in the heterogeneous integration of high-speed III–V compound semiconductors, such as GaN and InGaAs, onto silicon substrates. This facilitates the development of hybrid chips that combine the logic density of CMOS with the high-frequency and optoelectronic capabilities of III–V materials.

For emerging device structures, MOCVD enables the creation of advanced device architectures, such as TFETs that utilize band-to-band tunneling for ultra-low power consumption, and non-volatile memory by combining N-type MoS_2 and P-type WSe_2 for non-von Neumann computing frameworks. For manganite thin films, MOCVD enables the production of functional manganites for applications in spintronics and field-effect sensing.

Strategic Benefits of MOCVD over CVD: Although standard chemical vapor deposition (CVD) typically achieves higher material quality in laboratory environments, MOCVD is favored for industrial manufacturing owing to its wafer-scale consistency and demonstrated ability to produce uniform monolayer films across 8- and 12-inch wafers. Repeatability, enabled by enhanced control over precursor flow and reaction kinetics, is essential for reliable mass production. When it comes to managing thickness, MOCVD offers atomic-level control of layer thickness and stoichiometry, which is crucial for technology nodes below 5 nm.

8.3.4 Focused Electron Beam-Induced Deposition

Focused electron beam-induced deposition (FEBID) is a direct-write nanofabrication method that does not require masks and has become essential for advancing nanoelectronics. It facilitates the creation of intricate 2D and 3D nanostructures with resolutions below 10 nm by breaking down precursor gas molecules locally using a focused electron beam (Huth et al. 2012, Fedorov et al. 2014).

FEBID's contributions to nanoelectronics are significant. FEBID possesses the distinctive capability to fabricate 3D magnetic nano-architectures, including vertical nanowires and interconnected networks for 3D nanomagnetism and spintronics. These structures are crucial for low-power, high-density technologies such as NML and 3D domain-wall conduits. The FEBID technique is used to deposit superconducting materials for fluxtronics (vortex-based logic) and for achieving topological superconductivity relevant to Majorana-based quantum computing. FEBID offers an integrated approach to pattern, deposit, and dope sensitive materials such as graphene and carbon nanotubes while preserving the substrate, overcoming a significant drawback of ion beam techniques in carbon nanoelectronics. FEBID enables the production of gate electrodes on non-planar designs, e.g., Si-nanowire transistors, and facilitates 3D interconnects for the upcoming vertical chip architectures. Essentially, FEBID provides distinct advantages for the development of customized 3D nanostructures and specialized components, positioning it as critical for investigating new physics and achieving next-generation electronic, spintronic, and quantum technologies.

8.3.5 Wet-Chemical Synthesis

Wet-chemical synthesis is a flexible "bottom-up" method for advanced nanoelectronics beyond CMOS, enabling precise control of the structural and chemical properties of materials at the nanoscale (Cerofolini and Mascolo 2007, Zijian et al. 2022). In contrast to conventional vacuum-based deposition techniques, these solution-phase approaches allow the synthesis of various 1D and 2D nanomaterials, which are vital for future logic and memory technologies. Wet chemistry offers solutions to address the scaling limits of CMOS technology by creating functional nanomaterials and architectures such as 3D transistors and TFETs with customized properties, thus paving the way for efficient, multifunctional nanoelectronic systems.

Materials and Applications: Wet chemistry is employed to produce high-k amorphous metal-oxide thin films, e.g., copper calcium titanate, which act as superior dielectrics for 2D electronic devices, offering minimal hysteresis and low-power functionality. Techniques such as hydrothermal synthesis yield one-dimensional (1D) nanostructures, e.g., ZnO nanowires and BiSb-silica composites, essential for flexible electronics and active cooling in nanoscale applications. Sol-gel procedures are used to synthesize functional nanoparticles, such as plasmonic gold nanoparticles and magnetic materials (e.g., Sr_2FeReO_6) for innovative sensors, catalysts, and magnetic memory uses. The solution-processed oxide heterojunctions are being increasingly used for producing thin-film transistors (TFTs) with enhanced mobility and electrical property modulation.

Primary Synthesis Techniques: The sol-gel process is a well-established method that involves transforming a molecular precursor (typically a metal alkoxide) into a gel via hydrolysis and agitation, making it ideal for generating metal-oxide nanoparticles and thin films. Hydrothermal/solvothermal synthesis techniques are carried out in autoclaves at elevated temperature and pressure to grow crystalline nanostructures either directly on substrates or as templates. The chemical reduction process entails reducing metal ions in a stabilizing agent solution to create uniform nanoparticles. The reverse microemulsion method uses an oil-water phase to regulate the size and shape of nanoparticles at ambient temperatures.

Advantages and Challenges: Wet-chemical synthesis is a cost-effective process. It provides a more economical alternative to costly vacuum-based techniques such as CVD or MBE. This method offers scalability and is well-suited for the mass production of various types of materials. It also provides controllability by enabling precise adjustment of size, shape, and surface chemistry via varying reaction conditions (pH, temperature, surfactants).

Among challenges, purity is of great concern. Wet-chemical synthesis is often perceived as a "dirty" process compared to vacuum methods, usually necessitating significant purification to achieve electronic-grade quality. For integration, the logical assembly of nanoparticles from a liquid suspension onto solid device substrates poses technical challenges.

Despite its versatility, CMOS compatibility is limited because some wet processes might contaminate sensitive CMOS surfaces, frequently requiring dry etching or protective layers.

8.3.6 Self-Assembly

Self-assembly is a crucial method for nanoelectronics, enabling the transition from conventional "top-down" lithography to "bottom-up" molecular organization. As silicon transistors approach their atomic limits, self-assembly enables the fabrication of intricate, defect-free structures—such as arrays of carbon nanotubes, molecular logic gates, and 2D material heterostructures—that are challenging to pattern with light alone (Naik and Ramgopal Rao 2016, Jawad et al. 2026).

8.3.6.1 Mechanisms of Self-Assembly

In static self-assembly, components naturally arrange themselves into a configuration of minimal free energy. This category includes self-assembled monolayers (SAMs), which form single-molecule-thick films on surfaces and are vital for the functionalization of molecular electronics.

The DSA method employs external forces, such as electric or magnetic fields or lithographic templates, to position components at specific locations. The graphoepitaxy technique uses pre-patterned grooves to direct the alignment of BCPs into dense, ordered lines for advanced logic and memory applications. A DNA-guided assembly approach exploits the accuracy of DNA base pairing to serve as a "scaffold" for nanodevices. It can position carbon nanotubes or metal nanoparticles with sub-nanometer precision into circuit-like arrangements.

8.3.6.2 Emerging Applications of Self-Assembly Beyond CMOS

Self-assembly is applied in the patterning of 2D materials: Deterministic self-assembly enables 2D materials such as graphene or MoS_2 to be "wrinkled" or shaped into specific designs, facilitating the development of flexible and wearable sensors with enhanced sensitivity. In molecular and spintronic devices, self-assembly enables the formation of multiferroic nanocomposites that couple magnetic and electric properties at their interfaces, meeting a critical requirement for the next generation of low-power spintronic logic. In material computing, researchers are investigating "material reservoir computing," in which the self-assembled structure of a material, such as a network of nanowires, performs computations, potentially replacing conventional processor designs.

8.3.6.3 Benefits and Obstacles of Self-Assembly

Self-assembly offers scalability. It operates in parallel; millions of components can be organized simultaneously across an entire wafer, significantly reducing manufacturing costs compared to serial patterning. While self-assembly can achieve atomic-scale resolution, controlling random defects and ensuring long-range "registration" (alignment with other chip layers) remains the main challenge for industrial implementation. The focus is on managing precision in relation to defects.

Ultimately, self-assembly employs materials such as BCPs, DNA, and 2D materials that spontaneously generate precise nanoscale patterns for transistors and circuits, overcoming the limitations of traditional lithography. This technique enables the creation of intricate, regular structures, such as cylinders or lines, at the nanoscale, which is essential for developing smaller, more efficient devices for future computing. It also enables the integration of new materials, such as GaN and nickelates made from rare-earth elements ($RNiO_3$, where R represents elements such as Pr, Nd, Sm, or Y), as well as functionalities suitable for neuromorphic computing or advanced sensors, providing solid, cost-effective pathways for ongoing scaling. Self-assembly is a bottom-up method for constructing complex nanodevices, allowing natural principles to govern structure and propelling novelty beyond the current limitations of silicon technology.

Example 8.6 A promising alternative to conventional CMOS technology is the self-assembly of uniform metallic nanoparticles (NPs) into organized structures for potential use as interconnects or components in advanced devices. Determine the approximate number of gold (Au) atoms present in a single spherical gold nanoparticle with a diameter of $D = 10$ nm. Provided information includes: Diameter of the nanoparticle $(D) = 10$ nm, Density of bulk gold $\rho_{\text{Au}} = 19.3\ \text{g/cm}^3$, Molar mass of gold $(M_{\text{Au}}) = 197.0\ \text{g/mol}$, and Avogadro's number $(N_{\text{A}}) = 6.022 \times 10^{23}\ \text{atoms/mol}$.
Solution: First, we calculate the radius (r) of the nanoparticle in centimeters:

$$r = \frac{D}{2} = \frac{10}{2} = 5\ \text{nm} \tag{8.12}$$

Next, we convert this r value to centimeters:

$$r = 5\ \text{nm} = \frac{5}{10^7}\ \text{cm} = 5 \times 10^{-7}\ \text{cm} \tag{8.13}$$

We then find the volume V of the spherical nanoparticle:

$$V = \frac{4}{3}\pi r^3 = \frac{4}{3} \times \frac{22}{7} \times \left(5 \times 10^{-7}\right)^3\ \text{cm}^3 \tag{8.14}$$

This results in

$$V = \frac{4}{3} \times \frac{22}{7} \times 125 \times 10^{-21} = 523.8095 \times 10^{-21} \approx 5.238 \times 10^{-19}\ \text{cm}^3 \qquad (8.15)$$

Subsequently, we calculate the mass (m) of the nanoparticle:

$$m = \rho_{\text{Au}} V = 19.3\ \text{g/cm}^3 \times 5.238 \times 10^{-19}\ \text{cm}^3 = 101.09 \times 10^{-19}\ \text{g} = 1.01 \times 10^{-17}\ \text{g} \qquad (8.16)$$

Next, we determine the number of moles (n) of gold in the nanoparticle:

$$n = \frac{m}{M_{\text{Au}}} = \frac{1.01 \times 10^{-17}\ \text{g}}{197\ \text{g/mol}} = 0.0051269 \times 10^{-17}\ \text{mol} \approx 5.13 \times 10^{-20}\ \text{mol} \qquad (8.17)$$

Finally, we calculate the total number of atoms (N):

$$N = n \times N_{\text{A}} = 5.13 \times 10^{-20}\ \text{mol} \times 6.022 \times 10^{23}\ \text{atoms/mol} = 30.893 \times 10^{3}\text{atoms} = 30{,}893\ \text{atoms} \qquad (8.18)$$

Thus, a gold nanoparticle with a diameter of 10 nm contains roughly 30,893 atoms.

8.4 DISCUSSION AND CONCLUSIONS

Recent studies indicate a transition from conventional "top-down" strategies to a combination of hybrid and completely "bottom-up" techniques for achieving precision below 10 nm. It underscores a movement away from gradual enhancements in silicon processing toward a more revolutionary exploration of novel physics, materials, and entirely different manufacturing methods to satisfy future computing requirements.

If fabrication gives birth to the silicon brain, then packaging forms the skin, bones, and nervous system, making it vital to every modern electronic device. Packaging is the unsung hero that transforms fragile semiconductor dies into robust, reliable products. It connects the microscopic world of chips with our everyday reality, channeling power, signals, and protection. Without packaging, the marvels of modern electronics would simply not exist. It shields chips from corrosion, moisture, and physical harm. It forges the crucial electrical connections that link a chip's inner workings to the broader circuitry. It dissipates heat, ensuring chips stay cool even during intense performance. It coordinates the flow of signals and power, directing them precisely where they are needed. These are the classic roles of packaging. Yet, its influence stretches far "beyond CMOS technology." What more can packaging do? The next chapter holds the answer.

REFERENCES

Amashaev R. R., S. M. Isubgadzhiev, M. H. Rabadanov and I. M. Abdulagatov 2025 Development of atomic layer deposition technological platform for the synthesis of micro- and nanoelectronics materials, *Russian Microelectronics*, 54(1): 76–90.

Banqiu Wu B. and A. Kumar 2014 Extreme ultraviolet lithography and three dimensional integrated circuit—A review, *Applied Physics Reviews*, 1: 011104, https://doi.org/10.1063/1.4863412

Barcelo S. and Z. Li 2016 Nanoimprint lithography for nanodevice fabrication, *Nano Convergence*, 3: 21, https://doi.org/10.1186/s40580-016-0081-y

Biswas A., I. S. Bayer, A. S. Biris, T. Wang, E. Dervishi and F. Faupel 2012 Advances in top–down and bottom–up surface nanofabrication: Techniques, applications & future prospects, *Advances in Colloid and Interface Science*, 170 (1–2): 2–27, https://doi.org/10.1016/j.cis.2011.11.001

Cerofolini G. F. and D. Mascolo 2007 Chapter: A hybrid route from CMOS to nano and molecular electronics. In: Korkin A., E. Gusev, J. Labanowski and S. Luryi (Eds.) *Nanotechnology for Electronic Materials and Devices, Nanostructure Science and Technology*, Springer, Boston, MA, USA, pp. 1–65, https://doi.org/10.1007/978-0-387-49965-9_1

Chang P., W. C. Lee, T. D. Lin, C. H. Hsu, J. Kwo and M. Hong 2011 MBE—Enabling technology beyond Si CMOS, *Journal of Crystal Growth*, 323(1): 511–517, https://doi.org/10.1016/j.jcrysgro.2010.11.102

Cheng X., D. Liang, M. Jiang, Y. Sha, X. Liu, J. Liu, Q. Cao and J. Shi 2025 Review of directed self-assembly material, processing, and application in advanced lithography and patterning, *Micromachines (Basel)*, 16(6): 667. https://doi.org/10.3390/mi16060667

Fedorov A. G., S. Kim, M. Henry, D. Kulkarni and V. V. Tsukruk 2014 Focused-electron-beam-induced processing (FEBIP) for emerging applications in carbon nanoelectronics, *Applied Physics, A*, 117, 1659–1674, https://doi.org/10.1007/s00339-014-8628-4

Giannopoulos I., I. Mochi, M. Vockenhuber, Y. Ekinci and D. Kazazis 2024 Extreme ultraviolet lithography reaches 5 nm resolution, *Nanoscale*, 16: 15533–15543, https://doi.org/10.1039/D4NR01332H

Han W. and S. Xiao 2013 Focused ion beam (FIB) and its applications, *Materials China*, 32 (12): 716–727, https://doi.org/10.7502/j.issn.1674-3962.2013.12.02

Henini M. (Ed.) 2012 *Molecular Beam Epitaxy: From Research to Mass Production*, 2nd Edition, Elsevier Amsterdam, pp. 744.

Höflich K., G. Hobler, F. I. Allen, T. Wirtz, G. Rius, L. McElwee-White, A. V. Krasheninnikov, M. Schmidt, I. Utke, N. Klingner, M. Osenberg, R. Córdoba, F. Djurabekova, I. Manke, P. Moll, M. Manoccio, J. María De Teresa, L. Bischoff, J. Michler, O. De Castro, A. Delobbe, P. Dunne, O. V. Dobrovolskiy, N. Frese, A. Gölzhäuser, P. Mazarov, D. Koelle, W. Möller, F. Pérez-Murano, P. Philipp, F. Vollnhals and G. Hlawacek 2023 Roadmap for focused ion beam technologies, *Applied Physics Reviews*, 10: 041311, https://doi.org/10.1063/5.0162597

Hu X.-H. and S. Xiong 2022 Fabrication of nanodevices through block copolymer self-assembly, *Frontiers in Nanotechnology*, 4: 762996. https://doi.org/10.3389/fnano.2022.762996

Huff M. 2021 Recent advances in reactive ion etching and applications of high-aspect-ratio microfabrication, *Micromachines*, 12(8): 991, https://doi.org/10.3390/mi12080991

Huth M., F. Porrati, C. Schwalb, M. Winhold, R. Sachser, M. Dukic, J. Adams and G. Fantner 2012 Focused electron beam induced deposition: A perspective, *Beilstein Journal of Nanotechnology*, 3: 597–619, https://doi.org/10.3762/bjnano.3.70

Hwang C. S., and C. Y. Yoo (Eds.) 2014 *Atomic Layer Deposition for Semiconductors*, Springer Science+Business Media, New York, NY, USA, pp. 263, https://doi.org/10.1007/978-1-4614-8054-9

Jawad M. S., W. U. Rahman, M. Khan and A. M. Faysal 2026 Nanomaterial assembly pathways: Comparative insights into self-assembly and directed assembly techniques, *Next Materials*, 10: 101422, https://doi.org/10.1016/j.nxmate.2025.101422

Johnson K. C. 2019 Maskless EUV lithography, an alternative to e-beam, *Journal of Micro/Nanolithography, MEMS, and MOEMS*, 18(4): 043501, https://doi.org/10.1117/1.JMM.18.4.043501

Lin W.-K. and L. J. Guo 2025 30 years of nanoimprint: development, momentum and prospects, *Opto-Electronic Technology*, 1, 250001.

Lu W. and C. M. Lieber 2007 Nanoelectronics from the bottom up, *Nature Materials*, 6(11): 841–850, https://doi.org/10.1038/nmat2028

Naik T. and V. Ramgopal Rao 2016 Self assembled monolayer applications for nano-scale CMOS, *7th International Nanoelectronics Conference (INEC)*, Chengdu, China, 9–11 May, https://doi.org/10.1109/INEC.2016.7589436

Resnick D. 2014 9 – Nanoimprint lithography, In: Feldman M. (Ed.), *Nanolithography: The Art of Fabricating Nanoelectronic and Nanophotonic Devices and Systems*, Woodhead Publishing, Philadelphia, PA, USA, pp. 315–347, https://doi.org/10.1533/9780857098757.315

Sharma E., R. Rathi, J. Misharwal, B. Sinhmar, S. Kumari, J. Dalal and A. Kumar 2022 Evolution in lithography techniques: Microlithography to nanolithography, *Nanomaterials*, 12(16), 2754, https://doi.org/10.3390/nano12162754

Shen C., W. Zhan, M. Li, Z. Sun, J. Tang, Z. Wu, C. Xu, B. Xu, C. Zhao and Z. Wang 2024 Development of in situ characterization techniques in molecular beam epitaxy, *Journal of Semiconductors*, 45(3): 03130, https://doi.org/10.1088/1674-4926/45/3/031301

Tallegas S., T. Baron, G. Gay, C. Aggrafeil, B. Salhi, T. Chevolleau, G. Cunge, A. Bsiesy, R. Tiron, X. Chevalier, C. Navarro, K. Aissou, I. Otsuka, S. Halila, S. Fort and R. Borsali 2013 Block copolymer technology applied to nanoelectronics, *Physica Status Solidi (c)*, 10(9): 1195–1206, https://doi.org/10.1002/pssc.201300002

Tseng A. A., K. Chen, C. D. Chen and K. J. Ma 2003 Electron beam lithography in nanoscale fabrication: recent development, *IEEE Transactions on Electronics Packaging Manufacturing*, 26(2): 141–149, https://doi.org/10.1109/TEPM.2003.817714

van de Kerkhof M., H. Jasper, L. Levasier, R. Peeters, R. van Es, J.-W. Bosker, A. Zdravkov, E. Lenderink, F. Evangelista, P. Broman, B. Bilski and T. Last 2017 Enabling sub-10 nm node lithography: Presenting the NXE:3400B EUV scanner, *Proceedings of the SPIE 10143, Extreme Ultraviolet (EUV) Lithography VIII*, San Jose, CA, USA, 24 March, p. 101430D, https://doi.org/10.1117/12.2258025

Wiśniewski P., R. Mroczyński and B. Majkusiak 2016 Reactive ion etching (RIE) of silicon for the technology of nanoelectronic devices and structures, *Proceedings of SPIE 10175, Electron Technology Conference ELTE 2016*, Wisla, Poland, 11–14 September, p. 101750F, https://doi.org/10.1117/12.2261676

Yang F. H. 2014 Chapter 2: Modern metal-organic chemical vapor deposition (MOCVD) reactors and growing nitride-based materials. In Huang J. J., H.-C. Kuo and S.-C. Shen (Eds.), *Nitride Semiconductor Light-Emitting Diodes (LEDs): Materials, Technologies and Applications*, Woodhead Publishing (an imprint of Elsevier), Duxford, pp. 27–65, https://doi.org/10.1533/9780857099303.1.27

Zhang X., N. Trainor, T. V. McKnight, A. R. Graves, Z. Wu, L. Xu, X. Zheng, T. Zhang, J. Zhu, T. Palacios, J. Kong, B. Groven, B. Tian, C. Duan, J. Chu and J. M. Redwing 2025 Metal–organic chemical vapour deposition for 2D chalcogenides, *Nature Reviews Methods Primers*, 5(1): 57, https://doi.org/10.1038/s43586-025-00429-4

Zijian L., L. Zhai, Y. Ge, Z. Huang, Z. Shi, J. Liu, W. Zhai, J. Liang and H. Zhang 2022 Wet-chemical synthesis of two-dimensional metal nanomaterials for electrocatalysis, *National Science Review*, 9(5): nwab142.

9 Integration and Packaging

9.1 INTRODUCTION

Traditional packaging focused on protecting the chip, while modern advanced packaging drives system performance. Today, packaging serves as a system-level performance enabler and a platform for innovation, particularly beyond-CMOS technologies. It is now a critical, active component that integrates diverse chiplets, enhances electrical performance, manages significant thermal loads, and supports system-level functionality in AI, 5G, and high-performance computing. Advanced packaging is essential for post-Moore CMOS scaling, using 2.5D and 3D integration, chiplets, and hybrid bonding to improve performance, density, and power efficiency. This chapter will address the impact of integration and packaging on dimensional scaling because the focus has shifted from the chip itself to how we stack and connect chips at shorter distances to build systems.

In addition to conventional CMOS scaling, emerging integration methods emphasize combining multiple materials and functionalities (Gill and Ghoreishizadeh 2023, Katiyar et al. 2025). The primary integration strategies for post-CMOS nanoelectronics are monolithic integration (which involves building layers sequentially on a single substrate) and heterogeneous integration (which merges individually manufactured components).

Monolithic integration is the process of constructing various components, frequently utilizing innovative materials, directly on a single substrate, generally silicon (Si) (Muñoz-Gamarra et al. 2016, Zhou and Xie 2018, Datta et al. 2019). Monolithic integration techniques involve fabricating multiple device layers directly on top of one another through sequential, low-temperature processes, often using nanoscale interconnections. This approach achieves the highest levels of integration density and performance but encounters difficulties related to material compatibility and thermal constraints.

Heterogeneous integration brings together separately manufactured components, or "chiplets," produced through different processes or materials and assembles them into a compact, unified system, often involving advanced packaging methods (Kazior 2014, Mishra et al. 2024). This method provides flexibility in leveraging the best materials for each function and enhances yield management. Heterogeneous integration combines chiplets with unique talents—like CPUs, memory, and IO—much like a cricket team unites bowlers, batsmen, and fielders to play as one. A cricket team provides a real-world example of heterogeneous integration (Figure 9.1). The strategies of heterogeneous integration empower designers to utilize the optimal performance traits of various materials (e.g., the high frequency

DOI: 10.1201/9781003753469-9

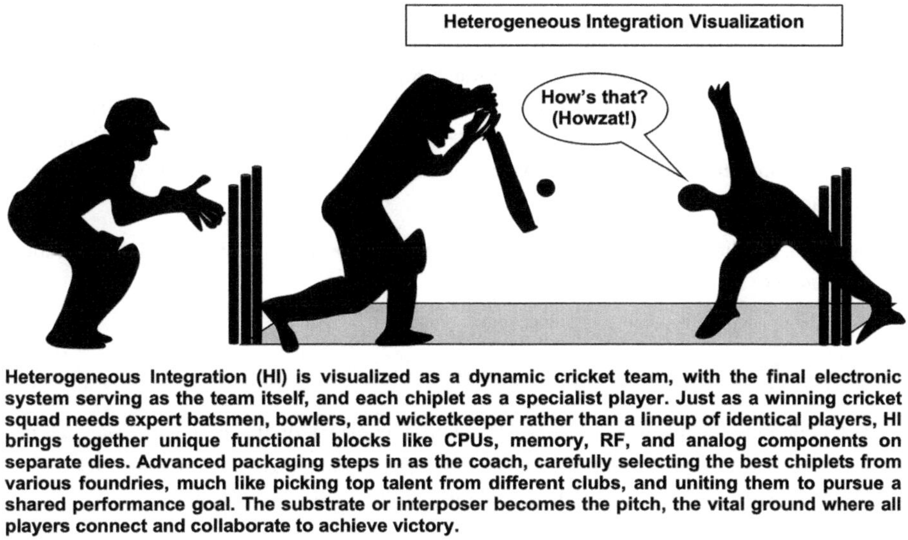

FIGURE 9.1 Picturing heterogeneous integration in semiconductors as a cricket team lineup. The processors, memory, and sensors are the star players, each bringing its own strengths to the field. This analogy highlights how these high-performance chiplets collaborate, just as cricket specialists combine their talents to boost the team's overall success.

of III–V devices, the exceptional sensing capabilities of 2D materials, and the high density of Si CMOS) to construct highly functional, integrated microsystems that surpass the capabilities of silicon alone.

9.2 HOMOGENEOUS (MONOLITHIC) INTEGRATION

9.2.1 Advanced Heteroepitaxy

Advanced heteroepitaxy/selective epitaxial growth (SEG) technique focuses on the direct growth of non-silicon materials such as III–V semiconductors like GaN and GaAs, or 2D materials atop a silicon substrate through methods like molecular beam epitaxy (MBE) or chemical vapor deposition (CVD) (Nakahata et al. 2002, Vescan et al. 2007). It can be selectively applied on patterned substrates (nanoheteroepitaxy) to control material variations and minimize defects. Therefore, this method is vital for the targeted growth of specific semiconductor materials, such as III–V compounds or silicon (Si), exclusively in designated "windows" on the substrate while preventing deposition on masking layers such as silicon dioxide. It plays a crucial role in forming high-performance channels for next-generation transistors, e.g., TFETs and merging photonics with existing CMOS technology.

Example 9.1 Heteroepitaxy

In a situation where a high-mobility III–V semiconductor channel material (e.g., InAs) is monolithically integrated onto a conventional Silicon (Si) substrate, the lattice constants of both materials play a vital role in assessing strain and defect density. The lattice constant for InAs is $a_{\mathrm{InAs}} = 6.058$ Å while that for silicon is $a_{\mathrm{Si}} = 5.431$ Å. Compute the lattice-mismatch percentage between the InAs epilayer and the silicon substrate, and comment on the value obtained.

Solution: To find the percentage of lattice mismatch ε between the InAs epilayer and the Si substrate, we use the formula:

$$\varepsilon = \frac{\left|a_{\mathrm{Epilayer}} - a_{\mathrm{Substrate}}\right|}{a_{\mathrm{Substrate}}} \times 100\% \tag{9.1}$$

Let's identify the provided values: The lattice constant of the epilayer $\left(a_{\mathrm{Epilayer}}\right)$ is $a_{\mathrm{InA}} = 6.058\text{Å}$, and the lattice constant of the substrate $\left(a_{\mathrm{Substrate}}\right)$ is $a_{\mathrm{Si}} = 5.431$ Å. We will use the lattice-mismatch formula to get the fractional mismatch as

$$\varepsilon = \frac{|6.058\,^{\circ}\mathrm{A} - 5.431\,^{\circ}\mathrm{A}|}{5.431\,^{\circ}\mathrm{A}} \times 100\% = \frac{0.627\,^{\circ}\mathrm{A}}{5.431\,^{\circ}\mathrm{A}} \times 100\% = 0.1154 \times 100\% \tag{9.2}$$

Finally, we compute the percentage leading to

$$\varepsilon = 11.54\% \tag{9.3}$$

Thus, the lattice mismatch percentage between the InAs epilayer and the Si substrate is roughly 11.54%. This considerable mismatch highlights the integration difficulties that must be addressed, such as a high density of threading dislocations, which can be tackled using advanced methods like nanoheteroepitaxy or buffer layers.

9.2.2 M3D Sequential Integration

Monolithic 3D (M3D) sequential integration method entails the stepwise fabrication of layers containing transistors and other elements on top of one another, featuring high-density, low-latency vertical connections (interlayer vias) (Batude et al. 2012, Mallik et al. 2017, Chaourani 2019). The primary challenge lies in executing subsequent fabrication processes at low temperatures to prevent damaging the already constructed lower layers. Imagine a processor built in vertical layers, each stacked neatly atop the other. This approach slashes wire delays, trims energy use, and shrinks the overall footprint. Monolithic 3D, or M3D, enables this by providing ultra-short wire connections, efficient heat management, and dense integration. The process begins with a base layer of transistors, created just like a traditional chip. Then, a fresh layer of active silicon is carefully grown above it using innovative, low-temperature methods. Finally, transistors are crafted on this upper layer through a specialized, gentle process (Gopireddy and Torrellas 2019). Figure 9.2 illustrates the concept.

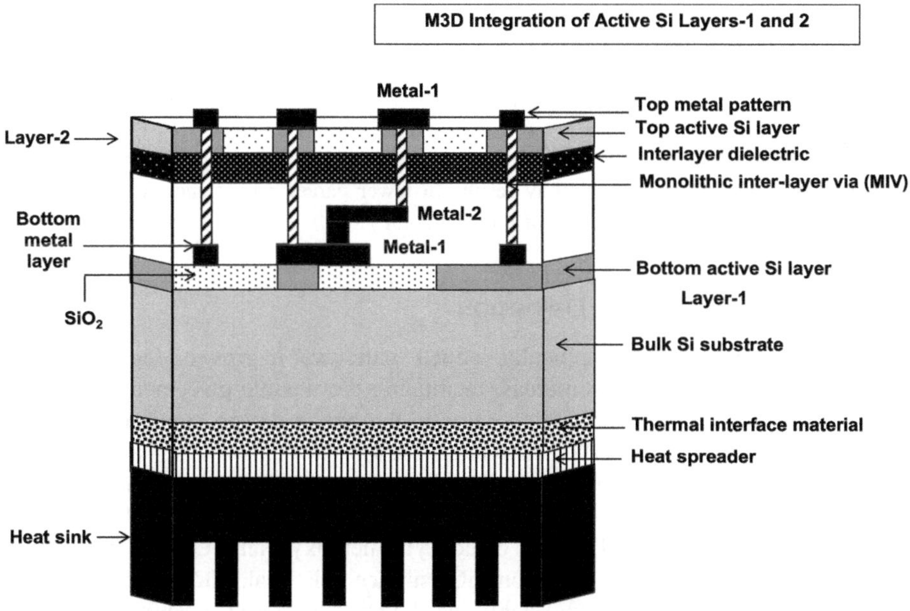

FIGURE 9.2 M3D integration of active layers 1 and 2. From bottom upwards: heat sink, heat spreader, Thermal interface material, bulk Si substrate, bottom active Si layer (layer-1), metal-1 and metal-2 in the bottom metal layer, monolithic interlayer via (MIV), interlayer dielectric, top active Si layer (layer-2), top metal pattern (metal-1).

Example 9.2 M3D Sequential Integration

A design team is evaluating a conventional 2D CMOS chip layout against a potential M3D sequential homogeneous integration variant of the same circuit, which utilizes two superimposed layers. In the 2D design, the average wire length for communication between two particular functional blocks measures 80 μm. In contrast, the M3D design benefits from ultra-fine pitch vertical interconnects, reducing the physical wire length necessary to connect the same two functional blocks (now situated in different vertical layers) to 16 μm. Determine the percentage decrease in wire length achieved by adopting the M3D integration method for this specific communication link.

Solution: The percentage decrease in wire length can be determined using the following formula:

$$\text{Percentage reduction} = \frac{\text{Original length} - \text{New length}}{\text{Original length}} \times 100\% \tag{9.4}$$

Provided values: Original Length (2D design) = 80 μm. New Length (M3D design) = 16 μm. Calculating the absolute decrease in length:

$$\text{Reduction} = 80\ \mu\text{m} - 16\ \mu\text{m} = 64\ \mu\text{m} \tag{9.5}$$

Computing the percentage decrease:

$$\text{Percentage reduction} = \frac{64\ \mu\text{m}}{80\ \mu\text{m}} \times 100\% = 0.8 \times 100\% = 80\% \quad (9.6)$$

The M3D integration method results in an 80% reduction in wire length for this communication link, which directly results in lower parasitic resistance, capacitance, and signal delay, a key advantage of this technology.

9.2.3 Selective Material Deposition

This method employs lithographically defined "windows" to grow or deposit specific materials exclusively in chosen areas, facilitating the flexible placement of various devices across a single substrate according to the circuit design requirements (Kafi et al. 2024, Karasulu et al. 2023, Kim et al. 2024).

AS-ALD of dielectric-on-metal (DoM) uses a two-step blocking strategy (Figure 9.3). First, dodecanethiol (DDT) selectively adsorbs on Cu to form a self-assembled monolayer (SAM). Next, octadecyltrimethoxysilane (OTMS) inhibitor is applied to form a SAM selectively on SiO_2 surface (Liu et al. 2023). Because DDT and OTMS have different thermal stabilities, the DDT layer on Cu can be removed at ALD process temperatures without affecting the OTMS inhibitor. This process exposes the Cu regions in a Cu/SiO_2 pattern for ALD. Thus, thermal removal of the DDT protector before AS-ALD allows ALD growth on Cu without interference, while ALD remains blocked on OTMS-covered SiO_2. This approach achieves DoM selectivity above 0.9 after depositing 5.6 nm of ZnO and 1.5 nm of Al_2O_3.

9.2.4 van der Waals Integration

For 2D materials (including MoS_2 or WSe_2), vdW integration utilizes dry-transfer techniques to stack atomically thin layers of distinct materials, establishing defect-free interfaces suitable for vertical complementary logic (Kang et al. 2025, Lee et al. 2025).

A scalable method uses a removable polymer buffer layer to integrate metal contacts with 2D semiconductors, enabling damage-free, wafer-scale integration of various metals (Ag, Au, Pd, etc.) (Kong et al. 2023). This approach prevents chemical contamination, reduces Fermi-level pinning, and improves device performance. WSe_2, a representative 2D semiconductor, is synthesized by chemical vapor deposition (CVD) on a silicon substrate with 300 nm of silicon oxide (Figure 9.4(a)). A 450-nm-thick PPC polymer is then spin-coated onto the WSe_2 layer (Figure 9.4 (b)). Source-drain electrodes (60 nm-thick Au) are deposited on the PPC using standard thermal evaporation (Figure 9.4(c)). The PPC layer serves as an effective barrier, protecting WSe_2 from deposition-induced damage and preventing chemical bonding between Au and WSe_2. The device is then heated to 250°C for 30 minutes under a nitrogen atmosphere (Figure 9.4(d)). During this step, the PPC decomposes into gases and dissipates (Figure 9.4(e)), allowing the Au to laminate onto WSe_2 through weak van der Waals interactions, as shown in Figure 9.4(f). This van der Waals integration process is compatible with industry standards and can be applied at the wafer scale.

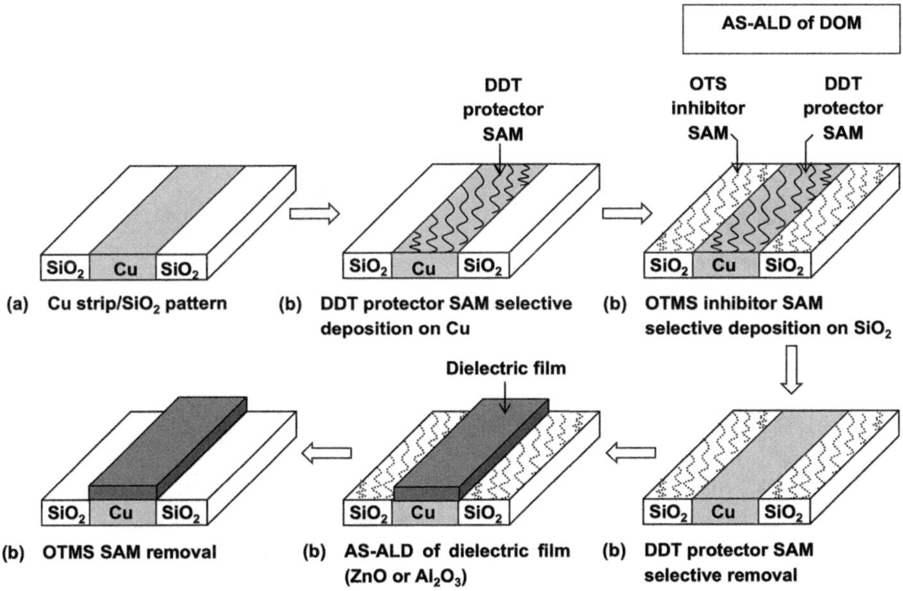

FIGURE 9.3 Schematic illustration of a generic area-selective ALD process using a patterned substrate with two materials: a growth area (Cu) for targeted deposition and a non-growth area (SiO_2), where deposition is undesired. Step (a) shows the patterned substrate with Cu and SiO_2 regions. In step (b), dodecanethiol (DDT) is applied, selectively chemisorbing to the non-growth area (SiO_2) and deactivating it from further precursor reactions. Step (c) demonstrates that the precursor surface reaction is inhibited on the deactivated non-growth area, allowing selective chemisorption only on the growth area (Cu). In step (d), the DDT self-assembled monolayer (SAM) is removed. Step (e) involves exposure to the co-reactant, which produces the first sub-monolayer of selectively deposited material (Cu) through surface reaction with the chemisorbed precursor. Repeating the ALD cycles results in selective deposition of the intended material (Cu) on the growth area until the desired thickness is achieved. In step (f), the OTMS is removed.

9.3 HETEROGENEOUS INTEGRATION

Heterogeneous integration is vital in nanoelectronics as it facilitates the merging of various, specialized materials and components into a single, compact system. This method transcends the physical and performance constraints associated with relying exclusively on silicon scaling, enabling enhanced features, improved performance, and lower power consumption that would otherwise be unattainable. With the deceleration of traditional CMOS scaling (Moore's Law), the traditional dimensional scaling of CMOS approaches its fundamental physical limits. Heterogeneous integration provides a pathway for ongoing innovation and performance enhancement by concentrating on new architectures and materials for information processing and memory via the lateral and 3D integration of new devices and materials, rather than merely reducing the size of existing ones. By integrating different components like logic chips, sensors, memory, and photonics, it becomes feasible to design systems with capabilities that are hard or impossible to achieve with a single technology.

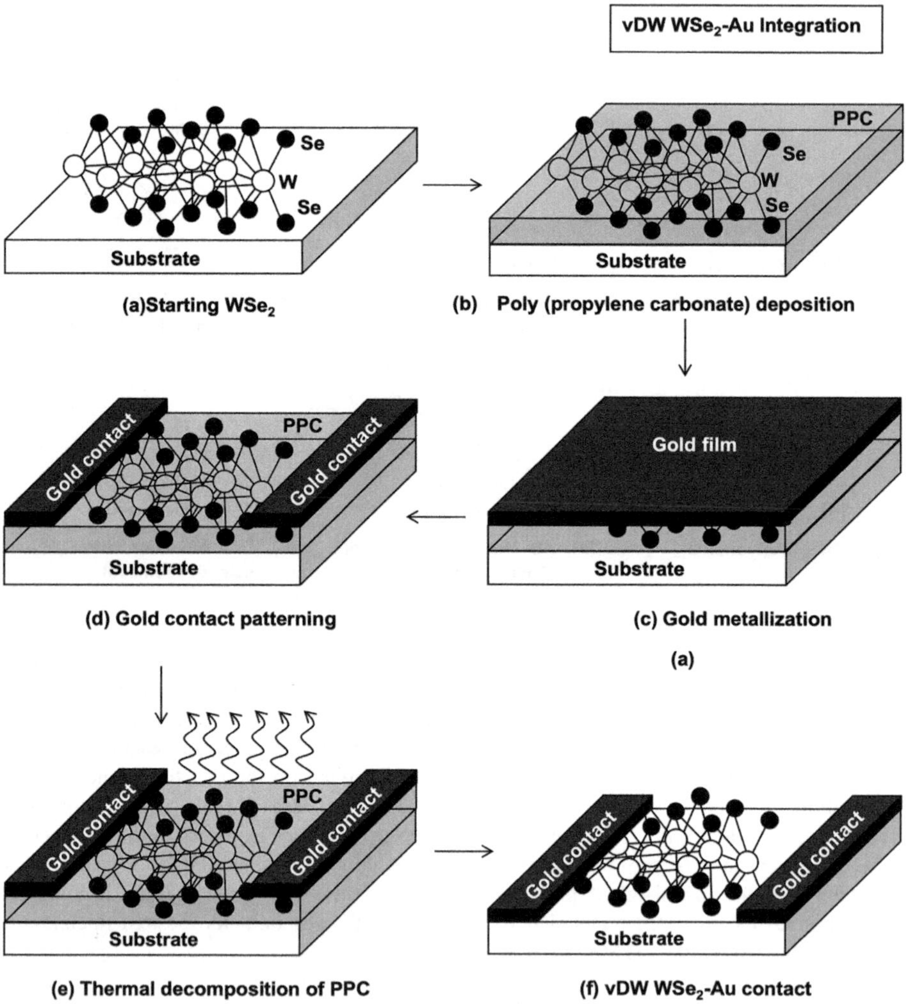

FIGURE 9.4 Wafer-scale Au–WSe_2 van der Waals (vdW) integration processes with six steps: (a)WSe_2 flake preparation on substrate, (b) poly(propylene carbonate) (PPC) buffer spin-coating, (c) metal deposition, (d) metal patterning, (e) PPC buffer dry-decomposition through annealing, and (f) intimate contact formation between Au and WSe_2.

This material diversification supports "More-than-Moore" applications, which aim to enhance functionality beyond merely increasing transistor density, surpassing physical limitations.

Extending CMOS Capabilities by Creating and Facilitating Novel Functionalities: Heterogeneous integration merges the high-density integration and established processing of silicon CMOS with the superior performance of alternative materials. It permits the fusion of materials with distinct properties, such as III–V semiconductors, 2D materials, optical elements, and microelectromechanical systems (MEMS) that are difficult or infeasible to produce on a single silicon substrate. III–V materials are utilized for high-frequency uses, while 2D materials can be employed for innovative logic and memory devices.

Optimizing Performance: The amalgamation of different technologies helps optimize performance characteristics, such as maximizing gain and efficiency for RF circuits or developing low-power, adaptable devices. By selecting the most appropriate material and manufacturing method for each specific function (e.g., the high-frequency capabilities of III–V devices combined with the density of Si CMOS), the overall performance of the system is improved, which includes enhancements in speed, energy efficiency, bandwidth, and signal processing abilities.

Constructing "Intelligent" Systems: The close, on-chip integration of control circuitry (CMOS) with specialized components (such as RF devices or sensors) enables the development of self-adapting, reconfigurable, and "intelligent" micro- and nano-systems. This allows for features like on-chip calibration to mitigate temperature changes or aging, thus increasing reliability and lifespan, and enabling new applications by delivering the required performance and integration levels, such as for the Internet of Things (IoT) and neuromorphic computing.

Miniaturization and Cost Efficiency: Heterogeneous integration enables the production of smaller, lighter, and more compact devices by merging multiple functions into a single package, rather than relying on larger, off-chip assemblies. This method also enhances cost-effectiveness in manufacturing by applying existing, established, low-cost silicon production processes for the base platform.

Thus, heterogeneous integration serves as a crucial bridge, prolonging the lifespan and augmenting the capabilities of the existing silicon CMOS platform by incorporating high-performance, beyond-CMOS materials and devices to address the increasing demands of advanced applications such as AI, 5G communications, and high-performance computing.

Example 9.3 Power saving by heterogeneous integration: A standard CMOS logic gate operates with a supply voltage $(V_{\mathrm{DD,CMOS}})$ of 0.8 V and has an average dynamic power usage (P_{CMOS}) of 10 nW during a particular operation. A new generation "beyond-CMOS" device, employing heterogeneous integration with innovative materials (such as 2D semiconductors or ferroelectrics), can achieve the same function with a lower supply voltage $(V_{\mathrm{DD,Beyond-CMOS}})$ of 0.4 V. Assuming that the dynamic power consumption of both types of devices adheres to the same relationship $(P \propto CV^2{}_{\mathrm{DD}}f$, where C represents capacitance and f denotes frequency), determine the percentage of power savings realized by utilizing the beyond-CMOS device for the same task.

Solution: Dynamic power (P) varies with the square of the supply voltage (V_{DD}^2).

$$P \propto V_{\mathrm{DD}}^2 \tag{9.7}$$

We can assess the power of the two devices by examining their voltage ratio:

$$\frac{P_{\text{Beyond-CMOS}}}{P_{\text{CMOS}}} = \left(\frac{V_{\text{DD,Beyond-CMOS}}}{V_{\text{DD,CMOS}}}\right)^2 \tag{9.8}$$

We plug in the values and determine $P_{\text{Beyond-CMOS}}$:

$$\begin{aligned} P_{\text{Beyond-CMOS}} &= P_{\text{CMOS}}\left(\frac{V_{\text{DD,Beyond-CMOS}}}{V_{\text{DD,CMOS}}}\right)^2 = 10\ \text{nW}\left(\frac{0.4\ \text{V}}{0.8\ \text{V}}\right)^2 = 10\ \text{nW} \times (0.5)^2 \\ &= 10\ \text{nW} \times 0.25 = 2.5\ \text{nW} \end{aligned} \tag{9.9}$$

Power saved is

$$P_{\text{Saved}} = P_{\text{CMOS}} - P_{\text{Beyond-CMOS}} = 10\ \text{nW} - 2.5\ \text{nW} = 7.5\ \text{nW} \tag{9.10}$$

The percentage of power saved is

$$\%P_{\text{Saved}} = \frac{P_{\text{Saved}}}{P_{\text{CMOS}}} \times 100\% = \frac{7.5\ \text{nW}}{10\ \text{nW}} \times 100\% = 75\% \tag{9.11}$$

Using a technology that surpasses traditional CMOS, an engineer can achieve a 50% decrease in operating voltage, resulting in a 75% reduction in power consumption for that particular operation.

9.3.1 BEOL Integration

The back-end-of-line (BEOL) integration strategy focuses on embedding innovative devices such as memristors, phase change materials (PCMs), and 2D material transistors into the interconnect layers of conventional CMOS chips (Sandrini et al. 2015, Srimani et al. 2020). This approach allows for advanced functionalities like in-memory computing while preserving the integrity of the front-end CMOS process. The development of thin-film transistors (TFTs) within BEOL layers enables them to perform roles such as signal buffering and power management. Key aspects of this strategy include:

Monolithic Silicon Photonics: The integration of both passive and active photonic components, such as waveguides and modulators, using CMOS-compatible methods.

Integration of Novel Materials: Incorporating materials like PCMs and 2D semiconductors, such as graphene and MoS_2, to enable applications in non-volatile memory and adaptable photonic devices. Graphene-enabled electronic and photonic devices boast remarkable properties, unlocking performance levels once thought unattainable. Successfully weaving this extraordinary material into mainstream semiconductor production marks a pivotal leap toward real-world applications (Neumaier et al. 2019). Figure 9.5 shows the steps of graphene integration.

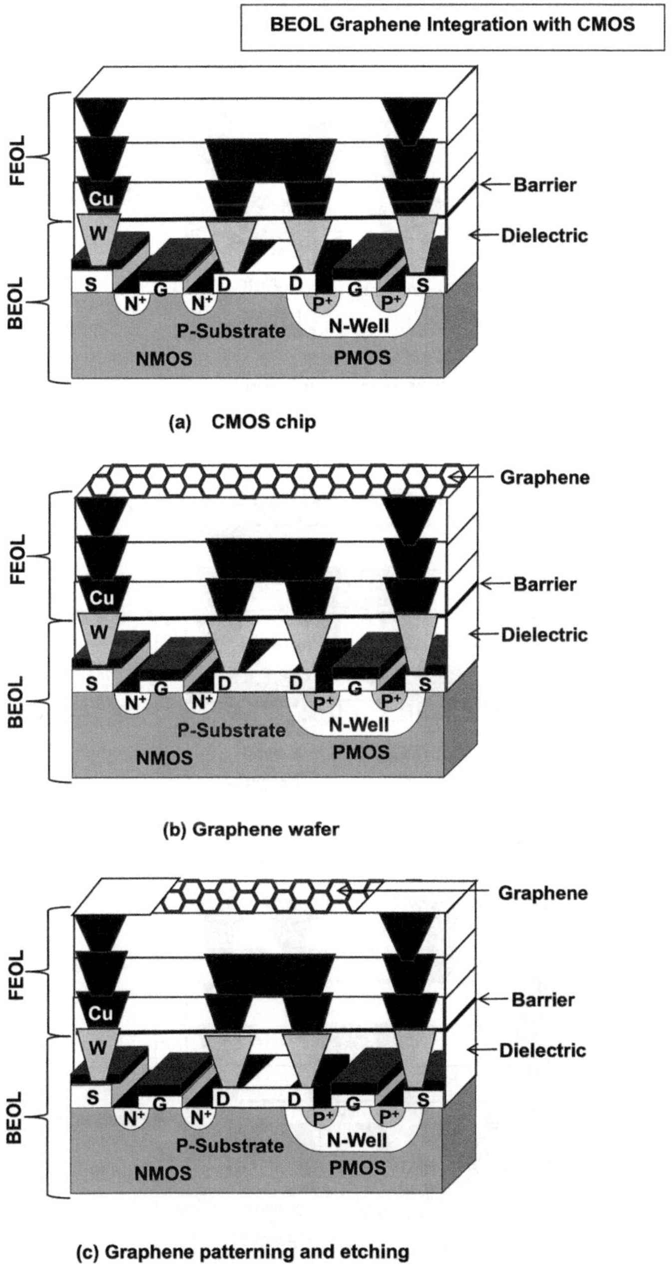

FIGURE 9.5 BEOL graphene integration with CMOS: (a) CMOS chip, (b) graphene transfer, (c) graphene patterning and etching, (d) graphene encapsulation, (e) via patterning and etching, and (f) via and contact metallization. This flow chart illustrates the journey of BEOL integration. It begins with a cross-sectional view of a CMOS chip, before graphene enters the scene. Next, a sheet of graphene is transferred across the entire wafer, ready for transformation. Oxygen plasma sculpts the graphene layer into precise patterns, which are then shielded with a

(Continued)

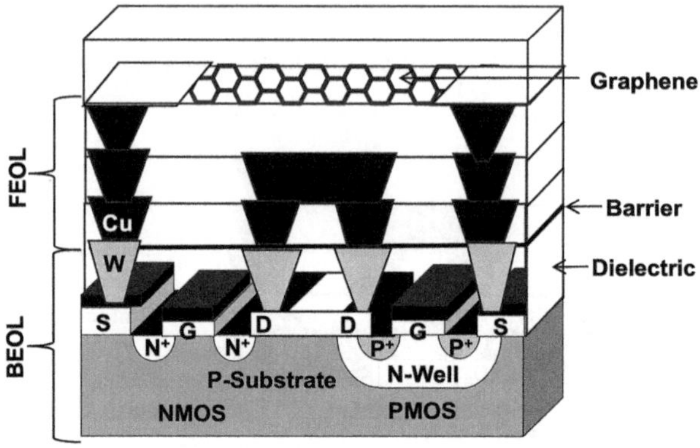

(d) Graphene encapsulation

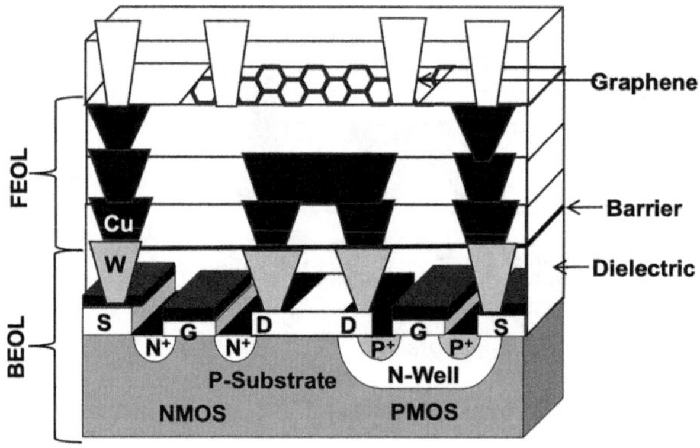

(e) Via patterning and etching

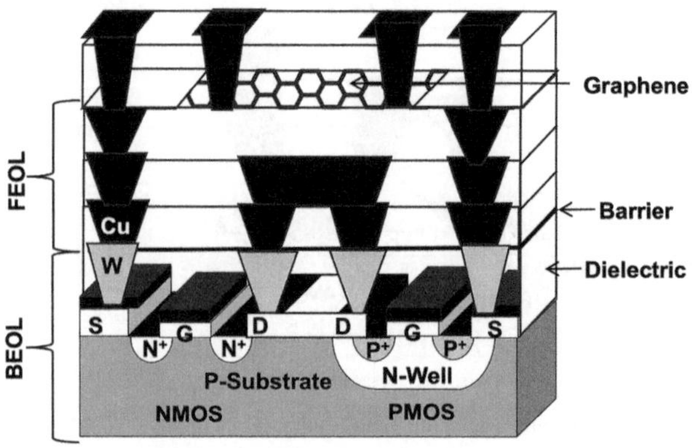

(f) Via and contact metallization

FIGURE 9.5 (Continued) protective dielectric layer. Dry etching carves out vias through the top dielectric, and these channels are filled with metal to create robust edge contacts to the graphene. With these steps complete, further interconnect layers can be added, seamlessly linking the graphene to the underlying Si CMOS devices. This integration approach is ideal for embedding graphene sensors onto wafers that already house Si CMOS logic.

Integration of Electronic/Photonic Devices: The ability to combine circuits, like III-nitride-based LEDs and HEMTs/MOSFETs, into a single chip to create optoelectronic integrated circuits.

Overall, this strategy enhances chip functionality while maintaining compatibility with existing technologies. The importance of BEOL integration in advancing beyond-CMOS nanoelectronics is vital for overcoming the constraints of conventional scaling. This process shifts from merely passive wiring to an active, functional platform that can accommodate new materials and devices, which supports heterogeneous integration, 3D architectures, and innovative functionalities such as neuromorphic computing. BEOL has transformed from a mere "wiring" system into a key active layer that greatly enhances the functionality and performance of next-generation nanoelectronics.

The contributions of BEOL integration in transcending CMOS nanoelectronics include the following:

Incorporating Innovative Materials: Modifications in the BEOL process are being made to include unconventional materials that are incompatible with the high-temperature requirements of the front-end-of-line (FEOL) procedure. These include: (i) 2D materials (like graphene, MoS_2, and WSe_2), which facilitate the development of high-performance transistors operating at low processing temperatures. (ii) Ferroelectric materials (such as HfO_2-based oxides) for energy-efficient, non-volatile memory devices like ferroelectric tunnel junctions (FTJs) used in neuromorphic computing and memory-in-logic applications. (iii) Alternative metals (including ruthenium, molybdenum, and cobalt) that can replace copper interconnects at advanced nodes, addressing resistance problems that occur as wire dimensions are reduced.

Integrating Varied Functionalities (Heterogeneous Integration): BEOL integration enables the seamless fusion of different technologies onto a single chip, including MEMS for on-chip sensing applications, III–V semiconductor components designed for high-speed radio-frequency (RF) functionalities, and energy harvesting devices that utilize pyroelectric materials.

Enhancing Interconnects Performance: As devices shrink, BEOL must evolve to manage increased interconnect density, which necessitates the development of new materials, process improvements, and scaling innovations for metal wiring and dielectric layers to diminish resistance and capacitance. To tackle the increasing resistance and capacitance (RC) delay issues associated with shrinking metal lines, BEOL emphasizes advanced metallization methods (e.g., hybrid metallization, semi-damascene) aimed at reducing resistance, new dielectric materials (low-*k* substances and air gaps) designed to lessen capacitance, and scaling enhancements such as self-aligned vias and buried power rails to improve circuit performance and maintain structural strength.

Facilitating New Computing Architectures: BEOL enables the exploration of new computing approaches that go beyond traditional Boolean logic, such as in-memory computing and neuromorphic computing, where processing and memory elements are co-located or integrated, significantly improving energy efficiency for AI applications.

Functionalization: The upper sections of the chip can be tailored for specific functions, allowing for new capabilities and improved connectivity without significantly expanding the chip's size.

Facilitating Heterogeneous Integration: Overall, BEOL offers the foundation for vertically stacking additional active components (transistors, memory, sensors) atop existing silicon (Si) CMOS circuitry, to connect new with established CMOS transistors, incorporating 2D materials, carbon nanotubes, and oxide semiconductors onto the silicon wafer. This increases device density and enhances functionality within a compact design—a crucial demand as traditional 2D scaling becomes less effective. BEOL methods are central to 3D integration plans, enabling the stacking of multiple device layers. This approach exceeds the limits of planar scaling, resulting in more compact memory and computing solutions.

Challenges in BEOL integration include the following:

Thermal Constraints: The low-temperature processing needed for BEOL (<400°C) poses challenges when integrating new materials that might need higher processing temperatures for crystallization or other procedures.

Material Compatibility: Ensuring that new materials and interconnects work with existing BEOL processes without causing damage to the underlying FEOL transistors is a significant challenge.

Reliability: It is essential to maintain the performance and reliability of both the novel devices and interconnects amid the stresses associated with the BEOL process.

Example 9.4 BEOL Integration: Imagine a typical copper (Cu) BEOL interconnect line at a current technology node having a standard resistivity of $\rho_{\mathrm{Nominal}} = 2.0 \times 10^{-8}\ \Omega\mathrm{m}$. For a prospective beyond-CMOS nanoelectronics integration approach, the required dimensions of the lines are considerably smaller. Due to effects at the nanoscale, the effective resistivity of the metal in these dimensions increases by a factor of 3. Given the dimensions of an interconnect line in this upcoming technology: Length (L): 10 μm, Width (W): 14 nm, and Height (H): 42 nm (with an aspect ratio of 3, used to mitigate some resistance growth), calculate the total electrical resistance (R) of this interconnect line.

Solution: Determining the effective resistivity $\rho_{\mathrm{Effective}}$: The problem indicates that the effective resistivity rises by a factor of 3 from the nominal value.

$$\rho_{\mathrm{Effective}} = \rho_{\mathrm{Nomimal}} \times 3 = 2 \times 10^{-8} \times 3 = 6 \times 10^{-8}\ \Omega\mathrm{m} \tag{9.12}$$

Converting all measurements to a consistent unit (meters):

$$L = 10\ \mu\mathrm{m} = 10 \times 10^{-6}\ \mathrm{m} = 1 \times 10^{-5}\ \mathrm{m} \tag{9.13}$$

$$W = 14\ \mathrm{nm} = 14 \times 10^{-9}\ \mathrm{m} = 1.4 \times 10^{-8}\ \mathrm{m} \tag{9.14}$$

$$H = 42 \text{ nm} = 42 \times 10^{-9} \text{ m} = 4.2 \times 10^{-8} \text{ m} \tag{9.15}$$

Determining the cross-sectional area (A) of the interconnect line:

$$A = W \times H = 1.4 \times 10^{-8} \text{ m} \times 4.2 \times 10^{-8} \text{ m} = 5.88 \times 10^{-16} \text{ m}^2 \tag{9.16}$$

Computing the overall resistance (R):

$$\begin{aligned} R = \rho_{\text{Effective}} \frac{L}{A} &= 6 \times 10^{-8} \, \Omega\text{m} \times \frac{1 \times 10^{-5} \text{ m}}{5.88 \times 10^{-16} \text{ m}^2} = \frac{6 \times 10^{-8-5+16}}{5.88} \frac{\Omega\text{m} \times \text{m}}{\text{m}^2} \\ &= \frac{6}{5.88} \times 10^3 \, \Omega = 1.0204 \times 10^3 \, \Omega = 1020.4 \, \Omega \end{aligned} \tag{9.17}$$

The overall electrical resistance of the interconnect line is roughly 1020.4 Ω. This elevated resistance value underscores a significant challenge in scaling interconnects, prompting research into alternative materials such as carbon nanotubes (CNTs) or innovative integration methods.

9.3.2 Wafer Bonding

Wafer bonding (die-to-wafer/wafer-to-wafer) is a commonly used technique involving physically bonding finished or partially processed wafers or individual dies using methods such as direct bonding or microbumps (Wernicke et al. 2024). Different material wafers (e.g., III–V and Si CMOS) that are fully or partially processed are bonded together (either face-to-face or using an interposer), and typically one substrate is removed to facilitate vertical interconnections employing through-silicon vias (TSVs).

Bao et al. 2021 demonstrate a method to integrate III–V compound semiconductor materials with SOI-CMOS on a common Si substrate. The steps in this method are delineated in Figure 9.6. Three types of 200 mm wafers were prepared for this experiment: Si (001) wafers, Si-CMOS/SOI wafers processed through FEOL steps in Si foundries, and InGaAs HEMT or AlInGaP LED epitaxial films grown on GaAs/Ge/Si (001) wafers with a 6° off-cut toward the [110] direction by MOCVD. A 500 nm PECVD SiO_2 layer was deposited on the Si-CMOS/SOI wafer, followed by densification in N_2 to remove residual gases and byproducts. The oxide surface was planarized using chemical-mechanical planarization (CMP). Prior to bonding, both the SOI and Si handle wafers were treated with O_2 plasma to enhance surface hydrophilicity, then rinsed with de-ionized water and spin-dried to ensure cleanliness and sufficient hydroxyl (–OH) group density for bonding. Post-bonding was conducted at 300°C in atmospheric-pressure N_2 for 3 hours to strengthen the bond. After post-bonding, the Si substrate from the SOI wafer was removed by mechanical grinding and wet-chemical etching in tetramethylammonium hydroxide (TMAH) solution, completing the first-layer transfer. For the second bonding and layer transfer, the BOX layer was removed in HF solution and replaced with a PECVD oxide, followed by densification, planarization, and deposition of a PECVD Si_3N_4 layer to address pinholes and outgassing. The III–V/Si wafers (InGaAs HEMT and AlGaInP LED epitaxial films on Si) underwent the same oxide and nitride deposition

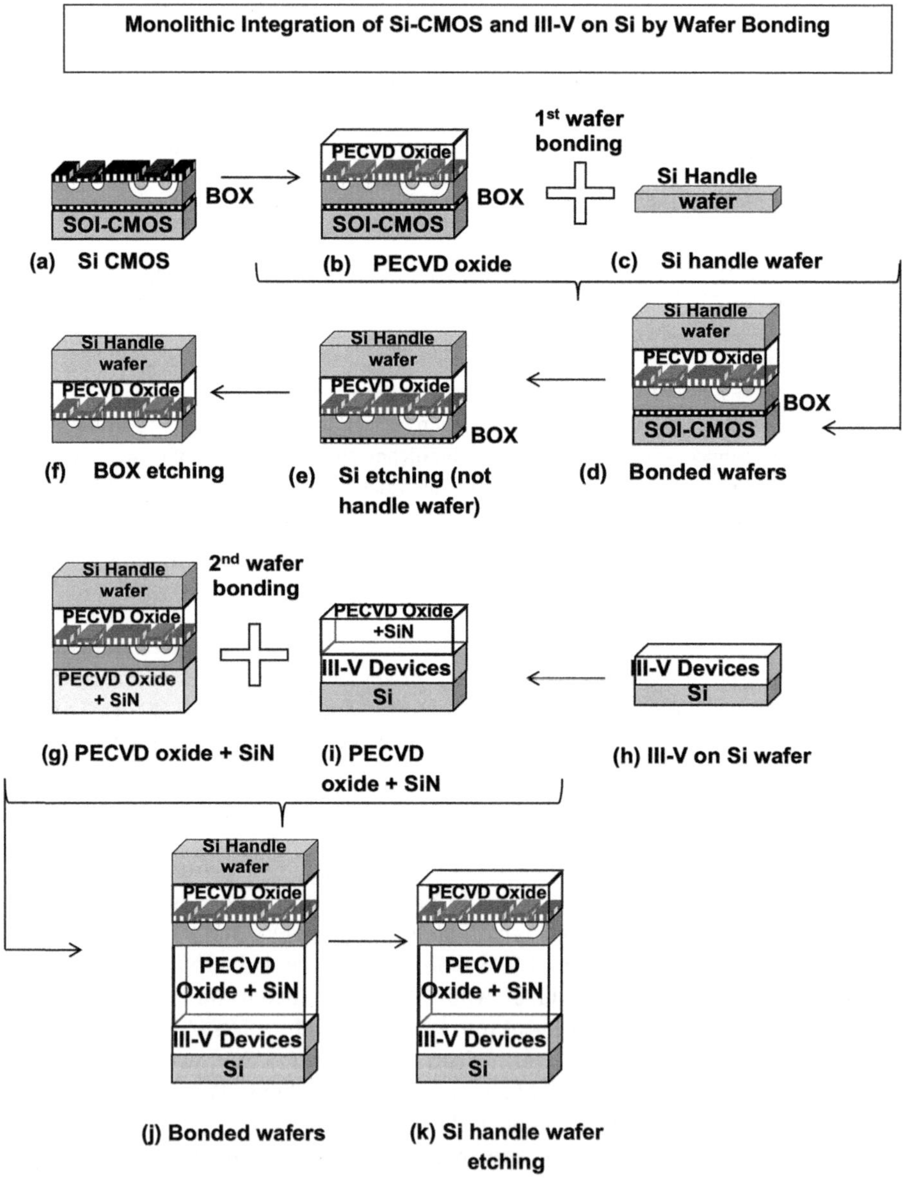

FIGURE 9.6 Monolithic integration of Si-CMOS and III–V on Si by double wafer bonding and layer transfer process. First, we temporarily bond the SOI-CMOS layer to a Si handle wafer. Next, we bond the III–V on Si wafer to the SOI-CMOS and Si handle wafer. Finally, we release the Si handle wafer to obtain the SOI-CMOS on III–V/Si hybrid structure on a common Si substrate.

processes. The wafer pairs were then bonded, and the Si handle was released to produce the Si-CMOS/III–V on Si wafer.

Wafer bonding is essential in advancing "beyond CMOS" nanoelectronics, as it supports heterogeneous integration and the development of 3D integrated circuits (3D ICs), which are crucial for addressing the physical and performance constraints of traditional 2D silicon scaling. This process allows for the creation of high-performance, multifunctional devices by physically connecting various materials and components at the wafer level. This approach overcomes the challenges of a single-material platform, enabling the development of more complex devices for multiple applications, including high-speed computing and optical systems. Significant functions of wafer bonding are:

Heterogeneous Integration: Wafer bonding, primarily via direct bonding, enables the integration of materials with different properties and lattice structures, such as III–V semiconductors with silicon-based CMOS. It permits combining materials (e.g., silicon, gallium arsenide (GaAs), indium phosphide (InP), gallium nitride (GaN), lithium niobate ($LiNbO_3$) that cannot be combined monolithically due to fundamental differences in crystal lattice constants or thermal expansion coefficients. This integration enables the construction of systems that harness the distinct properties of each material, such as combining high-speed electronics/optoelectronics (III–V materials) with established silicon CMOS platforms, high-power electronics (wide-bandgap semiconductors like SiC and Ga_2O_3) with substrates that efficiently dissipate heat, and photonic elements like high-speed electro-optical modulators alongside silicon waveguides.

3D Integration and 3D Integrated Circuits (3D ICs): This method enables vertically stacking and bonding multiple wafers or dies, allowing the creation of 3D integrated circuits. Such stacking significantly reduces interconnect lengths. Reducing the physical distance between components increases signal speed, reduces signal delays, and lowers power consumption, leading to faster, more efficient devices compared to traditional 2D configurations. This is a crucial approach for ongoing performance enhancement as CMOS dimensional scaling nears its limits. Hybrid bonding, a vital 3D technology, employs direct copper-to-copper connections to achieve extremely high interconnect density and bandwidth with minimal latency.

Advanced Substrate Engineering: Wafer bonding is used to fabricate specialized substrates, such as SOI and Germanium-On-Insulator (GeOI) wafers, which serve as foundational platforms for advanced CMOS and innovative devices, including high-performance field-effect transistors (FETs). The Smart Cut™ process, which has seen commercial success, incorporates ion implantation and wafer bonding to transfer high-quality thin films onto various substrates.

Micro-Electro-Mechanical Systems (MEMS) Integration and Advanced Packaging: It is a crucial technology for nanoelectronics, enabling the development of compact, high-density, high-performance systems. This encompasses the integration of sensors (such as MEMS and NEMS) with their control electronics on a single chip, resulting in smaller, more powerful devices like those found in smartphones and medical imaging systems. In the MEMS field, wafer bonding is used to create sealed cavities that safeguard delicate mechanical structures (e.g., pressure sensors, accelerometers, gyroscopes) or to integrate MEMS with CMOS circuitry on a single chip, resulting in compact, resilient microsystems.

Novel Device Capabilities: The ability to merge dissimilar materials paves the way for the development of entirely new device types and functionalities, such as lab-on-a-chip devices, implantable biosensors, and energy-harvesting systems, that extend beyond the limitations of standard CMOS technology.

Addressing Limitations: Traditional CMOS scaling encounters physical limitations. Wafer bonding enables the integration of new materials and capabilities beyond conventional scaling. It also allows combining different generations of technologies, as demonstrated by the integration of advanced memory with logic circuits.

Wafer-Level Testing: The ability to test devices at the wafer level improves efficiency and effectiveness in the development process.

On the whole, wafer bonding is a vital enabling technology, providing a versatile platform for integrating diverse, high-performance materials and intricate 3D architectures, thus fostering innovation in nanoelectronics that exceeds the boundaries of conventional CMOS scaling.

Example 9.5 Wafer Bonding

In a wafer-to-wafer bonding process aimed at heterogeneous integration, achieving an alignment accuracy of ± 50 nm in the $X - Y$ plane is essential for effective vertical interconnection. The wafers being bonded have a diameter of 300 mm. An alignment system offers a rotational accuracy of 3×10^{-4} degrees. Compute the maximum allowable linear misalignment (in nanometers) at the edge of the 300 mm wafer resulting from this angular error, and assess whether the system satisfies the necessary linear alignment criteria.

Solution: We will apply the arc length formula, which approximates linear displacement for small angles:

$$\Delta L = R \times \Delta\theta_{\text{Radians}} \tag{9.18}$$

where R represents the wafer radius and $\Delta\theta_{\text{Radians}}$ denotes the rotational error in radians. First, we need to convert the wafer diameter into radius (in mm): Radius $R = 300\ \text{mm}/2 = 150$ mm. Next, we convert the rotational error from degrees to radians: since π radians equals 180 degrees, we find

$$\begin{aligned}\Delta\theta_{\text{Radians}} &= 3 \times 10^{-4}\ \text{degrees} \times \frac{\pi}{180\ \text{degrees}} = 3 \times 10^{-4} \times \frac{22}{7} \times \frac{1}{180} \approx 0.05238 \times 10^{-4} \\ &= 5.238 \times 10^{-6}\ \text{radians}\end{aligned} \tag{9.19}$$

Now we can calculate the linear misalignment (in mm):

$$\Delta L = R \times \Delta\theta_{\text{Radians}} = 150\ \text{mm} \times 5.238 \times 10^{-6}\ \text{radians} = 785.7 \times 10^{-6}\ \text{mm} \tag{9.20}$$

After that, we will convert the linear misalignment into nanometers (nm):

$$\Delta L = 785.7 \times 10^{-6}\ \text{mm} = 785.7 \times 10^{-6} \times 10^{-3}\ \text{m} = \frac{785.7 \times 10^{-6} \times 10^{-3}}{10^{-9}}\ \text{nm}$$
$$= 785.7\ \text{nm} \tag{9.21}$$

Thus, the maximum linear misalignment at the wafer edge due to the identified rotational error is approximately 785.7 nm. The required alignment accuracy is ±50 nm. The calculated misalignment (785.7 nm) significantly exceeds the specified requirement. Therefore, the system fails to meet the alignment criteria for this particular nanoelectronic application, as the rotational error alone could lead to functional issues, e.g., improperly connected vertical interconnects.

9.3.3 Chiplet/Die-to-Wafer Assembly

System-in-package (SiP)/chiplets strategy consists of fabricating different components as separate small dies (chiplets) with various technologies (such as III–V power amplifiers, RF MEMS, Si CMOS) and then assembling these pre-fabricated "chiplets" into one compact package by utilizing advanced packaging technologies like flip chip bonding or 2.5D integration on an interposer after arranging them onto a shared silicon interposer or carrier substrate using microbumps (Gao et al. 2022, Lu 2026).

Example 9.6 Chiplet/Die-To-Wafer Assembly

A system-in-package (SiP) is developed utilizing a die-to-wafer (D2W) assembly technique. This technique combines two distinct types of chiplets onto a foundational wafer: Chiplet A (Logic), which is produced on its own wafer and has a die yield of 90% (or 0.90), and Chiplet B (Memory), fabricated on a separate wafer with a die yield of 85% (or 0.85). The assembly method features a bonding yield of 98% (or 0.98) for each chiplet placement during the D2W process. To achieve the final SiP design, one Chiplet A and two Chiplet B's must be successfully bonded onto a known good site on the base wafer. Assuming all yields function as independent probabilities, calculate the overall yield for a single, fully functional SiP (System-in-Package).

Solution: To determine the total system yield, one must multiply the independent probabilities of all necessary successful outcomes. The probability of obtaining a good Chiplet A is $P(\text{A}) = 0.90$. The probability of acquiring a good Chiplet B (first one) is $P(\text{B}_1) = 0.85$. The probability of securing a good Chiplet B (second one) is $P(\text{B}_2) = 0.85$. For Chiplet A, the probability of a successful bond is $P\ (\text{Bond A}) = 0.98$. The probability of achieving a successful bond for Chiplet B (first) is $P\ (\text{Bond B}_1) = 0.98$. Lastly, the probability of a successful bond for Chiplet B (second) is $P\ (\text{Bond B}_2) = 0.98$.
The overall system yield can be calculated by multiplying all these probabilities together:

$$\text{Total yield} = P(\text{A}) \times P(\text{B}_1) \times P(\text{B}_2) \times P(\text{Bond A}) \times P(\text{Bond B}_1) \times P(\text{Bond B}_2) \tag{9.22}$$

Therefore, the equation can be expressed as

$$\text{Total yield} = 0.90 \times 0.85 \times 0.85 \times 0.98 \times 0.98 \times 0.98 \quad (9.23)$$

This approximates to

$$\text{Total yield} = 0.612 \quad (9.24)$$

Ultimately, the total yield approximately equals 0.612. The final calculated yield for a single, complete, functional SiP is around 61.2%. This analysis emphasizes the importance of maintaining high individual die and assembly yields in chiplet-based production, as total yield can decline significantly when integrating multiple components.

9.3.4 Micro-Transfer Printing

Micro-Transfer Printing (MTP or μTP) is a scalable technique that involves the accurate transfer or direct growth of non-silicon materials (like 2D semiconductors or III–V materials) onto a completed or partially processed silicon wafer. It utilizes an elastomeric stamp to accurately pick up small "chiplets" (microstructures) from their original substrate and place them with precision onto a host wafer (Zhang et al. 2021, Downey et al. 2023). Using the stamping technique, a large number of small prefabricated material "coupons" (such as III–V lasers or 2D materials) is quickly and accurately transferred onto a target substrate. The method is highly adaptable for integrating a wide range of materials, including III–V lasers/detectors with Si photonics.

Figure 9.7 offers a closer look at the μTP process (Zhang et al. 2023). It all begins with a carefully constructed stack: the III–V device layer, a sacrificial release layer, and a substrate (Figure 9.7(a)). The release layer, such as InGaAs or InAlAs for InP-based epitaxy, is chosen for its selective etchability, making the source devices ready for printing. Next, devices are patterned onto the III–V wafer (Figure 9.7(b)) and shielded with photoresist or a dielectric layer, which is then opened locally to reveal the release layer beneath (Figure 9.7(c)). Ingeniously designed tethers support the device coupons, holding them in place until the moment of printing, when they are engineered to snap away cleanly. The release layer is then etched away, undercutting the devices (Figure 9.7(d))—often with a cold $FeCl_3$:H_2O solution for InP-based structures. At this stage, the devices are primed for transfer. A PDMS stamp is pressed onto the chosen device or array, then swiftly lifted (Figure 9.7(e)), snatching the coupon and breaking the tethers in one smooth motion. The harvested devices are then precisely aligned to their new home on the target wafer using fiducial markers and printed onto the silicon photonic integrated circuits through a series of careful steps: laminating the stamp (Figure 9.7(f)), applying a gentle shear, and slowly peeling the stamp away (Figure 9.7(g and h)). Although this study spotlights III–V devices, the μTP technique is versatile and can be applied to a broader range of materials.

Layer transfer techniques are comparable to transfer printing, but generally on a larger scale. This approach entails producing a material layer on a donor substrate and subsequently transferring it to a target substrate using release layers. Devices are then fabricated on the transferred layer in situ.

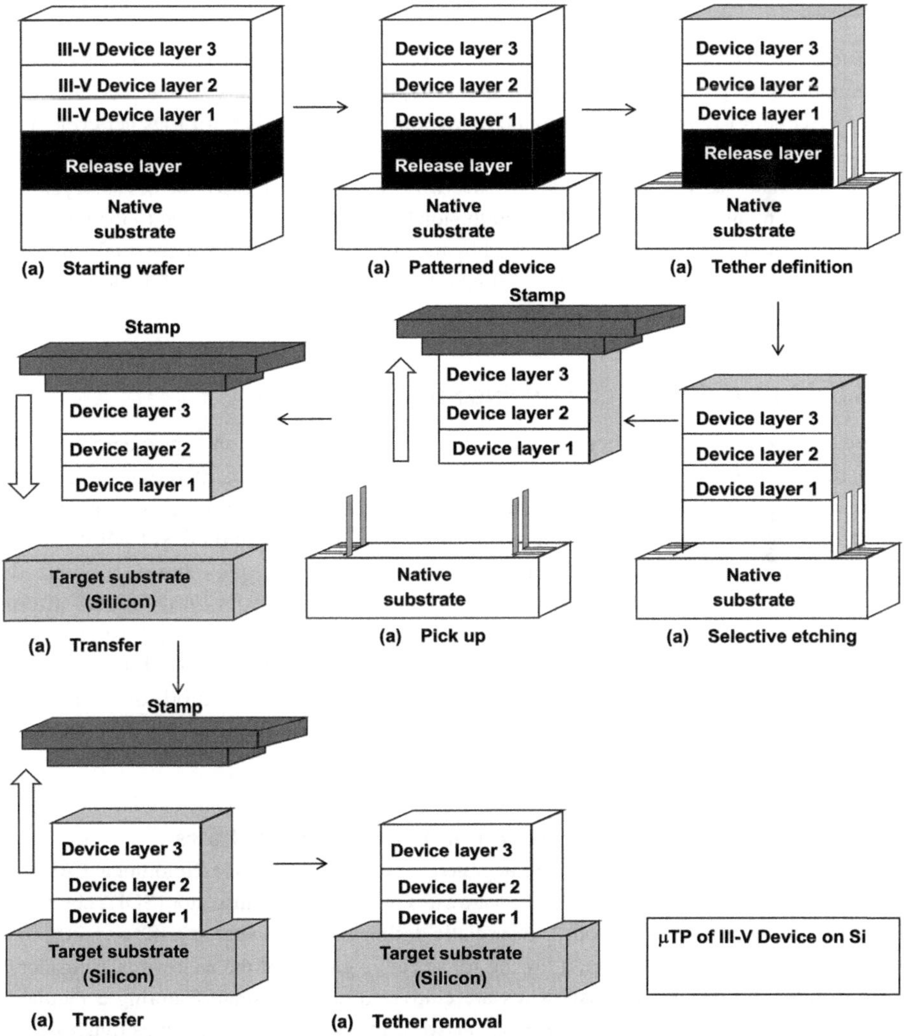

FIGURE 9.7 µTP of III–V device on Si: (a) Starting stack that includes the III–V device layer(s), a sacrificial release layer, and a substrate; (b) Patterning the devices on the III–V source wafer; (c) Patterning the photoresist or a dielectric layer with local openings to expose the release layer; (d) Selectively etching the release layer to undercut the devices; (e) Laminating a PDMS stamp to the device or device array of interest on the source wafer and moving the stamp upwards at a high retraction velocity; (f) Printing the devices onto the target Si PICs through a sequence of motions of laminating the stamp on the target; and (g) and (h) Applying shear displacement and slowly retracting the stamp.

Example 9.7 Micro-Transfer Printing

A micro-transfer printing method is utilized to position GaN micro-LEDs onto a silicon CMOS wafer. A single stamp is crafted to collect and arrange an array of 500 micro-LEDs in a single operation. The manufacturing team aims to achieve an effective yield of 99.8% per print cycle. Determine the total number of micro-LEDs that are successfully transferred after 100 print cycles.

Solution: First, we calculate the number of LEDs successfully transferred in each cycle. Number of LEDs in one cycle = 500, and yield per cycle = 99.8% or 0.998. The number of successful LEDs per cycle is found by multiplying the number of LEDs per cycle by the yield per cycle. Therefore, successful LEDs per cycle = $500 \times 0.998 = 499$. Next, we calculate the cumulative total of successfully transferred LEDs across all cycles. The total number of cycles = 100, and the total successful LEDs is equal to the number of successful LEDs per cycle multiplied by the number of cycles. Thus, the total number of successful LEDs = $499 \times 100 = 49{,}900$. Hence, a cumulative total of 49,900 micro-LEDs is successfully transferred after completing 100 cycles.

9.3.5 In Situ Integration on Engineered Substrates

Here, device layers are either grown or deposited in openings on "engineered" silicon substrates (such as silicon on lattice-engineered substrates (SOLES)) that have a template layer compatible with the crystal structure of the non-silicon material, promoting high-quality growth of materials like InP or GaN alongside standard Si CMOS.

Catalytic chemical vapor deposition (CCVD) techniques have enabled the direct creation of monolayer and bilayer graphene transistors on oxidized silicon, eliminating the need for transferring delicate graphene sheets (Wessely et al. 2012). Similarly, zinc oxide nanowires were grown directly on CMOS substrates at a remarkably low temperature of 90°C by hydrothermal method, demonstrating impressive sensitivity to trace amounts of toluene vapor even in humid conditions (Santra et al. 2013).

Example 9.8 In Situ Integration on Engineered Substrates

A research group is working on an integrated circuit that goes beyond traditional CMOS technology by cultivating silicon (Si) nanowires on a silicon-on-insulator (SOI) substrate using an in situ growth method. The specially designed substrate has specific openings in a silicon dioxide (SiO_2) layer, allowing for selective growth of the nanowires. Consider the following details: The nanowires are cultivated on a substrate featuring a regular square pattern of openings. The distance between the centers of the openings (pitch size) is 1 μm. Assume that one nanowire is produced per opening in this pattern. Task: Determine the maximum possible density of the nanowires on the substrate, expressed in nanowires per square micrometer (μm^{-2}).

Solution: The nanowire density can be assessed based on the pitch size of the array, which indicates the area designated for each nanowire.
Calculating the area per nanowire:

Given that the nanowires are arranged in a square pattern with a pitch of 1 μm, the area designated for each nanowire, along with its surrounding region, is:

$$\text{Area per nanowire} = \text{Pitch} \times \text{Pitch} = 1\ \mu\text{m} \times 1\ \mu\text{m} = 1\ \mu\text{m}^2 \tag{9.25}$$

Now, let us compute the nanowire density. The density represents the count of nanowires per unit area:

$$\text{Density} = \frac{\text{Number of nanowires}}{\text{Area}} \tag{9.26}$$

For a single unit cell in the array:

$$\text{Density} = \frac{1\ \text{nanowire}}{1\ \mu\text{m}^2} = 1\ \mu\text{m}^{-2} \tag{9.27}$$

The highest achievable density of the nanowires on this specially engineered substrate is 1 μm^{-2} (indicating one nanowire per square micrometer).

9.3.6 3D Integration with TSVs and 2.5D Interposers

3D Integration with TSVs: While a general integration technology, and sometimes categorized as advanced packaging rather than integration methods, TSVs are a vital enabling technology for 2.5D and 3D heterogeneous integration, and play a crucial role in enabling the vertical stacking of diverse components in heterogeneous systems, facilitating vertical connectivity among different chip layers (Shen and Chen 2017, Abrams et al. 2020, Abdullah and Lee 2024). They offer essential vertical electrical connections by passing through the silicon substrate to an interposer (such as silicon or glass) that directs signals between chiplets. Embedded multi-die interconnect bridge (EMIB)/advanced packaging methods involve embedding small silicon bridges or interposers into the package substrate to enable high-density, high-speed communication between neighboring chiplets without requiring complete wafer-scale bonding.

Let us distinguish between 2D, 2.5D, and 3D IC packaging schemes (Figure 9.8). In 2D IC packaging, individual chips or dies are placed side by side on a substrate and connected using wire bonding or flip-chip technology. As transistor counts increase, this approach results in longer interconnects, higher power consumption, and slower signal transmission. 2.5D packaging addresses these issues by placing multiple active dies on a silicon interposer, thereby reducing interconnect lengths and improving performance and bandwidth. 3D packaging advances integration by stacking multiple dies vertically within a single package, enabling faster communication and a more compact design. Thus, the primary distinction is the level of vertical integration. In 2D IC packaging, components are on a single plane. In 2.5D IC packaging, components remain on the

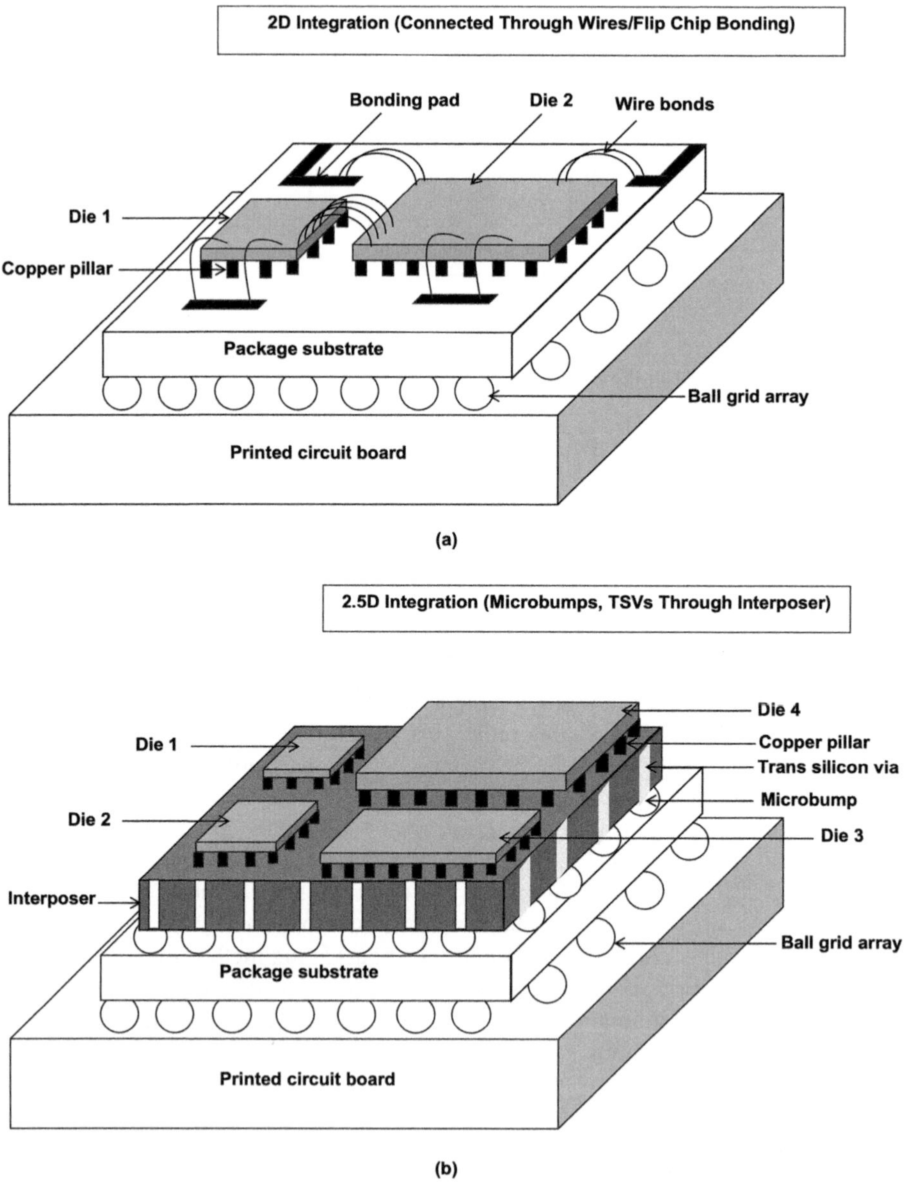

FIGURE 9.8 2D, 2.5D, and 3D integration schemes: (a) 2D integration (connected through wires/flip chip bonding), (b) 2.5D integration (microbumps, TSVs through interposer), and *(Continued)*

same plane, but an interposer enables greater integration. In 3D IC packaging, components are stacked vertically, allowing for higher integration and potentially better performance and power efficiency.

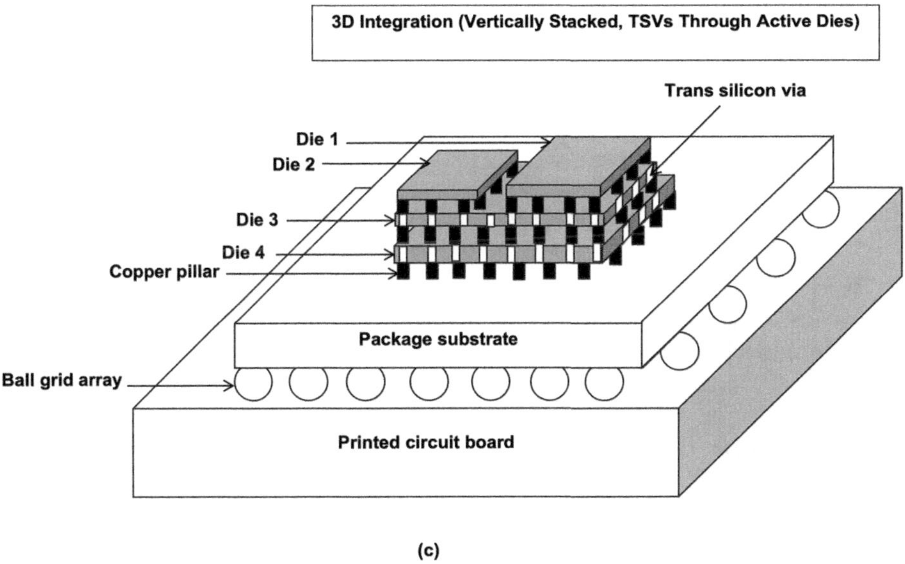

FIGURE 9.8 (Continued) (c) 3D integration (vertically stacked, TSVs through active dies).

Example 9.9 3D Integration with TSVs

A typical 2D integrated circuit (IC) features a critical interconnection wire with an average length of $L_{2D} = 100\ \mu m$. By utilizing 3D homogeneous integration with $N = 4$ stacked layers (tiers), this crucial wire length is reduced vertically through the use of through-silicon vias (TSVs), allowing components to connect directly in the third dimension. The average wire length for a 3D IC is estimated to decrease by a factor of $\sqrt{N}$ when compared to the corresponding 2D design, thanks to the enhanced routing efficiency in the vertical direction. Calculate the new average wire length L_{3D} in the 3D integrated circuit.

Solution: The equation for the diminished average wire length in a 3D IC with N layers is:

$$L_{3D} = \frac{L_{2D}}{\sqrt{N}} \tag{9.28}$$

where $L_{2D} = 100\ \mu\text{m}$ and $N = 4$. Therefore,

$$L_{3D} = \frac{100\ \mu\text{m}}{\sqrt{4}} = \frac{100\ \mu\text{m}}{2} = 50\ \mu\text{m} \tag{9.29}$$

Now, let us determine the percentage reduction in wire length: The percentage decrease is computed as:

$$\text{Percentage reduction} = \frac{L_{2D} - L_{3D}}{L_{2D}} \times 100\% \tag{9.30}$$

which simplifies to

$$\text{Percentage reduction} = \frac{100 - 50}{100} \times 100\% \tag{9.31}$$

This further resolves to

$$\text{Percentage reduction} = \frac{50}{100} \times 100\% = 50\% \tag{9.32}$$

The updated average wire length (L_{3D}) is 50 μm, and the percentage decrease in wire length is 50%. This notable reduction in wire length directly contributes to reduced power consumption (as less power is required to drive signals over shorter distances) and the potential for increased operating speed.

Example 9.10 3D Integration with TSVs

A standard 2D integrated circuit (IC) possesses a transistor density of $\rho_{2D} = 100$ million transistors per square millimeter (mm^2). In a beyond-CMOS approach, this technology is modified to use a uniform 3D integration method that stacks N identical device layers vertically. Assuming flawless vertical alignment and no extra area from through-silicon vias (TSVs) or inter-layer bonding (in an ideal scenario), how many layers, N, are necessary to attain a desired volumetric transistor density of $\rho_{3D} = 1$ billion transistors per cubic millimeter (mm^3)? What will be the resulting vertical pitch (thickness) P of each device layer in nanometers (nm)?

Solution: Here is the outline of the solution. We determine the required number of layers (N). The volumetric density is the planar density multiplied by the quantity of layers per unit height. If we consider a unit height (1 mm) for ease, the number of layers N required to reach the target 3D density can be calculated using the formula:

$$\rho_{3D} = N\rho_{2D} \tag{9.33}$$

or,

$$N = \frac{\rho_{3D}}{\rho_{2D}} = \frac{1 \text{ billion transistors/mm}^3}{100 \text{ million transistors/mm}^2} = \frac{1000 \text{ million}}{100 \text{ million}} = 10 \text{ layers} \tag{9.34}$$

Hence, 10 layers are needed.

To calculate the vertical pitch (P) for a single layer, we note that the total height of the 10 layers equals 1 mm (since ρ_{3D} is based on per cubic millimeter (mm^3). The pitch (thickness) P of each identical layer is the total height divided by the number of layers:

$$P = \frac{\text{Total height}}{\text{Number of layers}} = \frac{1 \text{ mm}}{10} = 0.1 \text{ mm} = 0.1 \times 10^{-3} \text{ m} = 0.1 \times 10^{-3} \times 10^{9} \text{ nm} = 100{,}000 \text{ nm} \tag{9.35}$$

The vertical pitch of an individual device layer is 100,000 nm.

Note: This is a comparatively large, "ideal" pitch relative to the actual sub-10 nm feature sizes seen in advanced nodes, underscoring the substantial overhead in practical 3D integration. This problem demonstrates how increasing the number of vertically stacked layers can significantly enhance the overall system density and performance, a primary motivation for developments beyond traditional CMOS technologies.

9.3.7 Comparative Analysis of Integration Modalities

Although 2D integration continues to dominate in cost-sensitive markets, 2.5D and 3D integration technologies offer new levels of interconnect density, power efficiency, and compactness. Table 9.1 highlights how these approaches mark a leap from traditional planar, single-die designs to sophisticated, stacked, heterogeneous, and chiplet-based architectures.

9.3.8 3.5D Integration

3.5D integration is a semiconductor packaging approach that bridges the gap between 2.5D and 3D-IC, blending the dense, horizontal connections of 2.5D with the vertical stacking of 3D-IC (Bobba et al. 2013). This technology lets powerful logic chips, such as GPUs or AI accelerators, sit shoulder to shoulder with high-bandwidth memory on a silicon interposer, while also enabling the stacking of select chiplets, like CPU-on-cache, directly on top of each other via copper-to-copper hybrid bonding.

9.3.9 Reticle Stitching as a Key Enabler for Advanced Packaging

Reticle stitching stands at the forefront of advanced packaging and heterogeneous integration, unlocking new possibilities in chip design. This sophisticated lithographic technique empowers manufacturers to craft expansive silicon interposers and chips that break free from the traditional boundaries of lithography scanners. By seamlessly aligning and merging exposure fields, engineers can build unified circuits that surpass the standard 26 mm × 33 mm (854 mm^2) size limit. Such large-format chips and interposers are essential for specialized applications demanding rapid, high-bandwidth connections among vast arrays of computing or memory cores (Xu et al. 2024; Bottoms et al. 2025).

9.3.10 Failure Modes of Advanced Packaging

Failure modes in semiconductor packaging often arise from the uneasy alliance of materials such as silicon, copper, and epoxy molding compounds, which react differently to heat (Singh and Viswanadham 2012, Chen et al. 2024). The main culprit is the mismatch in how these materials expand and contract, known as the coefficient of thermal expansion (CTE) mismatch. This tug-of-war creates internal stresses that twist and bend wafers, a phenomenon called warpage. Warpage is not just a cosmetic issue; it can trigger a cascade of mechanical failures, such as the epoxy peeling away from surfaces. Silicon barely budges with heat, while epoxy and

TABLE 9.1
Comparison Table of 2D, 2.5D and 3D Integration Techniques

Sl. No.	Point of Comparison	2D Integration	2.5D Integration	3D Integration
1	Feature	It has a planar structure.	It is based on an interposer and chiplets.	Vertical stacking of dies is done using TSVs.
2	Architecture	All the components are mounted side-by-side on a single planar silicon die or PCB.	Multiple chiplets are placed side-by-side on a shared silicon or organic interposer.	Multiple dies are stacked vertically on top of each other, and connected by TSVs and micro-bumps.
3	Interconnects	It uses standard wires, wire bonding, or flip-chip bumps.	It uses through-silicon vias (TSVs) in the interposer and micro-bumps.	It uses TSVs through the dies; hybrid bonding or micro-bumps.
4	Vertical connection	No vertical connection is used. There are only horizontal metal layers.	Vertical connections are made by creating TSVs within the silicon interposer to connect the upper metal layers to the lower package substrate.	TSVs are used to pass signals through the silicon die. The mechanism of vertical connection *between* the stacked dies has evolved from microbumps to copper-to-copper hybrid bonding.
4	Interconnect density	It yields lowest interconnect density being limited by die edge or PCB routing. and on-chip wiring.	It shows medium to high interconnect density enabled using fine-pitch interposer routing.	It gives the highest interconnect density due to direct vertical connectivity between layers.
5	Performance (speed/ bandwidth)	Baseline performance is observed.	Performance is high and better than 2D integration.	Highest performance is achieved by showing lowest latency.
6	Thermal management	It is easiest because a single chip dissipates heat from a single flat surface through a simple heat sink.	It is moderate to high because shortened inter-chip connections limit heat spreads across the interposer surface.	It is most challenging as vertical stacking creates high density, trapped heat-induced "thermal bottlenecks."

(Continued)

TABLE 9.1 (Continued)

Sl. No.	Point of Comparison	2D Integration	2.5D Integration	3D Integration
7	Power efficiency	Efficiency is low to moderate due to long on-chip or PCB wires.	High efficiency is enabled by shortened inter-chip connections than 2D but longer than 3D.	Highest efficiency is possible owing to shortest possible electrical paths and lowest parasitics.
8	Form factor	It has the largest footprint.	It has a medium footprint, which is smaller than 2D but bigger than 3D.	It provides the smallest footprint and most compact design.
9	Complexity of fabrication	Its fabrication is straightforward.	Its fabrication is moderately complex.	Highest fabrication complexity is faced.
10	Cost	It has the lowest cost due to mature and simple manufacturing.	The cost is moderate because additional expenses are incurred for interposer and assembly.	It is most expensive because of need of specialized processing.
11	Examples of usage	It is used in consumer electronics, sensors, and microcontrollers.	It is used in high-bandwidth memory.	It is used in AI accelerators.

organic substrates expand much more, so when temperatures swing during manufacturing, these materials strain against each other. This relentless push and pull can fatigue solder joints, causing cracks in micro-bumps and ball grid arrays, or snap delicate copper traces in redistribution layers. Warpage itself is the visible sign of these hidden stresses, bending wafers in unpredictable ways. The consequences ripple through manufacturing: warped wafers can throw off lithography alignment, cause handling mishaps like vacuum leaks and broken wafers, shift dies out of place before they are secured, and slash reliability, sometimes cutting mean time between failures by up to 60%. Epoxy peel-off, or delamination, happens when the bond between the epoxy and another surface gives way, usually at the die's sharp corners where stress is highest. Poor surface preparation, like skipping plasma cleaning, leaves the epoxy unable to grip tightly. Moisture trapped inside can vaporize during heating, popping the epoxy loose, while contaminants like oils or silicone act as invisible saboteurs, weakening the bond even further.

To tackle these challenges, we consider three key strategies. First, we match the CTE by selecting underfill materials and molding compounds that closely match the silicon die's CTE. Next, we design the package with structural symmetry in the thickness direction to evenly distribute stress. Finally, we optimize the process by lowering reflow or curing temperatures and using a gentle cooling rate, ideally below 10°C/min, to reduce thermal stress.

9.4 OPTICAL INTERCONNECTS

As devices relentlessly undergo miniaturization, copper interconnects struggle to keep up with this pace. Grain-boundary and surface-roughness effects, along with wider wire widths, drag down their performance (Singh et al. 2024). The result is a triple threat: higher latency, more power consumed, and a greater risk of electromigration. Crosstalk, sparked by capacitive and inductive coupling between neighboring wires, further complicates issues. In the deep submicrometer era, copper interconnects have hit a wall, devouring most of the available power budget. To overcome these limits and meet the demand for greater bandwidth and lower power, researchers are turning to new solutions. OIs are emerging as a promising alternative, already finding real-world use in supercomputing data centers, telecommunications, AI and machine learning hardware, automotive systems, and consumer electronics. OIs bring a host of advantages: broad bandwidth, high speed, low latency, and minimal crosstalk, whether off-chip or on-chip. Their architecture is simple yet powerful, built from a transmitter, a waveguide, and a receiver. Optical waveguides enable signals to travel as beams of light at the speed of light, operating at wavelengths between 1300 and 1550 nm. Unlike traditional electrical lines, OIs do not waste energy on wire charging and discharging, eliminating a major source of power loss.

In "Beyond CMOS" technology, OIs are breaking through the barriers that limit traditional copper wiring. These pathways use light or photon beams instead of electricity or electron streams to move data swiftly between and within chips. Silicon photonics is at the center of this revolution, merging optical and electronic components onto a single silicon platform. Most silicon photonic integrated circuits are built on SOI wafers, where key components such as waveguides, modulators, and photodetectors are fabricated directly in silicon. Since silicon itself cannot shine, lasers made from III–V materials like indium phosphide are added, either by blending technologies or assembling them together in a hybrid approach or through co-packaging. Thanks to standard CMOS manufacturing, these advanced circuits can be produced in large quantities at a low cost.

OIs can be integrated at three distinct levels, each opening a new possibility. Inter-chip integration links separate chips across a circuit board or interposer, enabling high-speed chiplet architectures. Intra-chip integration weaves data links between functional blocks on a single silicon substrate, often relying on TSVs to bridge different layers. The most advanced, 3D Integration, stacks photonic and electronic layers through hybrid bonding, building dense, multi-functional platforms.

9.5 LOW-TEMPERATURE FABRICATION PROCESSES

These processes play a significant role in "Beyond-CMOS" nanoelectronics because they facilitate the combination of innovative materials and three-dimensional device designs that cannot withstand the elevated thermal requirements of conventional CMOS manufacturing (Gowrisanker et al. 2009, Aasmundtveit et al. 2020, Naqi et al. 2023). Essential functions and key roles of low-temperature techniques in beyond-CMOS nanoelectronics are:

Supporting Heterogeneous and 3D Monolithic Integration: Conventional CMOS methodologies typically function at high temperatures (often exceeding 500°C), which risks damaging pre-fabricated devices and interconnect layers (Back-End-of-Line, or BEOL) within a three-dimensional stacked integrated circuit (IC). Low-temperature (LT) processes, which operate at temperatures below 400°C–500°C, enable the monolithic integration of new functional layers (such as memory, sensors, or diverse logic devices) atop existing CMOS circuits without degrading their performance. By employing low-temperature bonding and BEOL integration, sophisticated nanoelectronic devices such as single-electron transistors (SETs) can be stacked atop existing CMOS layers, resulting in three-dimensional monolithic systems.

Adapting to Innovative Materials: Various emerging "beyond-CMOS" materials (including metal oxides, 2D materials such as graphene, and select organic compounds) can be sensitive to elevated temperatures, potentially leading to unwanted diffusion, oxidation, or physical degradation. Low-temperature techniques facilitate new materials, permitting the direct deposition of substances like graphene without harming the underlying substrate, thereby supporting their integration into electronic devices. Low-temperature methods, such as specific types of CVD, atomic layer deposition (ALD), and solution-based processing, are crucial for depositing and annealing these materials while maintaining their distinctive properties. ALD is widely used to deposit high-quality, ultra-thin, and conformal layers of materials, particularly high-κ dielectrics, which are essential for effective gate control in deeply scaled nanoelectronic devices, including those made from 2D materials. The ability to precisely control thickness at the atomic level is a significant advantage.

Integration of Innovative Devices: These techniques are essential for merging new device architectures and materials, such as those found in emerging devices like FeFETs, which do not align with high-temperature processing. Techniques like nanosecond laser processing enable selective annealing of thin films, making it suitable for both 3D integrated and flexible devices.

Enhancing Device Efficiency and Consistency: Low-temperature processing can yield superior material quality in certain situations. For instance, specific LT annealing techniques may achieve contact resistivity and overall device performance equivalent to, or even better than, those of high-temperature approaches for particular materials. Additionally, these techniques help manage critical dimensions (CD) and reduce variability in nanoscale constructs, such as quantum dots, which are essential for reliable device operation.

Enabling Flexible Electronics: The advancement of flexible and transparent circuitry, a vital area of new nanoelectronics, often requires deposition on temperature-sensitive substrates, such as polymers. Low-temperature manufacturing is a key requirement for the success of these technologies. Low-temperature techniques are suitable for flexible electronics. They are vital for creating devices on flexible substrates, as high-temperature manufacturing steps would compromise them.

Illustrations of low-temperature techniques in beyond-CMOS nanoelectronics are as follows:

Microwave Annealing: Achieving optimal material quality at low temperatures can be challenging due to insufficient atomic diffusion, but novel techniques can address this challenge. For example, low-temperature annealing methods, such as microwave annealing, can achieve performance levels comparable to traditional high-temperature anneals for silicide contacts. Microwave annealing has emerged as a possible solution for low-temperature processing, showing effective silicidation and oxide densification at lower thermal budgets.

Combustion Processing: This technique, when combined with solution deposition, produces high-performance metal oxide TFTs at temperatures as low as 200°C.

Plasma Processing: Low-electron-temperature plasmas are commonly used for material processing, facilitating gas-surface chemical reactions and anisotropic etching at low temperatures, which is essential for introducing new materials and techniques without causing structural damage.

Metal-Induced Dopant Activation (MIDA): This technique enables the production of high-performance Germanium (Ge) N- and P-MOSFETs at notably low temperatures (below 380°C).

In essence, low-temperature fabrication processes eliminate the thermal barrier to integrating a wide range of materials and intricate architectures needed to advance beyond-CMOS restrictions, especially in the pursuit of 3D integration and innovative functional devices.

9.6 DISCUSSION AND CONCLUSIONS

This chapter presented crucial monolithic integration methods for creating beyond-CMOS nanoelectronics, mainly focusing on the in situ development of innovative materials and device structures directly on a single semiconductor substrate, typically a conventional silicon wafer. Collectively, these techniques aim to overcome the physical constraints of traditional silicon CMOS by integrating novel materials and 3D device architectures, often capitalizing on quantum-mechanical properties or steep sub-threshold slopes to enhance performance. Let us summarize the main takeaways from this chapter.

The most critical technique for monolithic integration in beyond-CMOS nanoelectronics is generally regarded as Monolithic 3D (M3D), also known as sequential 3D integration. This method is vital because it addresses the inherent limitations of 2D scaling by vertically stacking functional layers, thereby greatly enhancing integration density and performance. Monolithic 3D (M3D) integration method stacks layers of transistors or other devices with highly compact, nanoscale vertical interconnects (MIVs, with a pitch of about 100 nm), which are considerably smaller than standard TSVs. This enables the creation of highly compact, high-performance computing setups and denser memory-on-logic configurations. The main approaches utilized for homogeneous monolithic 3D integration in beyond-CMOS research consist of:

Vertical Transistor Stacking (CFETs): This technique involves vertically stacking N-type and p-type transistors (Complementary Field-Effect Transistors, or CFETs) on a single chip, often using silicon-based devices. Techniques to achieve this stacking include:

On-chip recrystallization of polycrystalline silicon (P-Si) films.

On-chip Wafer Bonding and Transfer (as a stage in monolithic integration): While specific transfer techniques are linked with heterogeneous integration, approaches such as "on-chip wafer bonding and transfer" or selective layer transfer can be applied within a single process flow to merge various material layers (e.g., a high-quality active channel layer) that might otherwise not tolerate the thermal constraints of the underlying CMOS layers.

Solid-State Epitaxy and Annealing: This technique entails the controlled crystallization of deposited amorphous or polycrystalline films into high-quality single-crystal layers directly on the chip. It serves as a means to achieve superior material quality without the elevated temperatures that could harm pre-fabricated devices in the lower layers of a monolithic 3D integration setup.

Nanoheteroepitaxy/Selective-Area Growth: This method entails the careful growth of new, lattice-mismatched materials (such as III–V semiconductor nanowires or GaN) directly on a pre-patterned silicon substrate. This approach enables integrating high-performance materials with current silicon CMOS platforms to leverage their unique features (e.g., optical characteristics for integrated photonics).

Advanced Lithography and Etching: Methods such as electron-beam lithography, reactive ion etching (RIE), and quasi-atomic layer etching

(QALE) are employed to define and fabricate the exceedingly small, three-dimensional structures required for nanoelectronics, including nanowires and nanosheets in Gate-All-Around FETs (GAAFETs).

Advanced Planar Techniques with Innovative Materials: For 2D materials (such as $MoTe_2$ or MoS_2), large-scale growth techniques that align with conventional silicon technology are being developed to produce homogeneous CMOS inverter arrays on a single wafer.

By utilizing these M3D techniques, the industry can shorten interconnect lengths to the nanoscale, reducing delays and power consumption while enabling new capabilities, such as integrated photonics and neuromorphic hardware, on a compact, high-density chip.

The predominant method for heterogeneous integration in post-CMOS nanoelectronics is seen as 3D integration, especially advanced techniques such as wafer bonding and hybrid bonding. This technique facilitates the vertical assembly and connection of various materials and components into a compact system. Rather than relying on a single approach, the field of heterogeneous integration employs a suite of sophisticated integration strategies to combine conventional silicon CMOS with innovative "beyond-CMOS" materials and devices.

Wafer Bonding/Hybrid Bonding: This fundamental technique is essential for building vertically stacked, 3D integrated systems. It involves face-to-face or face-to-back bonding of two or more wafers (or dies), often followed by substrate removal and the creation of high-density, low-parasitic vertical interconnects, such as TSVs.

Micro-Transfer Printing: This technique allows for the accurate positioning of prefabricated, ultra-thin layers of non-silicon materials (such as III–V semiconductors, 2D materials, or optical components) onto a finished or partially finished CMOS wafer. This is vital for incorporating materials that cannot withstand standard high-temperature silicon fabrication processes.

In Situ Fabrication/Epitaxy: In this technique, specific novel materials are directly grown or deposited on top of the CMOS wafer after standard processing is complete. This offers precise alignment and close on-chip connections, helping minimize parasitic effects.

Advanced Packaging (Chiplets): The chiplet approach employs advanced packaging technologies (such as fan-out wafer-level packaging) to integrate multiple smaller, specialized dies, horizontally (2D/2.5D integration) or vertically, into a single package. This provides the versatility to combine various functions (like memory, logic, sensors, and photonics) that may be manufactured using different optimal processes.

These techniques are essential because they enable the integration of new features (such as spintronics, photonics, novel memory devices, and quantum components) that cannot be realized with silicon alone, thereby addressing the physical and economic limitations of traditional CMOS scaling.

Having surveyed the latest breakthroughs in materials, logic, and memory devices, computing architectures, fabrication, and packaging, we are now well-versed in current achievements and state-of-the-art, and eager to enquire what the future holds in the next chapter.

REFERENCES

Aasmundtveit K. E., A. Roy and B. Q. Ta 2020 Direct integration of carbon nanotubes in CMOS, towards an industrially feasible process: A review, *IEEE Transactions on Nanotechnology*, 19: 113–122, https://doi.org/10.1109/TNANO.2019.2961415

Abdullah M. F. and H. W. Lee 2024 Technology review of CNTs TSV in 3D IC and 2.5D packaging: Progress and challenges from an electrical viewpoint, *Microelectronic Engineering*, 290: 112189, https://doi.org/10.1016/j.mee.2024.112189

Abrams N. C., Q. Cheng, M. Glick, M. Jezzini, P. Morrissey, P. O'Brien and K. Bergman 2020 Silicon photonic 2.5D Multi-chip module transceiver for high-performance data centers, *Journal of Lightwave Technology*, 38(13): 3346–3357, https://doi.org/10.1109/JLT.2020.2967235

Bao S., Y. Wang, K. Lina, L. Zhang, B. Wang, W. A. Sasangka, K. E. K. Lee, S. J. Chua, J. Michel, E. Fitzgerald, C. S. Tan and K. H. Lee 2021 A review of silicon-based wafer bonding processes, an approach to realize the monolithic integration of Si-CMOS and III–V-on-Si wafers, *Journal of Semiconductors*, 42(2): 023106, https://doi.org/10.1088/1674-4926/42/2/023106

Batude P., T. Ernst, J. Arcamone, G. Arndt, P. Coudrain and P.-E. Gaillardon 2012 3-D sequential integration: A key enabling technology for heterogeneous co-integration of new function with CMOS, *IEEE Journal on Emerging and Selected Topics in Circuits and Systems*, 2(4): 714–722, https://doi.org/10.1109/JETCAS.2012.2223593

Bobba S., P.-E. Gaillardon, C. Seiculescu, V. F. Pavlidis and G. De Micheli 2013 3.5-D integration: A case study, *2013 IEEE International Symposium on Circuits and Systems (ISCAS)*, Beijing, China, 19–23 May, pp. 2087–2090, https://doi.org/10.1109/ISCAS.2013.6572285

Bottoms C. M., R. C. Johnson, R. J. Lallement, C. W. Wong, A. R. Hubbard, S. N. Chowdhury and J. H. Rankin 2025 Multi-reticle stitching: Applications from packaging to high-NA EUV, *Proceedings SPIE 13424, Optical and EUV Nanolithography XXXVIII*, San Jose, CA, USA, 22 April, p. 1342400, https://doi.org/10.1117/12.3051542

Chaourani P. 2019 Sequential 3D Integration – Design Methodologies and Circuit Techniques, Doctoral dissertation, KTH Royal Institute of Technology, Stockholm, Sweden.

Chen Z., J. Zhang, S. Wang, C.-P. Wong 2024 Challenges and prospects for advanced packaging, *Fundamental Research*, 4(6): 1455–1458, https://doi.org/10.1016/j.fmre.2023.04.014

Datta S., S. Dutta, B. Grisafe, J. Smith, S. Srinivasa and H. Ye 2019 Back-end-of-line compatible transistors for monolithic 3-D integration, *IEEE Micro*, 39(6): 8–15, https://doi.org/10.1109/MM.2019.2942978

Downey B. P., S. Mack, A. Xie, D. Scott Katzer, A. C. Lang, J. G. Champlain, Y. Cao, N. Nepal, T. A. Growden, V. J. Gokhale, M. T. Hardy, E. Beam, C. Lee and D. J. Meyer 2023 Micro-transfer printing for heterogeneous integration of GaN and GaAs HEMTs, *IEEE Transactions on Electron Devices*, 70(6): 2994–3000, https://doi.org/10.1109/TED.2023.3269006

Gao G., L. Mirkarimi, G. Fountain, D. Suwito, J. Theil, T Workman, C. Uzoh, B. Lee, K. M. Bang and G. Guevara 2022 Die to wafer hybrid bonding for chiplet and heterogeneous integration: Die size effects evaluation-small die applications, *2022 IEEE 72nd*

Electronic Components and Technology Conference (ECTC), San Diego, CA, USA, 31 May–3 June, pp. 1975–1981, https://doi.org/10.1109/ECTC51906.2022.00310

Gill A. A. S. and S. S. Ghoreishizadeh 2023 Advances in nanomaterials integration in cmos-based electrochemical sensors: A review, *IEEE Sensors Journal*, 23(5): 4659–4671, https://doi.org/10.1109/JSEN.2023.3236068

Gopireddy B. and J. Torrellas 2019 Designing vertical processors in monolithic 3D, *2019 ACM/IEEE 46th Annual International Symposium on Computer Architecture (ISCA)*, Phoenix, AZ, USA, 22–26 June, pp. 643–656. https://doi.org/10.1145/3307650.3322233

Gowrisanker S., M. A. Quevedo-Lopez, H. N. Alshareef, B. E. Gnade, S. Venugopal, R. Krishna, K. Kaftanoglu and D. R. Allee 2009 A novel low temperature integration of hybrid CMOS devices on flexible substrates, *Organic Electronics*, 10(7): 1217–1222, https://doi.org/10.1016/j.orgel.2009.06.012

Kafi N., S. Kang, C. Golz, A. Rodrigues-Weisensee, L. Persichetti, D. Ryzhak, G. Capellini, D. Spirito, M. Schmidbauer, A. Kwasniewski, C. Netzel, O. Skibitzki and F. Hatami 2024 Selective growth of GaP crystals on CMOS-compatible Si nanotip wafers by gas source molecular beam epitaxy, *Crystal Growth & Design*, 24(7): 2724–2733, https://doi.org/10.1021/acs.cgd.3c01337

Kang T., J. Park, S. Y. Lee, H. Jung, J. Jeon, Y. H. Kim and S. Lee 2025 High-κ dielectric van der Waals integration on 2D semiconductors for three-dimensional complementary logic systems, *Nature Communications*, 16: 11648, https://doi.org/10.1038/s41467-025-66770-0

Karasulu B., F. Roozeboom and A. Mameli 2023 High-throughput area-selective spatial atomic layer deposition of SiO_2 with interleaved small molecule inhibitors and integrated back-etch correction for low defectivity, *Advanced Materials*, 35(25): 2301204, https://doi.org/10.1002/adma.202301204

Katiyar A.K., J. Choi and J. H. Ahn 2025 Recent advances in CMOS-compatible synthesis and integration of 2D materials, *Nano Convergence*, 12: 11, https://doi.org/10.1186/s40580-025-00478-1

Kazior T. E. 2014 Beyond CMOS: Heterogeneous integration of III–V devices, RF MEMS and other dissimilar materials/devices with Si CMOS to create intelligent microsystems, *Philosophical Transactions of the Royal Society A*, 372: 20130105, https://doi.org/10.1098/rsta.2013.0105

Kim H.-B., J.-M. Lee, D. Sung, J.-H. Ahn, W.-H. Kim 2024 Highly area-selective atomic layer deposition of device-quality $Hf_{1-x}Zr_xO_2$ thin films through catalytic local activation, *Chemical Engineering Journal*, 488: 150760, https://doi.org/10.1016/j.cej.2024.150760

Kong L., R. Wu, Y. Chen, Y. Huangfu, L. Liu, W. Li, D. Lu, Q. Tao, W. Song, W. Li, Z. Lu, X. Liu, Y. Li, Z. Li, W. Tong, S. Ding, S. Liu, L. Ma, L. Ren, Y. Wang, L. Liao, X. Duan and Y. Liu 2023 Wafer-scale and universal van der Waals metal semiconductor contact, *Nature Communications*, 14(1): 1014, https://doi.org/10.1038/s41467-023-36715-6

Lee G., Y. Oh, J. Hwang, S. W. Hong, S. Song, K. W. Shin, J. Lee, S. Yoon, D. Kyun Kim, H.-J. Kim-Lee, D. Kim, J. Kim, D.-C. Shin, M.-C. Yu, J. Lee, J.-Y. Park, H. M. Lee, S. W. Kim, B. Na, S. Yun, Y. Kim, J. Jeong, C. Park, K. Hwang and G. Yoo 2025 Scalable van Der Waals integration of III-N devices over 2D materials for CMOS-compatible architectures, *Advanced Materials*, 37(16): 2420060. https://doi.org/10.1002/adma.202420060

Liu T.-L., M. Harake and S. F. Bent 2023 Sequential use of orthogonal self-assembled monolayers for area-selective atomic layer deposition of dielectric on metal, *Advanced Materials*, 10(5): 2202134, https://doi.org/10.1002/admi.202202134

Lu M.-C. 2026 Manufacturing challenges of hybrid bonding for chiplets heterogenerous integration, *ASME, Journal of Electronic Packaging*, 148(1): 19, https://doi.org/10.1115/1.4068767

Mallik A., A. Vandooren, L. Witters, A. Walke, J. Franco, Y. Sherazi, P. Weckx, D. Yakimets, M. Bardon, B. Parvais, P. Debacker, B. W. Ku, S. K. Lim, A. Mocuta, D. Mocuta, J. Ryckaert, N. Collaert and P. Raghavan 2017 The impact of sequential-3D integration on semiconductor scaling roadmap, *2017 IEEE International Electron Devices Meeting (IEDM)*, San Francisco, CA, USA, 2–6 December, pp. 32.1.1–31.1.4, https://doi.org/10.1109/IEDM.2017.8268483

Mishra R.K., S. A. Thomas, D. S. Singh, J. Cherusseri, I. Chianella, H. Y. Nezhad and S. Goel 2024 Heterogeneous integration of 2D materials with silicon complementary metal oxide semiconductor (Si-CMOS) devices. In: Singh S., S. K. Sharma and D. Nandan (Eds.) *Beyond Si-Based CMOS Devices*. Springer Tracts in Electrical and Electronics Engineering. Springer, Singapore, pp. 149–179, https://doi.org/10.1007/978-981-97-4623-1_6

Muñoz-Gamarra J. L., A. Uranga and N. Barniol 2016 CMOS-NEMS Copper switches monolithically integrated using a 65 nm CMOS technology, *Micromachines (Basel)*, 7 (2): 30. https://doi.org/10.3390/mi7020030

Nakahata T., K. Sugihara, S. Maruno, Y. Abe and T. Ozekim 2002 Optimization of process conditions of selective epitaxial growth for elevated source/drain CMOS transistor, *Journal of Crystal Growth*, 243(1): 87–93, https://doi.org/10.1016/S0022-0248(02)01481-1

Naqi M., S. C. Jang, Y. Cho, J. M. Park, J. O. Oh, H. Y. Rho, H.-S. Kim and S. Kim 2023 Low temperature processed, highly stable CMOS inverter by integrating Zn-ON and tellurium thin-film transistors, *Journal of Information Display*, 24(3), 199–204. https://doi.org/10.1080/15980316.2023.2174195

Neumaier D., S. Pindl and M. C. Lemme 2019 Integrating graphene into semiconductor fabrication lines, *Nature Materials*, 18: 525–529. https://doi.org/10.1038/s41563-019-0359-7

Sandrini J., M. Thammasack, T. Demirci, P.-E. Gaillardon, D. Sacchetto, G. De Micheli and Y. Leblebici 2015 Heterogeneous integration of ReRAM crossbars in 180nm CMOS BEoL process, *Microelectronic Engineering*, 145: 62–65, https://doi.org/10.1016/j.mee.2015.03.011

Santra S., P. K. Guha, S. K. Ray, F. Udrea and J. W. Gardner 2013 SOI CMOS integrated zinc oxide nanowire for toluene detection, *2013 IEEE 5th International Nanoelectronics Conference (INEC 2013)*, Singapore, 2–4 January, pp. 119–121, https://doi.org/10.1109/INEC.2013.6465972

Shen W. W. and K. N. Chen 2017 Three-dimensional integrated circuit (3D IC) key technology: Through-silicon via (TSV), *Nanoscale Research Letters*, 12: 56, https://doi.org/10.1186/s11671-017-1831-4

Singh A., K. S. Sandha and M. K. Rai 2024 Investigation of optical interconnects for nanoscale VLSI applications, *Micro and Nanostructures*, 196: 207987, https://doi.org/10.1016/j.micrna.2024.207987

Singh P. and P. Viswanadham 2012 *Failure Modes and Mechanisms in Electronic Packages*, Springer US, Cham, pp. 370.

Srimani T., G. Hills, M. Bishop, C. Lau, P. Kanhaiya, R. Ho, A. Amer, C. Chao, A. Yu, A. Wright, A. Ratkovich, D. Aguilar, A. Bramer, C. Cecman, A. Chov, G. Clark, G. Michaelson, M. Johnson, K. Kelley, P. Manos, K. Mi, U. Suriono, S. Vuntangboon, H. Xue, J. Humes, S. Soares, B. Jones, S. Burack, Arvind, A. Chandrakasan, B. Ferguson, M. Nelson and M. M. Shulaker 2020 Heterogeneous integration of BEOL logic and memory in a commercial foundry: Multi-tier complementary carbon nanotube logic and resistive RAM at a 130 nm node, *2020 IEEE Symposium on VLSI Technology*, Honolulu, HI, USA, 16–19 June, pp. 1–2, https://doi.org/10.1109/VLSITechnology18217.2020.9265083

Vescan L., T. Stoica and E. Sutter 2007 Growth and properties of SiGe structures obtained by selective epitaxy on finite areas, *Applied Physics A*, 87(3): 485–490, https://doi.org/10.1007/s00339-007-3917-9

Wernicke T., B. Rebhan, V. Vuorinen, M. Paulasto-Kröckel, V. Dubey, K. Diex, D. Wünsch, M. Baum, M. Wiemer, S. Tanaka, J. Froemel, K. E. Aasmundtveit, H.-V. Nguyen and V. Dragoi 2024, Review—Recent developments in low temperature wafer level metal bonding for heterogenous integration, *ECS Journal of Solid State Science and Technology*, 13(10): 104005, https://doi.org/10.1149/2162-8777/ad8519

Wessely P. J., F. Wessely, E. Birinci, K. Beckmann, B. Riedinger and U. Schwalke 2012 Silicon-CMOS compatible in-situ CCVD grown graphene transistors with ultra-high on/off-current ratio, *Physica E: Low-dimensional Systems and Nanostructures*, 44(7–8): 1132–1135, https://doi.org/10.1016/j.physe.2011.12.022

Xu P., C. Marchese, G. Lepage, N. Golshani, R. Van Eenaeme, A. Mingardi, J. Van Ongeval, R. Magdziak, L. Halipre, D. Trivkovic, P. Verheyen, M. Chakrabarti, D. Velenis, A. Miller, F. Ferraro, Y. Ban and J. Van Campenhout 2024 Low-loss, multi-reticle stitched SiN waveguides for 300mm wafer-level optical interconnects, *2024 Optical Fiber Communications Conference and Exhibition (OFC)*, San Diego, CA, USA, 24–28 March, pp. 1–3.

Zhang J., L. Bogaert, C. Krückel, E. Soltanian, H. Deng, B. Haq, J. Rimböck, J. Van Kerrebrouck, G. Lepage, P. Verheyen, J. Van Campenhout, P. Ossieur, D. Van Thourhout, G. Morthier, W. Bogaerts and G. Roelkens 2023 Micro-transfer printing InP C-band SOAs on advanced silicon photonics platform for lossless MZI switch fabrics and high-speed integrated transmitters, *Optics Express*, 31(26): 42807–42821, https://doi.org/10.1364/OE.505112

Zhang L., C. Zhang, Z. Tan, J. Tang, C. Yao and B. Hao 2021 Research progress of microtransfer printing technology for flexible electronic integrated manufacturing, *Micromachines (Basel)*, 12(11): 1358, https://doi.org/10.3390/mi12111358

Zhou Y. and H. Xie 2018 Monolithic integration of carbon nanotubes and CMOS, In: Iniewski K. (Ed.), *Nano-Semiconductors: Devices and Technology*, CRC Press, Boca Raton, FL, USA, pp. 29–66, https://doi.org/10.1201/9781315217468-7

10 Future Outlook and Research Trends

10.1 INTRODUCTION

As we progressed through the chapters of this book, the holistic image of nanoelectronics that emerged was that researchers have explored alternative materials (such as high-mobility and wide bandgap semiconductors, graphene, and carbon nanotubes) and cutting-edge devices (like TFETs, NCFETs, FeFETs, spintronics, and quantum dots) in addition to CMOS nanoelectronics, to address the limitations of this technology, thereby facilitating quicker, smaller, and more energy-efficient computing by utilizing novel physical phenomena, including spin and tunneling, rather than solely relying on charge. The objective has been to create "complementary, not replacement," technologies to design solutions that enhance CMOS capabilities, establishing "Extended CMOS" for future generations of electronic systems. Now that we have gone through the achievements and possibilities of CMOS nanoelectronics, let's look ahead and imagine what the future might hold.

10.2 BOLTZMANN TYRANNY

At any given temperature T, thermal energy $k_B T$ leads to a broad energy distribution for electrons (the Boltzmann distribution), rather than a distinct cutoff. The expression $k_B T/q$ (where k_B represents Boltzmann's constant, T is the temperature, and q signifies the electron charge) denotes the thermal voltage, which is around 26 mV at room temperature. Subthreshold Swing (SS) quantifies the gate voltage V required to increase the drain current by a factor of 10 (1 decade).

The Constraint: Due to the thermal distribution, a voltage shift of roughly 2.3 $(k_B T/q)$ is necessary to surmount the barrier, resulting in approximately 60 mV/dec at room temperature; this phenomenon is referred to as "Boltzmann's Tyranny." (Nadeem et al. 2021, Chen et al. 2023, Dumé 2024).

For a typical transistor at room temperature (approximately 300 K), the essential Boltzmann limit—frequently known as "Boltzmann's Tyranny"—imposes two primary restrictions on switching energy:

Subthreshold Swing (Voltage Restriction): Standard field-effect transistors (FETs) operate via thermionic emission, where electrons must overcome an energy barrier to transition to the "on" state. Since the energy distribution of these electrons follows Boltzmann statistics, it requires only a slight change in gate voltage to alter the current by one order of magnitude (a "decade").

DOI: 10.1201/9781003753469-10

$$\mathrm{SS} = \left(\frac{k_{\mathrm{B}}T}{q}\right) \ln(10) = 60\ \mathrm{mV/dec} \tag{10.1}$$

The constraint is approximately 60 mV/dec at room temperature. Let us understand the energy implications: To achieve a substantial on/off ratio (typically 10^5), a transistor must utilize at least 300 mV of gate voltage. Given that switching energy is proportional to CV^2, this establishes a baseline for energy efficiency.

Information Erasure (Landauer Limit): At a more fundamental thermodynamic level, any irreversible logic operation (where information is lost or erased) is associated with a theoretical minimum energy dissipation known as the Landauer limit (Landauer 1991, Reeb and Wolf 2014, Bormashenko 2019): The constraint:

$$E_{\mathrm{Minimum}} = \left(\frac{k_{\mathrm{B}}T}{q}\right) \ln(2) \approx 2.9 \times 10^{-21}\mathrm{J} = 0.018\ \mathrm{eV} \tag{10.2}$$

at room temperature.

Current Situation: Contemporary transistors operate at energy levels approximately a million to a billion times higher than this theoretical limit.

Practical limits for "hyperscaling" in CMOS are estimated to be around 10 attojoules (aJ) per switch due to challenges in reducing voltage below the 500 mV Boltzmann threshold. Researchers are exploring alternative architectures to overcome "Boltzmann's Tyranny" by implementing switching mechanisms that do not depend on thermal injection across a barrier. Tunnel field-effect transistors (TFETs) utilize quantum tunneling to achieve an SS below 60 mV/dec. Negative-capacitance FETs (NC-FETs) integrate ferroelectric materials to enhance gate voltage internally. Cold-source FETs employ materials such as graphene to eliminate high-energy "hot" carriers.

Example 10.1 A next-generation nanoelectronic chip, designed to function beyond the limitations of CMOS technology, is targeted to achieve a power density requiring a heat dissipation rate of 1000 W/cm^2. Given that the chip has an area of 4 cm^2, we need to calculate the total power (in Watts) generated by the chip that must be dissipated as heat. Compute the total power produced.

Solution: One must multiply the power density by the chip's total area. Power density: 1000 W/cm^2, Chip area: 4 cm^2.

$$\text{Total power} = \text{Power density} \times \text{Chip area} = 1000\ \mathrm{Wcm^2} \times 4\ \mathrm{cm^2} = 4000\ \mathrm{W} = 4\ \mathrm{kW} \tag{10.3}$$

The total power generated by the chip that must be dissipated as heat is 4 kW (4000 W). This level of power dissipation (comparable to a small electric heater in a compact space) poses a significant challenge for thermal management in the future of nanoelectronics.

Example 10.2 A team of researchers is working on a new nanoelectronic switch that goes beyond traditional CMOS technology and works at 50°C. Their objective is to achieve a power consumption that is significantly lower than the fundamental Boltzmann limit for the energy required to switch a standard transistor, which is roughly $k_B T \ln(2)$ per switching event, where k_B is the Boltzmann constant $\left(1.381 \times 10^{-23}\ \mathrm{J/K}\right)$, and T is the temperature in Kelvin. Find the minimum energy (in Joules) that the researchers should strive to remain below per switching operation. Suppose a futuristic chip featuring one billion (1×10^9) beyond-CMOS devices could function exactly at the Landauer limit. What would be the total energy (in Joules) consumed during a single operation cycle in which every device is switched?

Solution: We must utilize the theoretical limit. First, we identify the values provided: Boltzmann constant k_B : 1.38×10^{-23} J/K, Temperature T : 273 + 50 = 323 K, and the Boltzmann limit formula:

$$E = k_B T \ln(2) = 1.38 \times 10^{-23}\ \mathrm{J/K} \times 323\ \mathrm{K} \times 0.69315 \tag{10.4}$$

which results in

$$E = 1.38 \times 323\ \mathrm{K} \times 0.69315 \times 10^{-23}\ \mathrm{J} = 308.9647 \times 10^{-23}\ \mathrm{J} \approx 3.09 \times 10^{-21}\ \mathrm{J} \tag{10.5}$$

Consequently, the minimum energy the researchers should target to stay below is about 3.09×10^{-21} J per switching event.
The total energy used in one operation cycle can be calculated as:

$$\begin{aligned}\text{Total energy} &= \text{Number of devices} \times \text{Energy per switch} = 1 \times 10^9 \times 3.09 \times 10^{-21}\ \mathrm{J} \\ &= 3.09 \times 10^{-12}\ \mathrm{J} = 3.09 \times 10^{-12} \times 10^{12}\ \mathrm{pJ} = 3.09\ \mathrm{pJ}\end{aligned} \tag{10.6}$$

This illustrates the significant disparity between the current energy consumption of devices and the theoretical minimum, which represents a key area of research.

Example 10.3 A new nanoelectronic device that goes beyond traditional CMOS technology is being developed to minimize the energy needed for each switching operation. One particular prototype features a gate capacitance C_G of 1×10^{-18} Farads (1 attofarad) and is designed to function with a target supply voltage V_{DD} of 0.2 V. Determine the minimum energy necessary to charge the gate capacitance during a single switching event.

Solution: We can use the energy stored in a capacitor formula:

$$E = \frac{1}{2} C_G V_{DD}^2 \tag{10.7}$$

In a typical switching cycle (when charging from a DC source), a comparable quantity of energy is lost as heat in the charging pathway. Consequently, the overall energy consumed from the power supply during one transition is:

$$E = \frac{1}{2} C_G V_{DD}^2 \times 2 = C_G V_{DD}^2 \tag{10.8}$$

The calculation proceeds as follows:

$$E = 1 \times 10^{-18}\ \mathrm{F}(0.2)^2\ \mathrm{V}^2 \tag{10.9}$$

resulting in

$$E = 1 \times 10^{-18}(0.2)^2\ \mathrm{FV}^2 = 0.04 \times 10^{-18}\ \frac{\mathrm{C}}{\mathrm{V}}\mathrm{V}^2 = 0.04 \times 10^{-18}\ \mathrm{CV} \tag{10.10}$$

leading to

$$E = 0.04 \times 10^{-18}\ \mathrm{C}\frac{\mathrm{J}}{\mathrm{C}} = 0.04 \times 10^{-18}\ \mathrm{J} \tag{10.11}$$

which equals 4×10^{-20} J. Thus, the minimum energy required for a single switching event in this prototype device is 40 zeptojoules (zJ). 1 zJ = 10^{-21} J
The room-temperature (300 K) thermal energy limit is

$$\begin{aligned} E &= k_B T \ln(2) = 1.38 \times 10^{-23}\ \mathrm{J/K} \times 300\ \mathrm{K} \times 0.69315 = 286.9641 \times 10^{-23}\ \mathrm{J} \\ &= 2.87 \times 10^{-21}\ \mathrm{J} \end{aligned} \tag{10.12}$$

The computed energy 4×10^{-20} J is approximately 4×10^{-20} J/2.87×10^{-21} J ~ 13.94 times greater than the room temperature thermal energy limit (2.87×10^{-21} J). The leading challenge researchers face in developing beyond-CMOS devices is creating a device capable of reliable switching at energy levels approaching or exceeding this thermal limit. This issue involves overcoming the effects of "Boltzmann's tyranny," which complicates maintaining a stable, logical state at very low voltages due to thermal fluctuations.

Example 10.4 A nanoelectronic device for beyond-conventional CMOS technology is being designed to take the place of a standard silicon-based transistor that operates at a 10 nm scale. The main objective of this new device is to reduce energy consumption per operation to no more than 1/10 that of the traditional device, as the significant heat generation poses a key challenge to conventional scaling. Assuming the leading-edge CMOS transistor at the 10 nm scale requires an average of 1 femtojoule (fJ) for each switching action, calculate the highest permissible energy consumption for the new beyond-CMOS device to achieve its target.

Solution: The goal for energy consumption is set at 1/10th of that of the traditional device, which consumes 1 fJ. This gives a target consumption of 1 fJ $(1/10) = 0.1$ fJ.

To convert this, we note that 1 fJ equals 1000 attojoules (aJ). Thus, the maximum permissible energy consumption (in aJ) is calculated as $0.1\ \text{fJ} \times 1000\ \text{aJ}/\text{fJ} = 100\ \text{aJ}$. Therefore, the new beyond-CMOS device must limit its energy consumption to no more than 100 aJ per switching operation to address the power efficiency challenge.

10.3 OVERLAY ERROR

An overlay error is the positional misalignment between two consecutive layers during semiconductor manufacturing. It represents the vector offset between the center of a pattern on the uppermost layer and the corresponding pattern on the layer below. It is a significant factor affecting yield, resulting from tool deficiencies (such as lithography scanners and masks), process fluctuations (including wafer stress and film effects), and handling. This error is assessed through optical or electronic measurements and corrected using advanced metrology techniques and machine learning to ensure device functionality. As device feature dimensions decrease to the nanometer range (e.g., 5 and 3 nm nodes), the acceptable limits for these errors tighten considerably, often requiring precision within 20%–30% of the critical dimension (CD). Mastering overlay control is crucial for producing modern, intricate chips, relying on enhanced tools, more intelligent software, and a deeper understanding of processes.

Example 10.5 For an upcoming nanoelectronic chip with a designated critical dimension (CD) of 5 nm, the industry norm for permissible overlay error (the accuracy of alignment between various patterned layers) is generally a fraction of the CD. If the maximum permissible overlay error is defined as 20% of the critical dimension, what is the upper limit of acceptable overlay error in nanometers (nm)?

Solution: Identifying the provided values: Target critical dimension (CD) = 5 nm, and Maximum permissible overlay error percentage = 20% (or 0.20 in decimal form) Determining the maximum allowable overlay error:

$$\begin{aligned}\text{Overlay error} &= \text{CD} \times \text{Maximum permissible overlay error in decimal form}\\ &= 5\ \text{nm} \times 0.20 = 1\ \text{nm}\end{aligned} \tag{10.13}$$

The upper limit for acceptable overlay error is 1 nm. This minimal tolerance underscores a significant challenge in advanced nanofabrication, as reliable nanometer-level alignment across an entire wafer requires highly advanced, precise lithography equipment.

Example 10.6 A key issue in metal-organic chemical vapor deposition (MOCVD) for next-generation materials such as GaN for high-power applications or 2D transition-metal dichalcogenides for ultra-thin layers is achieving consistent layer thickness across large wafers. If an MOCVD process aims to produce a monolayer film with an average thickness of 1 nm, and current studies indicate a typical thickness variation (standard deviation, 1σ) of 0.3 nm across a 300 mm wafer, what is the percentage variation (1σ as a percentage of the target thickness) in the film thickness? Does this satisfy the strict requirements for the reliability of nanoscale devices?

Solution: The percentage variation is determined as:

$$\%\text{Variation} = \frac{\text{Standard deviation}}{\text{Target thickness}} \times 100\% = \frac{0.3 \text{ nm}}{1 \text{ nm}} \times 100\% = 30\% \qquad (10.14)$$

Clarification of the Challenge: A 30% variation in thickness presents a significant hurdle for the production of reliable beyond-CMOS nanoelectronics. In traditional high-volume manufacturing, particularly as dimensions progress into the nanometer scale, device parameters become extremely sensitive to such variations. This degree of non-uniformity across the wafer can result in:

Considerable Device-to-Device Discrepancies: Transistors located in different areas of the same wafer may exhibit markedly different performance characteristics (for instance, variations in threshold voltage and inconsistent on/off currents).

Diminished Yield and Reliability: Variable properties complicate the integration of these materials into intricate, high-density circuits with satisfactory manufacturing yields.

Enhancing this uniformity to a significantly lower, industry-acceptable level (generally $<5\%$ or even $<2\%$) is a primary research focus in MOCVD to facilitate the practical realization of these promising future technologies.

10.4 POISSON YIELD MODEL

In CMOS nanoelectronics, the Poisson yield model serves as a key statistical framework for estimating the "die yield," which indicates the likelihood that a produced integrated circuit (IC) is operational and devoid of defects (Yuan 2007, Praepattharapisut et al. 2014). The model postulates that defects are randomly and independently scattered across a semiconductor wafer. The probability Y of a die being defect-free (the yield) is defined as:

$$Y = Y_0 \exp\left(-AD_0\right) \qquad (10.15)$$

where the ultimate die yield Y_0 is a non-random (systematic) yield component that reflects losses unrelated to the chip's area (e.g., process-wide failures). A is the critical area of the die—the particular space where a defect would lead to a functional failure. D_0 is the average defect density, generally quantified in defects per unit area (for instance, cm^2).

The Poisson model assumes that defects are "points" distributed according to a Poisson distribution. It is most effective when defects are evenly distributed, and their sizes are minor compared to the wafer's size. Sensitivity to scale: As CMOS technology progresses into the nanometer range, the "critical area" becomes increasingly intricate. Even the displacement of a single atom or the presence of a minuscule particle can lead to a short or an open circuit in densely packed nano-scale layouts. The Poisson model serves as the statistical foundation, yet contemporary semiconductor manufacturing employs more advanced models to better address the complex defect patterns in advanced CMOS technologies.

Example 10.7 A new nanoelectronic device, classified as "beyond-CMOS," shows the potential for enhanced performance but encounters production issues tied to defect density. During an initial fabrication attempt, a specific defect appears randomly on a 100 mm by 100 mm test wafer. The average defect density is recorded at 2 defects per square centimeter (defects/cm^2). To achieve a manufacturing yield of at least 80% for operational chips, it is essential to determine the maximum permissible area for each functional chip (in mm^2). Determine this area using the Poisson yield model.

Solution: The formula of the Poisson yield model is given as:

$$\text{Yield} = \exp\left(-\text{Area} \times \text{Defect density}\right) \tag{10.16}$$

The first step is to identify the provided values: Target yield: 80% = 0.8, Defect count: 2 defects/cm^2.
Next, we need to rearrange the formula to solve for Area (A). Taking the natural logarithm of the yield, we get

$$\ln\left(\text{Yield}\right) = -\text{Area} \times \text{Defect density} \tag{10.17}$$

After substituting the values into the formula, it is found that the upper limit for a single functional chip's area to satisfy the 80% yield requirement is approximately 11.157 mm^2.

$$\begin{aligned}\text{Area} &= -\frac{\ln\left(\text{Yield}\right)}{\text{Defect density}} = -\frac{\ln\left(0.8\right)}{2\ \text{defects/cm}^2} = -\frac{-0.22314}{2}\ \text{cm}^2 = 0.11157\ \text{cm}^2 \\ &= 0.11157 \times 100\ \text{mm}^2 = 11.157\ \text{mm}^2\end{aligned} \tag{10.18}$$

The problem underscores the difficulty of achieving smaller chip dimensions to reduce the impact of high defect rates typically associated with early phases of beyond-CMOS fabrication techniques.

Example 10.8 Atomic Layer Deposition
A significant challenge in the incorporation of new materials for devices beyond CMOS technology is attaining uniformity in film thickness. If an ALD (atomic layer deposition) process targets a growth rate of 0.04 nm per cycle and the necessary thickness for a specific gate dielectric layer is 2.0 nm, what is the theoretical number of ALD cycles required?

Solution: To determine the number of cycles, the required film thickness must be divided by the growth rate per cycle:

$$\text{Number of cycles} = \frac{\text{Required thickness}}{\text{Growth rate per cycle}} = \frac{2.0\ \text{nm}}{0.04\ \text{nm per cycle}} = 50\ \text{cycles} \tag{10.19}$$

Therefore, 50 cycles are required to achieve the desired thickness of 2.0 nm. The related research obstacle: In practice, achieving this target uniformly across a complex, high-aspect-ratio 3D structure (such as a FinFET) is challenging due to variations in precursor diffusion, leading to issues like edge-placement inaccuracies and uneven film thickness across the chip.

Example 10.9 A 3D stacked integrated circuit (IC) is suggested for use in beyond-CMOS nanoelectronics, utilizing a process that involves the stacking and bonding of several individual dies (chips). If every individual chip has a known good die (KGD) yield of 95% (0.95), what is the overall yield of a module incorporating four such chips in a single stack, given that the failure of any one chip leads to the failure of the entire module?
Solution: To calculate the overall yield of the stacked module, we need to multiply the individual yields of all the chips:

$$\begin{aligned}\text{Overall yield} &= \text{Yield of chip 1} \times \text{Yield of chip 2} \times \text{Yield of chip 3} \times \text{Yield of chip 4} \\ &= 0.95 \times 0.95 \times 0.95 \times 0.95 = (0.95)^4 = 0.8145\end{aligned} \tag{10.20}$$

The total final yield of the four-chip stacked module is roughly 81.45%. This calculation demonstrates how rapidly the final yield can decrease as more components are combined, underscoring a crucial manufacturing challenge in advanced packaging. So, a significant challenge lies in the manufacturing yield of the completed assembly.

10.5 FURTHER EXPLORATION OF BEYOND-CMOS TECHNOLOGICAL ALTERNATIVES

10.5.1 Photonics and Optical Computing as a Beyond-CMOS Technology

Photonics and optical computing stand at the edge of a new technological frontier, swapping out electrons for the speed and elegance of light (Kazanskiy et al. 2022). For tasks such as AI matrix multiplication, photons offer blazing-fast performance and efficiency. Yet, despite their brilliance, these technologies still wrestle with tough challenges in memory and miniaturization, especially when compared to the well-oiled CMOS industry.

10.5.1.1 Merits of Optical Computing

(i) Extremely High Speed and Low Latency: Because light races far ahead of electrical signals in copper, photonic systems can achieve sub-nanosecond latency and potentially process data up to a 1000 times faster than today's silicon chips.

(ii) Gigantic Parallelism: Photons glide past each other without interference, letting countless beams of light share the same space at once through different wavelengths. This unlocks a new level of parallel computing power.

(iii) Thermal Efficiency Restriction: Since photons encounter almost no resistance and produce little heat as they travel, photonic systems can sidestep the elaborate cooling systems that traditional CPUs require.

(iv) High Bandwidth: Optical channels act as superhighways for data, carrying far more information than their electrical counterparts. This makes them perfect for connecting data centers and powering massive AI training.

(v) Imperviousness to Interference: Photonic circuits shrug off electromagnetic interference and crosstalk, two obstacles that often trip up electronic circuits as they shrink to nanometer scales.

10.5.1.2 Demerits of Optical Computing

(i) The Memory Bottlenecks: Storing light efficiently remains out of reach. As a result, optical systems constantly convert signals back to electricity for memory, creating energy-hungry bottlenecks that erase much of their speed advantage.
(ii) Miniaturization Constraints: The wavelength of visible light (~400–700 nm) is much larger than the dimensions of modern transistors (currently at 3 nm or less). This makes optical components up to 1000 times larger than their electronic counterparts, limiting chip density.
(iii) High Signal Conversion Costs: In hybrid systems, flipping signals between light and electricity can gulp down as much as 30% of the system's energy, becoming a major drain on efficiency.
(iv) Flexibility Crunch: While CMOS chips can be easily reprogrammed with software, early optical processors were obstinately rigid, needing physical hardware swaps to change their calculations.
(v) Manufacturability and Economics: Photonic hardware still lacks a mature supply chain. Building these components demands extraordinary precision, driving up costs and keeping yields lower than in the seasoned silicon industry.

10.5.2 Superconducting Rapid Single Flux Quantum (RSFQ) Logic as a Beyond-CMOS Technology

Superconducting logic stands at the forefront of "beyond-CMOS" innovation, utilizing Josephson junctions (JJs) in place of traditional transistors to execute operations (Bairamkulov and De Micheli 2024). This technology promises dramatic leaps in speed and energy efficiency, yet it must overcome daunting challenges in scaling and the need for extreme cooling.

10.5.2.1 Positive Aspects of Superconducting Logic

(i) Blazing Speed: These circuits race at frequencies close to 770 GHz, switching in mere picoseconds. One flip-flop has already shattered records by hitting 750 GHz.
(ii) Unmatched Energy Efficiency: Harnessing materials with almost no electrical resistance, superconducting systems can be up to 100 times more efficient than cutting-edge CMOS. For exascale supercomputers, this means reduction of power use by as much as 500 times.
(iii) Lossless Connections: Superconducting microstrip lines transmit data across chips at nearly the speed of light, covering long distances with virtually zero distortion or loss.
(iv) True 3D Scalability: While CMOS struggles with heat in dense stacks, superconducting logic stays cool, allowing advanced 3D integration for next-generation chips.
(v) Quantum Compatibility: This technology serves as the backbone for controlling and reading qubits in today's superconducting quantum computers.

10.5.2.2 Negative Aspects and Challenges

(i) Cryogenic Requirements: Achieving superconductivity requires chilling these systems to near-absolute zero, a feat that demands costly, power-hungry cooling equipment.
(ii) Low Integration Density: Compared to CMOS, these systems pack far fewer components onto a chip. Bulky inductors in logic gates make miniaturization a stubborn challenge.
(iii) Frail Driving Ability: Josephson junctions generate only faint currents and voltages, often falling short of powering memory chips or connecting to devices outside the cryogenic realm.
(iv) Extreme Magnetic Sensitivity: Even a stray magnetic field can disrupt these circuits, threatening performance or causing total breakdown. Elaborate shielding and protective barriers become essential.
(v) Lack of Manufacturing Maturity: While CMOS technology races ahead, superconducting fabrication is still finding its footing. Feature sizes remain much larger, trailing far behind the cutting edge of silicon.

10.5.3 Quantum-dot Cellular Automata as a Beyond-CMOS Technology

Quantum-dot Cellular Automata (QCA) is a computational approach that abandons transistors entirely, using the precise arrangement of electrons to encode binary data rather than electrical current (Chen and Zhao 2022). As a leading contender in the race to move beyond CMOS technology, QCA sidesteps the barriers of traditional silicon, such as runaway heat and unwanted capacitance, that threaten to halt further miniaturization.

10.5.3.1 Core Mechanism and Architecture

In contrast to traditional transistors, which shuffle charges around, QCA stores information in the unique polarization of electrons within each cell. Picture a tiny square made of four quantum dots, each holding two extra electrons. These electrons, repelling each other, settle into opposite corners, forming two distinct polarization states: $P = -1$ for logic "0" and $P = +1$ for logic "1." Instead of current, QCA relies on the subtle dance of neighboring cells influencing one another to pass information. Its logic circuits are crafted from majority gates and inverters, the building blocks from which complex operations like addition and multiplication emerge.

10.5.3.2 Advantages Over CMOS

(i) Ultra-Low Power Consumption: With no current flowing between source and drain, these devices consume power so sparingly that their energy use is measured in mere millielectron volts.
(ii) Extreme Density of Integration: Each cell is just a few nanometers across, packing far more devices into a chip than traditional CMOS ever could.
(iii) High Operational Speed: In theory, QCA circuits could blaze along at terahertz frequencies, leaving today's technologies far behind.

10.5.3.3 Implementation and Research Status

Researchers are racing to refine layouts and tackle the physical challenges that stand between QCA and real-world applications:

(i) Clocking Schemes: Synchronization is achieved through a four-phase clocking mechanism (Switch, Hold, Release, Relax) that controls the potential barriers within cells to ensure unidirectional data flow.
(ii) Design Tools: Engineers rely on advanced simulation software to test and perfect logic designs long before any hardware is built.
(iii) Obstacles Faced: Despite its promise, QCA remains mostly experimental, held back by its sensitivity to manufacturing flaws, latency issues, and the absence of reliable quantum error correction in today's prototypes. At present, experimental metal-dot devices must operate at cryogenic temperatures, usually between 300 mK and 1 K. This chilling requirement stems from the relatively large size of the quantum dots in today's fabricated devices. But this technology aims to develop molecular designs that could one day operate seamlessly at room temperature.

10.5.4 Molecular Electronics as a Beyond-CMOS Technology

Molecular electronics, which uses single molecules as the building blocks of electronic devices, stands at the forefront of "Beyond CMOS" innovation as silicon nears its ultimate limits. Researchers have mapped out three inventive strategies for assembling molecular circuits: directly connecting a molecule to electrodes, physically linking two molecule–electrode units, and chemically joining these components (Li et al. 2023).

While molecular electronics dazzles with its potential for radical miniaturization and a spectrum of new functions, it must overcome significant challenges in reliability and large-scale production. For now, its most promising role may be in hybrid systems, where molecules deliver specialized capabilities like ultra-dense memory or advanced sensing, rather than fully supplanting silicon.

10.5.4.1 Advantages of Molecular Electronics

(i) Realizing Ultimate Miniaturization: Molecules mark the edge of what is possible in shrinking technology, promising device densities so high they could pack a thousand times more components than today's silicon chips into the same space.
(ii) High Energy Efficiency: Molecular switches consume energy at astonishingly low levels, decreasing the heat and wasted power that haunt current chip technology.
(iii) Capability for Bottom-Up Fabrication: Instead of costly top-down methods, molecular devices can build themselves from the ground up through self-assembly and chemistry, promising lower costs and gentler resource use.
(iv) Offering Functional Diversity: By customizing a molecule's shape and makeup, scientists unlock smart features like sensing, quantum tricks, and spin control that traditional semiconductors can only dream of.

(v) Biocompatibility Benefit: Many molecular systems, including DNA and protein circuits, blend seamlessly with living tissue, paving the way for next-generation medical implants and biosensors.

10.5.4.2 Shortcomings of Molecular Electronics

(i) Unpredictability of Performance and Fragility of Connections: Creating a dependable link between a single molecule and large-scale electrodes is like trying to balance a marble on a pinhead. Even the tiniest change at the contact point can send device performance swinging wildly from one attempt to the next.
(ii) Dealing with Interconnection Challenges (The 5Cs): Imagine trying to choreograph billions of tiny dancers, each with their own quirks. Bringing them together into a seamless performance means wrestling with the "5Cs": concatenation, connectivity, crosstalk, compatibility, and cost-effectiveness.
(iii) Undesirably High Environmental Sensitivity: Molecules act like delicate instruments, easily thrown off tune by their environment. Shifts in temperature, humidity, oxygen, or light can quickly degrade their electronic properties or trigger unwanted reactions.
(iv) Imposition of Fan-out and Signal Gain Limits: Many molecular circuits stumble when asked to juggle multiple outputs or maintain crisp, clear signals. These hurdles make it difficult to implement the complex logic required for advanced computing.
(v) Difficulty in Integration with Existing Technology: Connecting the tiny world of molecular devices to the much larger realm of CMOS technology is like building a bridge between two distant islands. The engineering challenges at this interface are still far from solved.

10.5.5 Valleytronics Logic as a Beyond-CMOS Technology

Valleytronics logic is a futuristic alternative to silicon, applying the hidden valleys of electrons in 2D materials, such as TMDs, to enable fast, energy-efficient, and ultra-compact data processing (Schaibley et al. 2016). It uses the valley index—the unique energy peaks and valleys within a material's band structure—to encode and process information. By tapping into the momentum of electrons, not just their charge or spin, valleytronics promises to usher a new era of ultra-efficient computing.

10.5.5.1 Upsides of Valleytronics Logic

(i) High Energy Efficiency: Valleytronics can potentially eliminate the power-hungry nature of traditional electrical interconnects. Valleytronic operations could reduce energy use, requiring as little as 30 meV, compared with roughly 900 meV demanded by today's technology.
(ii) Ultrafast Processing Ability: Valleytronics delivers fast data speeds, rivaling spintronics but with the potential edge of greater stability, thanks to its reliance on quantum waves rather than fragile magnetic spins.
(iii) Reversible Computing Possibility: By storing extra input information in valley states, valleytronics paves the way for logically reversible gates, sidestepping the need for complex, energy-intensive circuits.

(iv) Superior Stability: While spintronics can falter under external magnetic fields, valleytronics draws its strength from electron momentum, making it less prone to interference and more robust overall.
(v) Multi-Valued Logic Facilitation: Valleytronics breaks free from binary limits. By leveraging two distinct valleys and a "null" state, it can enable quaternary or even more complex logic systems.

10.5.5.2 Downsides of Valleytronics Logic

(i) Short Coherence Time Limitations: One of the biggest obstacles is the fleeting valley-polarization lifetime at room temperature, which often lasts only picoseconds or nanoseconds, making reliable information processing a real challenge.
(ii) Requirement of Extreme Operating Conditions: Most experiments demand frigid, ultralow temperatures to keep delicate valley signals from being overwhelmed by bulk currents.
(iii) Manufacturing Difficulties: Scaling up the production of flawless, atomically thin materials like transition-metal dichalcogenides is a major challenge. Even tiny defects or grain boundaries can seriously undermine performance.
(iv) Control Challenges: Steering electrons into specific valleys is no easy feat, often requiring intricate triggers such as circularly polarized lasers or finely tuned mechanical strain.
(v) Integration Concerns: Building reliable electrical contacts and integrating valleytronic materials with standard silicon electronics remain stubborn obstacles.

10.5.6 Field-Emission Vacuum Nanoelectronics as a Beyond-CMOS Technology

Field-emission vacuum nanoelectronics (FEVN) is a cutting-edge "Beyond CMOS" technology that replaces the solid-state channel of conventional transistors with a nanoscale vacuum gap (Han et al. 2020). Applying the power of quantum tunneling, electrons leap from a razor-sharp cathode into a tiny vacuum or air-filled gap, yielding performance that pushes past the boundaries of traditional silicon semiconductors.

10.5.6.1 Core Advantages Over CMOS

(i) Ballistic Transport of Carriers: While CMOS electrons are slowed by constant collisions with the crystal lattice, electrons in a vacuum channel race forward without obstacles. This enables astonishing speeds from tens to hundreds of GHz, with the tantalizing possibility of reaching petahertz-scale performance.
(ii) Resilience Toward Harsh Environments: Vacuum shrugs off the threats that plague solid-state devices, resisting aging, extreme heat, and ionizing radiation. FEVN thrives where CMOS falters, making it a natural fit for aerospace, nuclear, and military missions.
(iii) High Breakdown Fields: Without a solid-state barrier, these devices can withstand much higher voltages, enabling powerful, high-frequency performance in a remarkably compact form.

(iv) Capability of Low Voltage Operation: Thanks to nanofabrication, these devices feature gaps smaller than 50 nm, allowing them to run on drive voltages as low as 0.24–2 V—matching the efficiency of the most advanced CMOS technology.

10.5.6.2 Key Implementation Challenges

(i) Lack of Complementary Logic: One of the biggest obstacles is that vacuum devices operate only with electrons, making them inherently unipolar. In contrast, traditional CMOS technology depends on both N-type and P-type transistors to keep power usage low. Inventing a true "P-type" counterpart for vacuum systems remains an unsolved puzzle, though some inventive designs are experimenting with nanoelectromechanical actuation to emulate this crucial function.
(ii) Reliability and Lifetime Issues: Maintaining field-emission stability over time is a real challenge. Cathodes wear down, current can become noisy, and even tiny traces of leftover gases or water can throw the whole system off balance.
(iii) Fabrication Scalability Problems: Researchers are testing CMOS-compatible methods with advanced materials such as graphene and gallium nitride, but getting every emitter in a large array to perform consistently remains a tough technical challenge.

10.6 DISCUSSION AND CONCLUSIONS

10.6.1 CMOS Remains the Mainstream Technology

The outlook for nanoelectronics beyond CMOS is promising, emphasizing innovative materials such as graphene and quantum dots, new devices including memristors and quantum dots, and unconventional computing architectures like neuromorphic and quantum systems to surpass traditional scaling limitations for improved efficiency and additional functionalities, facilitating advancements ranging from devices to sophisticated AI hardware (Dragoman and Dragoman 2021, Cresti 2023, Dhanaselvam and Balasubramanian 2025). However, CMOS is expected to maintain its lead in cost-effective general computing, while these specialized beyond-CMOS innovations supplement it. The trajectory of "Beyond CMOS nanoelectronics" is marked by a shift from merely miniaturizing silicon transistors to an integrated approach that incorporates novel materials and non-von Neumann architectures. While conventional CMOS continues to be the predominant framework, "Beyond CMOS" technologies are increasingly viewed as crucial enhancers for specific applications such as AI and neuromorphic computing.

The present environment for these emerging technologies encompasses a broad field. Steep-slope switches such as tunnel FETs (TFETs) and negative capacitance FETs (NC-FETs) are engineered to overcome the "60 mV/dec" physical limitation of CMOS, enabling much lower power consumption. Revolutionary next-generation devices are diverting beyond traditional charge-based logic to leverage new information carriers such as electron spin, photon polarization, and molecular states.

Spintronic approach harnesses the spin of electrons, rather than solely their charge, to manage and store data, providing non-volatile and energy-efficient solutions for memory and logic. Advancements in 1D carbon nanotubes (CNT) devices and 2D material devices based on graphene, MoS_2, and WSe_2 and are being pursued for their exceptionally high charge carrier mobility and potential replacement of silicon in high-performance logic applications.

Developments in beyond-CMOS nanoelectronics are promoting a departure from the conventional "von Neumann" design, which separates memory and processing toward more effective computing paradigms. Neuromorphic computing method employs memristors and other synaptic devices to emulate the structure of the human brain, delivering substantial efficiency improvements for AI applications. By leveraging quantum dots and superconducting devices, quantum computing paradigm aims to tackle complex problems that classical computers cannot solve. Stochastic computing technique exploits the inherent randomness of nanoscale devices such as low-barrier magnetic tunnel junctions to implement specialized probabilistic algorithms.

10.6.2 Advanced Packaging as a Feasible Route to Moore's Law Scaling for Dimensional Reduction

Instead of completely replacing CMOS, the "Bright Future" lies in strategic homogeneous and heterogeneous integration and "more than Moore" approaches. Advanced packaging enables chip size reduction, offering a fresh path beyond traditional Moore's law scaling, enabling smaller chip footprints, and increasing functional density. By stacking chips vertically, manufacturers reclaim valuable circuit board space. This approach also lets different chip types—logic, memory like HBM, and even optical components—sit side by side in a single package. Large chips can be split into smaller, mix-and-match chiplets, making them easier to produce. Technologies such as hybrid bonding and micro-bumps bring chips closer together, shortening signal paths and enabling more compact layouts. Wafer-level and panel-level packaging further miniaturize systems by spreading connections across larger areas or stacking more dies together.

It is re-emphasized that traditional packaging relies on large solder balls spaced far apart, while advanced methods use tiny micro-bumps and silicon interposers to squeeze more connections into tighter spaces. These miniature bumps shrink the chip's footprint and reduce unwanted electrical effects, boosting both speed and power efficiency. By using micro-bumps with pitches as small as 10–20 μm, advanced packaging dramatically increases connection density, enabling more compact chips.

Small chip sizes improve yield and lower manufacturing costs. Smaller chip sizes are a game-changer for the semiconductor industry, driving up manufacturing yield while slashing costs. By embracing chiplet architectures, manufacturers can sidestep the risk of defects that can ruin entire chips, since smaller dies are less likely to harbor critical flaws. This means more chips can be carved from each wafer, with fewer losses to defects or wasted silicon at the edges. The result is a higher yield and a lower cost per chip, as more usable units emerge from every fabrication run. Although manufacturing smaller nodes requires greater expense and complexity, it enables the creation of far more working chips per wafer, driving down the cost per chip. By reducing chip sizes, manufacturers can pack more functional units onto a single wafer, considerably increasing efficiency. This cost reduction comes from two main advantages: fitting more chips onto each wafer and achieving a higher percentage of usable chips. Instead of relying on a single, sprawling chip that can be ruined by a single tiny flaw, advanced packaging takes a modular approach, assembling smaller, more reliable components to improve overall yield.

We have noted under homogeneous integration, advanced heteroepitaxy enables the growth of pristine, single-crystal layers even when their atomic lattices do not match, using 2D interlayers or specialized growth methods to sidestep defects. M3D sequential integration improves chip stacking by building multiple device layers one atop another on the same wafer, using gentle temperatures to create ultra-dense, tightly connected structures. Selective material deposition offers pinpoint precision, laying down materials only where needed, which is crucial for crafting features smaller than 10 nm and streamlining the manufacturing process. Van der Waals integration assembles ultra-thin materials without the need for chemical bonds, relying on subtle atomic forces to unite even the most mismatched materials in a gentle, low-temperature embrace.

Heterogenous integration too is helpful in multifarious ways. BEOL integration weaves diverse components directly onto finished wafers during the final metallization, creating seamless connections. Wafer bonding uses oxygen plasma or hybrid copper bonding to unite entire wafers of different materials at low temperatures, forging dense vertical links. Chiplet and die-to-wafer assembly orchestrate independently crafted dies onto a single wafer, sidestepping reticle limits and boosting yields for intricate AI architectures. Micro-transfer printing employs a flexible stamp to lift thousands of tiny "coupons" from their source and accurately place them on unconventional substrates such as glass, plastic, or silicon. *In situ* integration on engineered substrates grows or constructs functional elements right within a custom platform, fine-tuning the interplay between varied materials. 2.5D interposers arrange chips side by side on a shared silicon base to bridge connections, while 3D integration stacks them vertically using through-silicon vias for ultra-short interconnects and unmatched density.

Co-optimization involves integrating advanced CMOS logic with specialized "Beyond CMOS" components such as non-volatile memory or sensors on the same chip to enhance overall system performance. On the whole, vertical scaling approaches focused on 3D spatial architectures and "more than Moore" strategies seek to boost functionality without further miniaturizing individual transistors. Figure 10.1 pictorially depicts the emerging futuristic landscape of this technology.

10.6.3 Silicon Photonics Facilitating the Miniaturization of Photonic Integrated Circuits

Silicon photonics is revolutionizing advanced packaging by merging optical and electrical components onto a single platform, harnessing the speed of light to shatter bandwidth barriers and slash power usage in data centers and AI systems

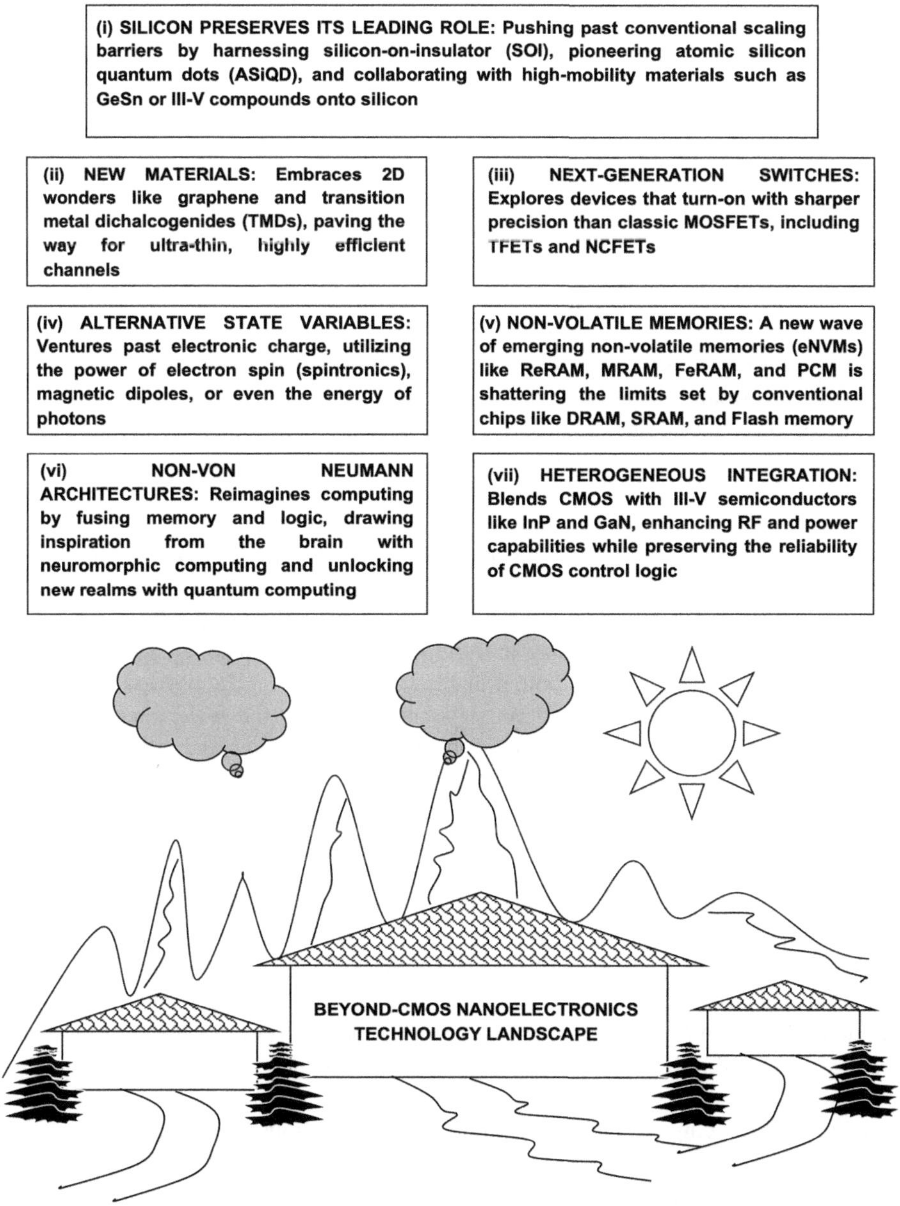

FIGURE 10.1 Beyond-CMOS technology landscape preserves silicon in its leading role and focuses on new materials, next-generation switches, alternative state variables, non-volatile memories, non-von Neumann architectures, and heterogeneous integration.

(Jalali and Fathpour 2006, Liu et al. 2025). As copper wiring reaches its limits for heat and speed, the industry is turning to photons rather than electrons to fuel the next wave of high-performance computing. This technology seamlessly links optical fibers to chips and connects electronic drivers to photonic circuits for lightning-fast data transfer. By enabling high-speed optical input and output, as well as co-packaged optics (CPO), silicon photonics is the backbone of dense, high-speed data transmission, where traditional copper simply cannot keep up. With CPO, photonics chips are now placed right beside GPUs or switches, dramatically cutting power draw and signal delays. As data traffic surges, silicon photonics can cut power consumption for data transmission by about half compared to electronics, easing bandwidth bottlenecks. It serves as a vital bridge between silicon-based electronic circuits and photonics, packing a multitude of optical and electro-optical functions into a compact footprint. However, challenges remain, such as integrating lasers directly onto silicon, managing high manufacturing costs, and achieving precise fiber-to-chip alignment. Advanced thermal management is also crucial, since silicon photonics is highly sensitive to temperature swings that can cause crosstalk and degrade performance in densely packed photonic integrated circuits (PICs).

Silicon photonics comes to life by integrating lasers, waveguides, and modulators directly onto a silicon canvas, using the same tried-and-true processes that build our everyday computer chips. This fusion lets engineers sculpt light at the tiniest scales, sending data racing along at speeds electricity can only dream of. The foundation of these chips lies in the silicon-on-insulator wafer, a sandwich of thin silicon, a glassy silicon dioxide layer, and a sturdy silicon base. Because silicon bends light more than silicon dioxide, it forms a perfect trap, guiding photons through hair-thin wires with total internal reflection. By tapping into the vast world of semiconductor manufacturing, silicon photonics becomes both scalable and affordable. The process begins with lithography, where light-sensitive coatings are applied to the wafer and exposed to intricate patterns that map out the future paths of light. Plasma etching then removes unwanted silicon, revealing the delicate rib-and-wire waveguides beneath. Implantation of elements like boron or phosphorus creates P–N junctions, the building blocks for active devices such as modulators. Protective layers of silicon nitride are added, shielding the circuit and keeping light on track. Since silicon struggles to detect or generate fast infrared signals, engineers get creative: lasers made from exotic materials like indium phosphide are bonded to the chip, while modulators act as lightning-fast shutters, flicking light on and off at breathtaking speeds. To convert light back into electrical signals, a thin layer of germanium is grown atop the silicon, absorbing infrared photons with ease. The final act is a feat of precision, as optical fibers must be aligned with the chip's microscopic waveguides to within a fraction of a nanometer. Ingenious techniques such as edge coupling and grating couplers help bridge the tiny gap between the chip and the fiber, completing the photonic masterpiece.

10.6.4 Beyond 3D Integration

Beyond 3D integration signals a leap into the future of chip packaging, where performance, density, and thermal efficiency reach new heights. This frontier includes techniques such as monolithic 3D (M3D), 3D-ICs enhanced with 2D materials, PICs and CPO, 4D heterogeneous integration, and cutting-edge substrate technologies. While traditional 3D-ICs stack finished chips, M3D builds multiple active silicon layers one after another on a single wafer, producing an unprecedented interconnect density at the transistor level. However, this approach brings its own set of manufacturing and thermal hurdles. By layering atomically thin 2D materials, such as MoS_2, instead of silicon, engineers can push scaling further, reduce power consumption, and introduce new functionalities that break free from silicon's constraints. Merging electronic ICs with photonic circuits in a single package promises to shatter the bandwidth barriers of electrical connections.

4D heterogeneous integration, still in its infancy, imagines packages that entwine advanced materials, sensors, and even time as a functional element, enabling chips that can reconfigure or adapt. In 4D printing, time becomes the fourth dimension, as printed objects morph their shape in response to external triggers, thanks to smart materials (Maher et al. 2025). While 3D printing builds objects that stay put, 4D printing brings them to life with movement and transformation. The real marvel lies in the material's memory, which lets it shift shape when it encounters triggers such as heat, cold, water, light, electricity, or magnetism. Imagine smart materials—shape-memory polymers, hydrogels, or liquid crystal elastomers—springing into action when the environment changes. This breakthrough lets us create electronics that adapt and evolve, something 3D printing could never do. Circuits can now fold or bloom into new forms, packing in more components or morphing antennas to tune into different frequencies. Wearable and implantable sensors flex and stretch with the body, always keeping a perfect connection, no matter how much we move. In the world of nano-robotics, tiny untethered robots powered by 4D-printed parts can swim through the body's smallest vessels or deliver medicine exactly where it's needed, triggered by a change in temperature or chemistry. Even batteries and supercapacitors are getting smarter, healing themselves or expanding and contracting as they charge and discharge, making them last longer and work better. Glass substrates are increasingly chosen over organic or silicon interposers to improve signal integrity and accommodate larger, more ambitious package designs.

In beyond-3D CMOS integration, time emerges as a vital manufacturing parameter, shaping the thermal, physical, and reliability boundaries of stacked devices. While the architecture rises in three dimensions, the temporal aspect acts as an invisible thread, interlacing through design and operation to influence system performance. Spatial stacking may lead the way, but time-dependent factors like reliability, aging, and lifecycle are equally essential. Advanced digital twin concepts bring this fourth dimension to life, using real-time sensors to create living, virtual replicas of semiconductor packages. These digital twins continuously mirror the physical state, simulating intricate interactions, diagnosing issues, predicting failures, and fine-tuning performance throughout the package's life from design to deployment. In reconfigurable circuits, time becomes a tool for transformation, allowing circuits to shift their function as needs evolve. Some beyond-CMOS innovations blend MEMS switches or III-V devices to craft intelligent circuits that adapt on the fly, e.g., 3D integration enables packaging timing modules, such as MEMS resonators, with ASICs, ensuring robust clock generation and long-term stability. Through these advances, engineers are building systems that can reshape themselves in real time to meet changing demands.

10.6.5 Simple Analogy to "Beyond CMOS Evolution"

The evolution of transportation is an easy-to-understand analogy to the progress of CMOS technology beyond CMOS status (Figure 10.2). CMOS is the sleek,

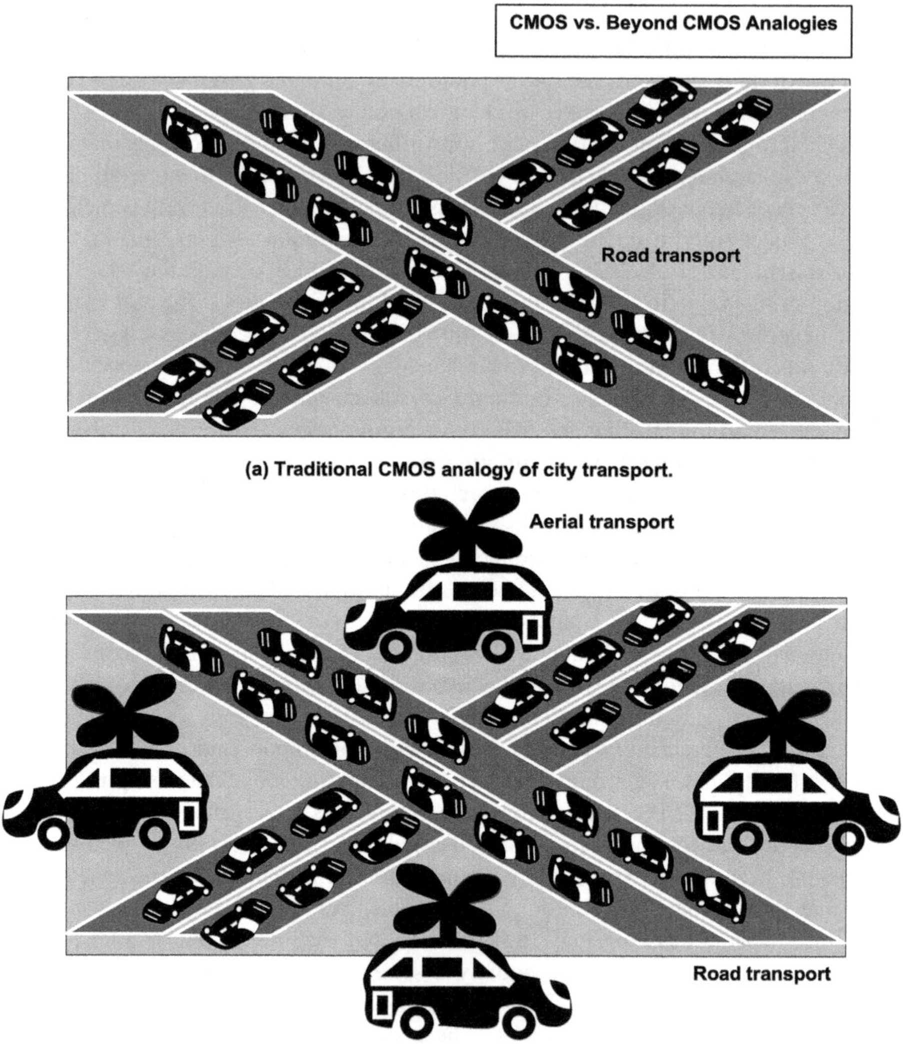

(a) Traditional CMOS analogy of city transport.

(b) Beyond CMOS analogy of city transport.

FIGURE 10.2 Simple analogies to explain the CMOS to beyond-CMOS transition: (a) CMOS is pictured as the optimized ground car, the ultimate modern car—sleek, swift, and efficient. Yet, no matter how advanced, it remains bound to the road, confined by the rules of physics. The leap from CMOS to beyond-CMOS occurs the moment cars take flight, breaking free from the limits of the road and soaring into a new realm of performance. (b) Beyond-CMOS is the true flying car, not just a faster car on the same road; it lifts off entirely, soaring above the traffic. By tapping into new realms of physics, such as spin or quantum states, it escapes the bottlenecks of heat and space that limit traditional designs. The future is hybrid. In tomorrow's world, we might drive our finely tuned ground car to a launchpad, then take to the skies. Standard CMOS will handle everyday tasks, while beyond-CMOS takes over for demanding jobs like AI or complex simulations, where the old road simply cannot keep up.

meticulously engineered car of today's highways. Decade after decade, we have tuned these cars to be swifter, lighter, and astonishingly efficient. At some point, shrinking the engine or slimming the tires any further makes the car wobble or overheat. This is the so-called scaling limit, where classical silicon technology runs up against the hard boundaries of physics and heat. Beyond-CMOS is the leap to wild new vehicles, like flying cars. The difference is that rather than simply souping up the car moving on the ground, it reinvents the journey itself. Instead of rolling on wheels powered by electricity, it might glide on magnetic spin (spintronics), criss-cross through brain-inspired circuits (neuromorphic computing), or jump across quantum dimensions. However, the challenge faced is just as flying cars demand fresh air traffic systems and new infrastructure, beyond CMOS calls for a complete reimagining of how we build and program our machines.

10.6.6 Overcoming Challenges En Route and Marching Forward

Despite the tremendous potential of beyond-CMOS technology, this field confronts significant obstacles to commercial viability. The process of transferring non-charge variables, such as spin, through conventional wires currently lags behind the movement of electrical charge, creating interconnect bottlenecks. Ensuring reliability, stability, and longevity at the atomic level remains a challenge for mass fabrication and manufacturing. Numerous theoretical models for atomic-scale devices still require extensive long-term statistical evidence for industrial-scale production due to a lack of experimental data. Optical computing, superconducting logic, quantum-dot-cellular automata, molecular electronics, valleytronics, and vacuum nanoelectronics promise impressive performance theoretically, but at present, they fall short in scalability, reliability, and manufacturing compatibility needed to dethrone CMOS at the sub-nanometer scale. Nonetheless, their importance as subsidiary technologies is laudable and cannot be overlooked.

Pushing through obstacles and forging ahead requires resilience, smart planning, and a mindset that turns hurdles into stepping stones for growth. To weather tough times, let us break daunting problems into manageable tasks, keep our eyes on the bigger picture, and embrace a growth mindset that turns setbacks into valuable lessons.

The March of Nanoelectronics
Undeterred by challenges, daunting and formidable,
Never in history has there been such upheaval.
As impossible goals become achievable,
With changes happening at an unprecedented pace,
Computers that once occupied a large space
Now need a very small place.
Running towards miniaturization,
Enabling new functionalities and applications,
Secure communication and reliable instrumentation,
At high speed and low power consumption,
Inexpensive nanoelectronics for all
Will make life comfortable and enjoyable —
More charming and joyful!

CMOS nanoelectronics, together with emerging non-CMOS technologies, are poised to shape a vibrant and enduring future, powering digital intelligence with groundbreaking integration and inventive packaging. This marks leapfrogging from the limitations of a one-size-fits-all chip design to the dynamic possibilities of a modular, high-density ecosystem.

REFERENCES

Bairamkulov R. and G. De Micheli 2024 Superconductive electronics: A 25-year review [feature], *IEEE Circuits and Systems Magazine*, 24(2): 16–33, https://doi.org/10.1109/MCAS.2024.3376492

Bormashenko E. 2019 Generalization of the Landauer principle for computing devices based on many-valued logic, *Entropy (Basel)*, 21(12): 1150, https://doi.org/10.3390/e21121150

Chen H. and L. Zhao, 2022 Quantum-dot cellular automata as a potential technology for designing nano-scale computers: Exploring the state-of-the-art techniques and suggesting the opportunities for the future, *Optik*, 265: 169431, https://doi.org/10.1016/j.ijleo.2022.169431

Chen J., Y.-Q. Zhu, X.-C. Zhao, Z.-H. Wang, K. Zhang, Z. Zhang, M.-Y. Sun, S. Wang, Y. Zhang, L. Han, X. Wu and T.-L. Ren 2023 PZT-enabled MoS_2 floating gate transistors: Overcoming Boltzmann tyranny and achieving ultralow energy consumption for high-accuracy neuromorphic computing, Nano Letters, 23 (22), 10196–10204, https://doi.org/10.1021/acs.nanolett.3c02721

Cresti A. (Ed.). 2023 *Beyond-CMOS: State of the Art and Trends*. ISTE Ltd and John Wiley & Sons, Inc., London, UK and Hoboken, NJ, USA, 448pp, https://doi.org/10.1002/9781394228713

Dhanaselvam P. S. and K. Balasubramanian 2025 Future prospective beyond CMOS technology design, In: Dhanaselvam, P. S., K. S. Rao, S. B. Rahi and D. S. Yadav (Eds.) *Field Effect Transistors*, Wiley, Hoboken, NJ, USA, pp. 73–80, https://doi.org/10.1002/9781394248506.ch5

Dragoman M. and D. Dragoman 2021 *Atomic-Scale Electronics Beyond CMOS*. Springer Nature Switzerland AG, Cham, 221pp, https://doi.org/10.1007/978-3-030-60563-6

Dumé, I. 2024 Vertical-nanowire transistors defeat the Boltzmann tyranny, *Physics World*, https://physicsworld.com/a/vertical-nanowire-transistors-defeat-the-boltzmann-tyranny/

Han J.-W., M.-L. Seol, J. Kim and M. Meyyappan 2020 Nanoscale complementary vacuum field emission transistor, *ACS Applied Nano Materials*, 3(11): 11481–11488, https://doi.org/10.1021/acsanm.0c02587

Jalali B. and S. Fathpour 2006 Silicon photonics, *Journal of Lightwave Technology*, 24(12): 4600–4615, https://doi.org/10.1109/JLT.2006.885782

Kazanskiy N. L., M. A. Butt and S. N. Khonina 2022 Optical computing: Status and perspectives, *Nanomaterials*, 12(13): 2171, https://doi.org/10.3390/nano12132171

Landauer R. 1991 Information is physical, *Physics Today*, 44(5): 23–29, https://doi.org/10.1063/1.881299

Li T., V. K. Bandari and O. G. Schmidt 2023 Molecular electronics: Creating and bridging molecular junctions and promoting its commercialization, *Advanced Materials*, 35(22): 2209088, https://doi.org/10.1002/adma.202209088

Liu Q., Y. Bian and J. Xiong 2025 Progress in passive silicon photonic devices: A review, *Photonics*, 12(9): 928, https://doi.org/10.3390/photonics12090928

Maher D., M. Piekarczyk, B. Rezaei and A.-I. Bunea 2025 Beyond 3D: Smart material systems for microscale 4D printing, *Advanced Materials Technologies*, 10(10): 2401950, https://doi.org/10.1002/admt.202401950

Nadeem M., I. Di Bernardo, X. Wang, M.S. Fuhrer and D. Culcer 2021 Overcoming Boltzmann's tyranny in a transistor via the topological quantum field effect, *Nano Letters*, 21(7): 3155–3161, https://doi.org/10.1021/acs.nanolett.1c00378

Praepattharapisut W., W. Pengchan, T. Phetchakul and A. Poyai, 2014 Yield analysis by Poisson yield model based on the defect analysis with derivative method," *2014 11th International Conference on Electrical Engineering/Electronics, Computer, Telecommunications and Information Technology (ECTI-CON)*, Nakhon Ratchasima, Thailand, 14–17 May, pp. 1–4, https://doi.org/10.1109/ECTICon.2014.6839864

Reeb D. and M. M. Wolf 2014 An improved Landauer principle with finite-size corrections, *New Journal of Physics*, 16(10): 103011, https://doi.org/10.1088/1367-2630/16/10/103011

Schaibley J., H. Yu, G. Clark, P. Rivera, J. S. Ross, K. L. Seyler, W. Yao and X. Xu 2016 Valleytronics in 2D materials, *Nature Reviews Materials*, 1: 16055, https://doi.org/10.1038/natrevmats.2016.55

Yuan T. 2007 Yield and Reliability Analysis for Nanoelectronics, PhD dissertation, University of Tennessee, https://trace.tennessee.edu/utk_graddiss/204

Index

Pages in *italics* refer to figures and pages in **bold** refer to tables.

W

Z

For Product Safety Concerns and Information please contact our EU representative GPSR@taylorandfrancis.com Taylor & Francis Verlag GmbH, Kaufingerstraße 24, 80331 München, Germany

Batch number: 10725800

Printed by Printforce, the Netherlands